S0-AYW-803

Quantum Mechanics and Its Emergent Macrophysics

Quantum Mechanics and Its Emergent Macrophysics

Geoffrey L. Sewell

PRINCETON UNIVERSITY PRESS

PRINCETON AND OXFORD

Copyright © 2002 by Princeton University Press

Published by Princeton University Press,
41 William Street, Princeton, New Jersey 08540

In the United Kingdom: Princeton University Press,
3 Market Place, Woodstock, Oxfordshire OX20 1SY

All Rights Reserved
Library of Congress Cataloging-in-Publication Data applied for.
Sewell, Geoffrey, L.
Quantum Mechanics and Its Emergent Macrophysics/
Geoffrey L. Sewell
p. cm.
Includes bibliographical references and index.
ISBN 0-691-05832-6 (alk. paper)

British Library Cataloging-in-Publication Data is available

This book has been composed in Times and Abadi

Printed on acid-free paper

www.pup.princeton.edu

Printed in the United States of America

Contents

Preface

The quantum theory of macroscopic systems has become a vast interdisciplinary subject, which extends far beyond the traditional area of condensed matter physics to general questions of complexity, self-organisation and chaos. It is evident that the subject must be based on conceptual structures quite different from those of quantum microphysics, since the macroscopic properties of complex systems are expressible only in terms of concepts, such as entropy and various kinds of order, that have no relevance to the microscopic world. Moreover, the empirical fact that systems of different microscopic constitutions exhibit similar macroscopic behavior provides grounds for suspecting that macrophysics is governed by very general features of the quantum properties of many-particle systems. Accordingly, it appears natural to pursue an approach to the theory of its emergence from quantum mechanics that is centred on macroscopic observables and certain general features of quantum structures, independently of microscopic details.

This book is devoted to precisely such an approach. It is thus in line with such classic works as Onsager's nonequilibrium thermodynamics, Landau's fluctuating hydrodynamics and the renormalisation theory of critical phenomena. However, thanks to developments in mathematical physics over the last three decades, we now have a framework, that was not available to Onsager and Landau, for a general approach to the problem of the passage from microphysics to macrophysics.

The developments in question are in the areas of operator algebraic quantum theory, quantum stochastic processes and classical dynamical systems. The relevance of these three areas to our project may be briefly summarised as follows. Firstly, operator algebraic quantum theory is the generalisation of quantum mechanics to systems that may have infinite numbers of degrees of freedom. It therefore provides the natural framework for the generic model of macroscopic systems, idealised as infinitely extended assemblies of particles of finite mean number density. This idealisation is essential for the precise specification of different levels of macroscopicality and for the sharp mathematical characterisations of quintessentially macroscopic properties of matter, such as phase transitions and irreversible evolutions, whose description would otherwise be blurred by finite size effects. Secondly, the theory of quantum stochastic processes, which is naturally cast within the operator algebraic framework, provides structures pertinent to the dynamics of both open and closed quantum systems, whether near to or far from thermal equilibrium. Thirdly, the theory of classical dynamical systems, with its various scenarios of regular and chaotic motion, becomes essential for the treatment of the classical evolutions that generally emerge at the macroscopic level.

This book represents a further development of the approach to the theory of collective phenomena presented in my OUP monograph "Quantum theory of collective phenomena", although it is designed to be readable as a self-contained work. Its essentially new features are embodied in the central roles of macroscopic observables and quantum stochastic processes in the treatment of collective phenomena, both near to and far from thermal equilibrium.

The book is divided into four parts. Part I consists of a concise presentation of the operator algebraic framework for quantum mechanics, and of a formulation, within this framework, of the concepts of symmetry, order and disorder, together with an approach to the theory of irreversibility. Part II provides a general quantum statistical treatment of both equilibrium and nonequilibrium thermodynamics. The former contains a new characterisation of a complete set of thermodynamical variables and the latter provides a nonlinear generalisation of the Onsager theory. Parts III and IV are concerned with ordered and chaotic structures that arise in some key areas of condensed matter physics. Specifically, Part III is devoted to a derivation of macroscopic superconductive electrodynamics from the basic assumptions of off-diagonal long range order, gauge covariance and thermodynamical stability by methods that circumvent the enormously complicated quantum many-body problem. Part IV, on the other hand, presents a framework for the theory of phase transitions in dissipative systems far from thermal equilibrium, as illustrated by a quantum stochastic treatment of a laser model that supports normal, coherent and chaotic radiation.

The choice of topics here, although inevitably dependent on my personal interests, has been made with the aim of presenting a coherent approach to the vast problem of the emergence of macroscopic phenomena from quantum mechanics.

The book is addressed to research physicists, mathematicians and scientists interested in interdisciplinary studies. I have endeavoured to keep the mathematics as simple as possible, without sacrificing rigour. The book is thus designed to be readable on the basis of a knowledge of standard quantum mechanics, mathematical analysis and vector space theory. Any additional mathematics required here is expounded in self-contained form, either in the main text or in appendices.

Geoffrey L. Sewell

Notation

We employ the following standard notation throughout the book. Other symbols are defined in the text.

$:=$ signifies "is defined to be equal to".

$X \backslash Y$ denotes the complement, in a set X, of a subset Y.

C denotes the set of all complex numbers: the complex conjugate of a complex number c will be denoted by $\bar{c}$.

R denotes the set of all real numbers.

$\mathbf{R}_+$ denotes the set of nonnegative real numbers.

Z denotes the set of all integers.

N denotes the set of all nonnegative integers.

Part I

_____

The algebraic quantum mechanical framework and the
description of order, disorder and irreversiblity in
macroscopic systems: prospectus

Part I is devoted to a concise operator algebraic formulation of the quantum
theory of macroscopic systems and to a description, within the framework
thereby provided, of symmetry, order, disorder and irreversibility.

We start, in Chapter 1, with a brief general discussion of the application of
quantum theory to macroscopic specimens of matter, arguing that these may
naturally be idealised as systems with infinite numbers of degrees of freedom
and thus described within a framework, due originally to Segal [Seg] and to
Haag and Kastler [HK], that is based on the algebraic structure of their observables rather than on any one of the infinity of possible underlying Hilbert
space representations thereof. We formulate this framework in Chapter 2, and
demonstrate how it serves to provide qualitative distinctions between the
descriptions of matter at different levels of macroscopicality and a natural
characterisation of its pure phases.

We then proceed, in Chapter 3, to a general formulation of entropy, symmetry, order and disorder, within the operator algebraic framework, and to a
discussion of the limitations of our present conception of these entities. In
particular, we demonstrate that, on the one hand, entropy does not necessarily
provide a measure of disorder, since some pure states can have spatially
chaotic structures; while, on the other hand, order, or structural organisation,
can carry relatively high entropy, just as coherent signals can coexist with
intense background noise.

Chapter 4 is devoted to the theory of irreversibility and macroscopic causality in conservative quantum systems. Here we formulate the concepts of
microscopic reversibility and macroscopic irreversibility within the operator
algebraic framework, and demonstrate how the latter ensues from the former,
subject to appropriate initial conditions, in rather simple concrete examples.

Chapter 1

Introductory discussion of quantum macrophysics

Quantum theory began with Planck's [Pl] derivation of the thermodynamics of black body radiation from the hypothesis that the action of his oscillator model of matter was quantised in integral multiples of a fundamental constant, h. This result provided a microscopic theory of a macroscopic phenomenon that was incompatible with the assumption of underlying classical laws. In the century following Planck's discovery, it became abundantly clear that quantum theory is essential to natural phenomena on both the microscopic and macroscopic scales. Its crucial role in determining the gross properties of matter is evident from the following considerations.

(1) The stability of matter against electromagnetic collapse is effected only by the combined action of the Heisenberg and Pauli principles [DL, LT, LLS, BFG].

(2) The third law of thermodynamics is quintessentially quantum mechanical and, arguably, so too is the second law.[1]

(3) The mechanisms governing a vast variety of cooperative phenomena, including magnetic ordering [Ma], superfluidity [La1, BCS] and optical and biological coherence [Ha1, Fr1], are of quantum origin.

As a first step towards contemplating the quantum mechanical basis of macrophysics, we note the empirical fact that macroscopic systems enjoy properties that are radically different from those of their constituent particles. Thus, unlike systems of few particles, they exhibit irreversible dynamics, phase transitions and various ordered structures, including those characteristic of life. These and other macroscopic phenomena signify that complex systems, that is, ones consisting of enormous numbers of interacting particles, are qualitatively different from the sums of their constituent parts. Correspondingly, theories of such phenomena must be based not only on quantum mechanics *per se* but also on conceptual structures that serve to represent the characteristic features of highly complex systems. Among the key general

---

[1] The essential point here is that the classical statistical mechanical formulation of entropy depends on an arbitrary subdivision of phase space into microcells (cf. [Fe, Chapter 8]).

concepts involved here are ones representing various types of order, or organisation, disorder, or chaos, and different levels of macroscopicality. Moreover, the particular concepts required to describe the ordered structures of
superfluids and laser light are represented by macroscopic wave functions
[PO, Ya, GH, Se1] that are strictly quantum mechanical, although radically
different from the Schrödinger wave functions of microphysics.

To provide a mathematical framework for the conceptual structures
required for quantum macrophysics, it is clear that one needs to go beyond
the traditional form of quantum mechanics [Di, VN1], since that does not
discriminate *qualitatively* between microscopic and macroscopic systems.
This may be seen from the fact that the traditional theory serves to represent
a system of N particles within the standard Hilbert space scheme, which takes
the same form regardless of whether N is 'small' or 'large'. In fact, it was this
very lack of a sharp characterisation of macroscopicality that forced Bohr [Bo]
into a dualistic treatment of the measuring process, in which the microscopic
system under observation was taken to be quantum mechanical, whereas the
macroscopic measuring apparatus was treated as classical, even though it too
was presumably subject to quantum laws.

However, a generalised version of quantum mechanics that provides the
required qualitative distinctions between different grades of macroscopicality
has been devised over the last three decades, on the basis of an *idealisation* of
macroscopic systems as ones possessing infinite numbers of degrees of freedom. This kind of idealisation has, of course, long been essential to statistical
thermodynamics, where, for example, the characterisation of phase transitions
by singularities in thermodynamical potentials necessitates a passage to the
mathematical limit in which both the volume and the number of particles of a
system tend to infinity in such a way that the density remains finite [YL, LY,
Ru1]. Its extension to the full description of the observables and states of
macroscopic systems [AW, HHW, Ru1, Em1] has served to replace the merely
quantitative difference between systems of 'few' and 'many' (typically 10^{24})
particles by the qualitative distinction between finite and infinite ones, and has
thereby brought new, physically relevant structures into the theory of collective phenomena [Th, Se2].

The key element of the generalisation of quantum mechanics to infinite
systems is that it is based on the algebraic structure of the observables, rather
than on the underlying Hilbert space [Seg, HK]. The radical significance of
this is that, whereas the algebra of observables of a finite system, as governed
by the canonical commutation relations, admits only one irreducible Hilbert
space representation [VN2], that of an infinite system has infinitely many
inequivalent such representations [GW]. Thus, for a finite system, the algebraic and Hilbert space descriptions are equivalent, while, for an infinite one,
the algebraic picture is richer than that provided by any irreducible representation of its observables.

Moreover, the algebraic quantum theory of infinite systems, as cast in a form designed for the treatment of fundamental problems in statistical mechanics and quantum field theory [Em1, BR, Th, Se2, Haa1], admits just the structures required for the treatment of macroscopic phenomena. In particular, it permits clear definitions of various kinds of order, as well as sharp distinctions between global and local variables, which may naturally be identified with macroscopic and microscopic ones. Furthermore, the wealth of inequivalent representations of the observables permits a natural classification of the states in both microscopic and macroscopic terms. To be specific, the vectors in a representation space[2] correspond to states that are macroscopically equivalent but microscopically different, while those carried by different representations are macroscopically distinct. Hence, the macrostate corresponds to a representation and the microstate to a vector in the representation space. This is of crucial significance not only for the description of the various phases of matter, but also for the quantum theory of measurement. The specification of the states of a measuring apparatus in microscopic and macroscopic terms has provided a key element of a fully quantum treatment [He, WE] of the measurement process that liberates the theory from Bohr's dualism.

Our approach to the basic problem of how macrophysics emerges from quantum mechanics will be centred on macroscopic observables, our main objective being to obtain the properties imposed on them by general demands of quantum theory and many-particle statistics. This approach has classic precedents in Onsager's [On] irreversible thermodynamics and Landau's fluctuating hydrodynamics [LL1], and is at the opposite pole from the many-body-theoretic computations of condensed matter physics [Pi, Tho]. Our motivation for pursuing this approach stems from the following two considerations. Firstly, since the observed laws of macrophysics have relatively simple structures, which do not depend on microscopic details, it is natural to seek derivations of these laws that are based on general quantum macrostatistical arguments. Secondly, by contrast, the microscopic properties of complex systems are dominated by the molecular chaos that is at the heart of statistical physics; and presumably, this chaos would render unintelligible any solutions of the microscopic equations of motion of realistic models of such systems, even if these could be obtained with the aid of supercomputers.

Thus, we base this treatise on macroscopic observables and certain general structures of complex systems, as formulated within the terms of the algebraic framework of quantum theory. The next three chapters are devoted to a concise formulation of this framework, for both conservative and open systems (Chapter 2), and of the descriptions that it admits of symmetry, order and disorder (Chapter 3), and of irreversibility (Chapter 4).

---

[2] To be precise (cf. Section 2.6.3), this is true for primary representations.

Chapter 2

The generalised quantum mechanical framework

The traditional form of quantum theory is based on the model in which the pure states and observables of a system correspond to the normalised vectors, ψ, and the self-adjoint operators, A, respectively, in a certain Hilbert space, $\mathcal{H}$, with the interpretation that the expectation value of an observable A for the state ψ is $(\psi, A\psi)$ (cf. [VN1]). To be precise, this model is designed for systems with finite numbers of degrees of freedom, since for any such system, the canonical commutation relations, which govern the algebraic structure of its observables, admit only one irreducible Hilbert space representation [VN2]. The picture of the system provided by this representation therefore fully captures its algebraic properties.

The situation is radically different, however, for a system with an infinite number of degrees of freedom, since the algebraic structure of its observables generally admits infinitely many inequivalent irreducible representations [Haa2, GW]. Hence, as first appreciated by Segal [Seg] and Haag and Kastler [HK], the picture based on the algebraic structure of the observables of such a system is much richer than that provided by any particular irreducible representation.

This last observation has led to a reformulation of quantum theory, in which the primary objects are the observables, endowed with algebraic structures stemming from the canonical commutation relations, rather than any particular representation thereof. This provides a natural generalisation of traditional quantum theory to systems with infinite numbers of degrees of freedom, such as those arising in both statistical mechanics and quantum field theory.

This chapter is devoted to a formulation of the algebraic framework of this generalised quantum theory. We have endeavoured to keep the mathematics as simple as possible here without sacrificing rigour. In the appendix to this chapter, we provide a rudimentary account of the theory of Hilbert spaces, which suffices for the purposes of this book. Readers who are not conversant with Hilbert spaces may find it helpful to begin the chapter by reading the appendix.

2.1 OBSERVABLES, STATES, DYNAMICS

The generic model of a physical system, Σ, as expressed in the most basic terms, consists of three essential components, $(\mathcal{O}, S, \mathcal{D})$, representing its observables, states and dynamics, respectively. Here, the observables are the variables, such as functions of positions and momenta, that can, in principle, be measured. The states, on the other hand, are the functionals on the observables that serve to specify their expectation values. Thus, if $A \in \mathcal{O}$ and $\rho \in S$, then $\rho(A)$ is the expectation value of the observable A when the state of Σ is ρ. We sometimes denote $\rho(A)$ by $\langle \rho; A \rangle$. Finally, $\mathcal{D}$ is a dynamical law that specifies the expectation value, $\langle \rho; A \rangle_t$, of the observables A at time t, given that the initial state of the system is ρ.

Evidently, the model of Σ comprises the structures of $\mathcal{O}$, S and $\mathcal{D}$. Here, we formulate these structures for nonrelativistic quantum systems of both finite and infinite numbers of degrees of freedom.

2.2 FINITE QUANTUM SYSTEMS

We start by examining both the algebraic and the Hilbert space structures of the standard model of finite quantum systems, as formulated, for example, in Von Neumann's book [VN1].

2.2.1 Uniqueness of the Representation

Let us consider first a single particle that is confined to move along a straight line. According to quantum theory, its position and momentum correspond to self-adjoint operators $\hat{q}$ and $\hat{p}$, respectively, in a separable Hilbert space $\mathcal{H}$, that satisfy the canonical commutation relation (CCR)

$$[\hat{q}, \hat{p}] = i\hbar I, \qquad (2.2.1)$$

where $\hbar = h/2\pi$ and $[A, B] = AB - BA$. This form of the CCR carries some domain problems, since it implies that $\hat{q}$ and $\hat{p}$ cannot both be bounded (cf. [Em1, Section 2b]). To avoid such problems, we adopt Weyl's scheme [We], in which the CCR are re-expressed in terms of the unitary operators

$$U(a) = \exp(ia\hat{q}), \quad V(b) = \exp(ib\hat{p}) \quad \forall a, b \in \mathbf{R}, \qquad (2.2.2)$$

and the relation (2.2.1) is sharpened to the form

$$U(a)V(b) = V(b)U(a)\exp(-i\hbar ab). \qquad (2.2.3)$$

Thus, in Weyl's picture, a representation of the CCR comprises a triple, $(U, V, \mathcal{H})$, such that U and V are strongly continuous unitary representations of $\mathbf{R}$ in a separable Hilbert space $\mathcal{H}$ that satisfy the algebraic relation (2.2.3). In this picture, the position and momentum observables are $-i$ times the infinitesimal generators of the unitary groups $U(\mathbf{R})$ and $V(\mathbf{R})$, respectively.

The problem of classifying the representations of the CCR was resolved by Von Neumann [VN2], who proved that all *irreducible*[1] ones are unitarily equivalent to that of Schrödinger, $(U_S, V_S, \mathcal{H}_S = L^2(\mathbf{R}))$, as defined by the equations

$$(U_S(a)f)(q) = f(q)\exp(iaq), \quad (V_S(b)f)(q) = f(q + \hbar b), \quad \forall f \in L^2(\mathbf{R}).$$
$$(2.2.4)$$

In other words, if $(U, V, \mathcal{H})$ is an irreducible representation of the CCR, then there is a unitary mapping, W, of $\mathcal{H}$ onto $\mathcal{H}_S$, such that $U(a) = W^{-1}U_S(a)W$ and $V(b) = W^{-1}V_S(b)W$. Note that it follows from Eqs. (2.2.4) that the position and momentum operators, $\hat{q}_S$ and $\hat{p}_S$, for the Schrödinger representation, defined as $-i$ times the infinitesimal generators of $U_S(\mathbf{R})$ and $V_S(\mathbf{R})$, respectively, are given by the standard formulae

$$(\hat{q}_S f)(q) = qf(q), \quad (\hat{p}_S f)(q) = -i\hbar \frac{df(q)}{dq}.$$
$$(2.2.5)$$

To summarise, the essentially unique irreducible representation of the CCR is determined by its algebraic structure. Further, no more physically relevant information can be encoded in the reducible representations, since these are merely direct sums of copies of the irreducible one.

The same situation prevails for a Pauli spin, $S = (S_x, S_y, S_z)$, whose algebraic properties are given by the angular momentum commutation relations

$$[S_x, S_y] = i\hbar S_z, \quad [S_y, S_z] = i\hbar S_x, \quad [S_z, S_x] = i\hbar S_y, \quad (2.2.6a)$$

together with the spin one half condition that

$$S^2 = \frac{3}{4}\hbar^2 I.$$
$$(2.2.6b)$$

To be specific, all irreducible representations of these relations are unitarily equivalent [Wi] to the two-dimensional one,

$$S = \frac{1}{2}\hbar\sigma,$$
$$(2.2.7)$$

where the components $(\sigma_x, \sigma_y, \sigma_x)$ of σ are the Pauli matrices

$$\sigma_x = \begin{pmatrix} 0 & 1 \\ 1 & 0 \end{pmatrix}, \quad \sigma_y = \begin{pmatrix} 0 & -i \\ i & 0 \end{pmatrix}, \quad \sigma_z = \begin{pmatrix} 1 & 0 \\ 0 & -1 \end{pmatrix}. \quad (2.2.8)$$

This representation may be conveniently expressed in terms of the basis vectors

---

[1] The representation $(U, V, \mathcal{H})$ is irreducible if $\mathcal{H}$ has no proper subspaces that are stable under $U(\mathbf{R})$ and $V(\mathbf{R})$.

$$\phi_1 = \begin{pmatrix} 1 \\ 0 \end{pmatrix}, \quad \phi_{-1} = \begin{pmatrix} 0 \\ 1 \end{pmatrix}, \tag{2.2.9}$$

via the formula

$$\sigma\phi_s = (\phi_{-s}, is\phi_{-s}, s\phi_s), \quad \text{for } s = \pm 1. \tag{2.2.10}$$

The above results concerning the essential uniqueness of the irreducible representation of the CCR and the spin algebra may readily be generalised to systems of finite numbers of degrees of freedom. These include systems of identical particles for which the demand that the observables be invariant under particle permutations implies that the vectors of the representation space are either symmetric or antisymmetric with respect to these permutations, according to whether the particles are bosons or fermions (cf. Chapter 15 of [Ja]). Further, in the case of systems confined to bounded spatial regions, the Schrödinger representation of the position and momentum operators has to be modified according to the boundary conditions, for example, those of Dirichlet if the confinement is affected by hard walls.

To summarise, the observables of a finite system have but one irreducible representation, and this is determined by their algebraic structure and by the statistics of the constituent particles.

2.2.2 The Generic Model

This last remark leads naturally to the following standard formulation [VN1] of the model of a finite system, Σ, in terms of the essentially unique irreducible representation of the algebraic structure of its observables in a separable Hilbert space $\mathcal{H}$.

The observables are represented by the self-adjoint operators in $\mathcal{H}$. To avoid domain problems, we restrict ourselves to the bounded ones. This entails no loss of generality, since any self-adjoint operator may be expressed in terms of its spectral projectors [VN1],[2] which are, of course, bounded; and furthermore, it is justified on the physical grounds that measurements provide evaluations only of bounded quantities. Accordingly, we represent the (bounded) observables of the system by the set, $\mathcal{O}$, of bounded, self-adjoint operators in $\mathcal{H}$.

The *pure* states of Σ correspond to the normalised vectors, ψ, in $\mathcal{H}$, with the interpretation that the expectation value of an observable A, for the state represented by ψ, is

$$\rho_\psi(A) = (\psi, A\psi) \equiv \text{Tr}(P(\psi)A), \tag{2.2.11}$$

where $P(\psi)$ is the projection operator for ψ. Thus, the functional ρ_ψ on $\mathcal{O}$, as

---

[2] See the appendix, item (20).

defined by this formula, is a state of Σ, in the sense of Section 2.1. Note that the states ρ_ψ and $\rho_{\psi'}$ are the same if and only if $\psi' = \psi \exp(i\alpha)$, where the phase α is a real constant.

The *mixed* states of the system are the weighted combinations of different pure ones, $\rho_n = \rho_{\psi_n}$, that is, they are the functionals of the form

$$\rho = \sum_n c_n \rho_n,$$

acting on $\mathcal{O}$, where $\{c_n\}$ is a denumerable set of positive numbers, whose sum is unity. Hence, by Eq. (2.2.11), ρ may be represented by the density matrix $\hat{\rho} = \sum_n c_n P(\psi_n)$, according to the formula

$$\rho(A) = \mathrm{Tr}(\hat{\rho}A), \quad \forall A \in \mathcal{O}. \tag{2.2.12}$$

In fact, since $P(\psi)$ is a density matrix, it follows from Eqs. (2.2.11) and (2.2.12) that the latter formula, when extended to *all* density matrices in $\mathcal{H}$, covers both the pure and the mixed states. Moreover, the correspondence between states and density matrices given by that formula is one-to-one, since the condition that $\mathrm{Tr}((\hat{\rho} - \hat{\rho}')A) = 0$, for all observables A, implies that $\hat{\rho} = \hat{\rho}'$.[3]

Thus, the states, S, comprise the functionals ρ of the observables, that correspond to the density matrices $\hat{\rho}$ according to Eq. (2.2.12). The pure states are the ones whose density matrices are one-dimensional projectors.

Note on Quantum Interference

Suppose that ψ is a linear combination, $c_1\psi_1 + c_2\psi_2$, of two orthogonal, normalised vectors ψ_1 and ψ_2, and that ρ is the mixed state corresponding to the density matrix $|c_1|^2 \rho_{\psi_1} + |c_2|^2 \rho_{\psi_2}$. Then it follows from Eqs. (2.2.11) and (2.2.12) that

$$\rho_\psi(A) = \rho(A) + 2\,\mathrm{Re}\,\bar{c}_1 c_2(\psi_1, A\psi_2).$$

Here, the last term represents a *quantum interference* effect that renders the expectation value of A for the state ψ different from the sum of its expectation values for ψ_1 and ψ_2, as weighted by the probabilities $|c_1|^2$ and $|c_2|^2$, respectively. This interference is characteristic of *quantum probability* theory, and stems from the fact that quantum expectation values are given by *probability amplitudes*, as represented by vectors in a Hilbert space, rather than by classical probabilities.

---

[3] This may be seen by putting $A = P(\psi)$, and thereby inferring that $(\psi, (\hat{\rho} - \hat{\rho}')\,\psi) = 0$ for all vectors ψ in $\mathcal{H}$.

The dynamics of the model is governed by the Hamiltonian, H, of the system. This is a self-adjoint operator[4] in $\mathcal{H}$, which is generally given by the sum of the kinetic and potential energy operators of the system. Time-translations are represented by the one-parameter group, $\{U_t \mid t \in \mathbf{R}\}$, of unitary transformations of $\mathcal{H}$, whose infinitesimal generator is $iH/\hbar$. Thus,

$$U_t = \exp(iHt/\hbar), \quad \forall t \in \mathbf{R}. \tag{2.2.13}$$

The dynamics may be equivalently formulated in the Heisenberg and Schrödinger pictures. In the former, it is carried by the observables, according to the principle that the evolute of $A(\in \mathcal{O})$ at time t is

$$A_t = U_t A U_t^{-1}, \tag{2.2.14}$$

while the states remain fixed. Thus, the time-dependent expectation value of an observable A at time t, given that the system is prepared in the state ρ, is

$$\langle \rho; A \rangle_t = \rho(A_t) \equiv \mathrm{Tr}(\hat{\rho} U_t A U_t^{-1}). \tag{2.2.15}$$

In the Schrödinger picture, on the other hand, the dynamics is carried by the states, according to the prescription that the evolute of the density matrix $\hat{\rho}$ at time t is

$$\hat{\rho}_t = U_t^{-1} \hat{\rho} U_t, \tag{2.2.16}$$

while the observables remain fixed. Thus, the time-dependent expectation value of the observable A at time t, for evolution from an initial state ρ, is $\mathrm{Tr}(U_t^{-1} \hat{\rho} U_t A)$, which is identical to the right-hand side of Eq. (2.1.15), by the cyclicity property of the Trace.

We note here that it follows from equations (2.2.13), (2.2.14) and (2.2.16) that the equations of motion for the Heisenberg and Schrödinger pictures are

$$\frac{dA_t}{dt} = \frac{i}{\hbar} [H, A_t] \tag{2.2.17}$$

and

$$\frac{d\hat{\rho}_t}{dt} = -\frac{i}{\hbar} [H, \hat{\rho}_t], \tag{2.2.18}$$

these being the equations of Heisenberg and Von Neumann, respectively.[5]

---

[4] It is generally unbounded, and therefore not an element of $\mathcal{O}$, although it is affiliated to this set, in that its spectral projectors belong to $\mathcal{O}$ [Se3].

[5] Since H is generally unbounded, these equations should be interpreted as pertaining to their matrix elements between vectors in the domain of H. Thus, Eq. (2.2.17) should be taken to signify that $d(f, A_t g)/dt = i/\hbar((Hf, A_t g) - (A_t^\star f, Hg))$ for f, g in this domain.

This completes our specification of the model, within the terms of Section 2.1. We now discuss its algebraic structure in a form that can be generalised to infinite systems.

2.2.3 THE ALGEBRAIC PICTURE

The above-defined set, $\mathcal{O}$, of observables possesses the following simple algebraic properties. If $A, B \in \mathcal{O}$ and $k \in \mathbf{R}$, then kA, $(A + B)$, $(AB + BA)$ and $i(AB - BA)$ all belong to $\mathcal{O}$. In other words, $\mathcal{O}$ is closed with respect to (a) multiplication by real numbers, (b) binary addition, (c) the binary symmetrised and antisymmetrised multiplications that send the pair of elements (A, B) to $(AB + BA)$ and $i(AB - BA)$, respectively. However, it is *not* closed with respect to binary multiplication, since, if A, B are noncommuting elements of $\mathcal{O}$, then $(AB)^\star = BA \neq AB$, and so AB is not self-adjoint. In view of this situation, it is simpler to express the algebraic properties of the observables in terms of those of the set, $\mathcal{A}$, of all bounded operators in $\mathcal{H}$, than to work directly with $\mathcal{O}$.

Thus, we note that $\mathcal{O}$ comprises the self-adjoint elements of $\mathcal{A}$, and that the latter set is closed with respect to binary addition, binary multiplication, multiplication by complex numbers and the adjoint mapping $A \to A^\star$. Accordingly, $\mathcal{A}$ is termed the *algebra of observables* of the system.

The states, S, may now be formulated as the functionals ρ on this algebra that correspond to the density matrices according to the Eq. (2.2.12), as extended to $\mathcal{A}$. Thus,

$$\rho(A) = \mathrm{Tr}(\hat{\rho}A) \quad \forall A \in \mathcal{A}. \tag{2.2.19}$$

It follows easily from this formula that the functional ρ possesses the following properties.

$$\rho(A^\star A) \geq 0 \quad \forall A \in \mathcal{A} \text{ (positivity)}, \tag{2.2.20}$$

$$\rho(\lambda_1 A_1 + \lambda_2 A_2) = \lambda_1 \rho(A_1) + \lambda_2 \rho(A_2) \quad \forall A_1, A_2 \in \mathcal{A},$$

$$\lambda_1, \lambda_2 \in \mathbf{C} \text{ (linearity)} \tag{2.2.21}$$

and

$$\rho(I) = 1 \text{ (normalisation)}. \tag{2.2.22}$$

The states are therefore the *positive linear normalised* functionals on the algebra of observables that correspond to the density matrices according to the formula (2.2.19). This latter condition, which is indeed restrictive, is equivalent to the following intrinsic condition [Em2]. For any sequence $\{E_n\}$ of orthogonal projectors in $\mathcal{A}$, whose sum is the identity,

$$\sum_n \rho(E_n) = 1. \qquad (2.2.23)$$

Note that this condition is nontrivial when the sequence $\{E_n\}$ is an infinite one, since it then implies a continuity condition, to the effect that the sum of a limit is equal to the limit of the sum. The functionals ρ that satisfy this condition are termed *normal*.

We now note that it follows from Eq. (2.2.19), or equivalently from Eqs. (2.2.20–2.2.23), that S is a *convex set*, that is, if $\rho_1, \rho_2 \in S$ and λ is a real number between 0 and 1, then $\lambda\rho_1 + (1 - \lambda)\rho_2 \in S$. Furthermore, the *extremal* elements of S, that is, those that cannot be expressed as weighted sums of other elements, are precisely the pure states, as defined by Eq. (2.2.11).

As regards the dynamics, we see from Eq. (2.2.14) that it corresponds to transformations α_t of $\mathcal{A}$, defined by the formula

$$\alpha_t A = U_t A U_t^{-1} \quad \forall A \in \mathcal{A}, t \in \mathbf{R}. \qquad (2.2.24)$$

These transformations possess the group property that $\alpha_t\alpha_s = \alpha_{t+s}$, Furthermore, they preserve the algebraic structure of $\mathcal{A}$, that is,

$$\alpha_t(\lambda A + \mu B) = \lambda\alpha_t A + \mu\alpha_t B,$$

$$\alpha_t(AB) = (\alpha_t A)(\alpha_t B) \quad \text{and} \quad \alpha_t(A^\star) = (\alpha_t A)^\star,$$

and so are termed *automorphisms* of this algebra. Thus, in the Heisenberg picture, the dynamics corresponds to a one-parameter group, $\alpha = \{\alpha_t \mid t \in R\}$, of automorphisms of $\mathcal{A}$.

To summarise, the model of Σ comprises the triple $(\mathcal{A}, S, \alpha)$, where $\mathcal{A}$ is its algebra of observables; S, its state space, is the set of positive, linear, normalised functionals on $\mathcal{A}$ that satisfy the normality condition (2.2.23); and α is the one-parameter group of automorphisms of $\mathcal{A}$, implemented by the unitaries U_t according to Eq. (2.2.24).

Finally, we note that the thermal equilibrium state, ρ_β, of the system at inverse temperature β is given by the Gibbs density matrix $\exp(-\beta H))/\mathrm{Tr}(idem)$; and, by Eq. (2.2.12) and the cyclicity of the Trace, this state is completely characterised [HHW] by the Kubo–Martin–Schwinger (KMS) conditions [Ku, MS], namely,

$$\rho(A(t)B) = \rho(BA(t + i\hbar\beta)) \quad \forall A, B \in \mathcal{A}, \qquad (2.2.25)$$

where $A(t) \equiv \alpha_t A$. More precisely [HHW],[6] these conditions may be expressed in the following form.

---

[6]   The point here is that Eq. (2.2.25) is formal, since α does not neccessarily have an analytic continuation to complex values of its argument t.

(KMS) For each pair of elements A, B of $\mathcal{A}$, there is a function, F_{AB}, on the strip $S_\beta = \{z \in \mathbf{C} \mid \text{Im}(z) \in [0, \hbar\beta]\}$, such that

(i) F_{AB} is analytic in the interior of S_β and continuous on its boundaries;

(ii) $F_{AB}(t) = \rho(B\alpha(t)A)$ $\forall t \in \mathbf{R}$, and

(iii) $F_{AB}(t + i\hbar\beta) = \rho([\alpha(t)A]B)$ $\forall t \in \mathbf{R}$.

2.3 INFINITE SYSTEMS: INEQUIVALENT REPRESENTATIONS

As we have already mentioned at the beginning of this chapter, systems of infinite numbers of degree of freedom differ from finite ones in the crucial respect that the algebraic relations governing their observables generally admit inequivalent irreducible representations. We now present a simple example which illustrates the basic reasons for this.

The model that we consider is that of a chain, Σ, of Pauli spins, located on the sites of the one dimensional lattice $\mathbf{Z}$. We represent these spins, in units of $\hbar/2$, by operators $\{\sigma_n = (\sigma_{n,x}, \sigma_{n,y}, \sigma_{n,z}) \mid n \in \mathbf{Z}\}$ in a Hilbert space $\mathcal{H}$, which satisfy the algebraic relations corresponding to Eqs. (2.2.6) and (2.2.7), namely

$$[\sigma_{n,x}, \sigma_{n,y}] = 2i\sigma_{n,z}, \text{ etc.} \qquad (2.3.1a)$$

and

$$\sigma_n^2 = 3I, \qquad (2.3.1b)$$

together with the condition that spins on different sites intercommute, that is,

$$[\sigma_{m,u}, \sigma_{n,v}] = 0 \quad \text{for } m \neq n, \ u, v = x, y, z. \qquad (2.3.1c)$$

In order to construct explicit representations of these relations, we introduce the set, S, of doubly infinite sequences, $s = \{s_n \mid n \in \mathbf{Z}\}$, each s_n taking the value ± 1. Thus, by Eq. (2.2.8), S is the set of all configurations of the eigenvalues, ± 1, of σ_z, on the lattice $\mathbf{Z}$, that is, it is the set of mappings $n \to s_n$ of $\mathbf{Z}$ into $\{-1, 1\}$. For each $n \in \mathbf{Z}$, we define θ_n to be the transformation of S, whose action on a configuration s reverses its nth component and leaves the rest unchanged, that is,

$$(\theta_n s)_m = s_m(1 - \delta_{mn}) - s_n \delta_{mn}. \qquad (2.3.4)$$

2.3.1 The Representation $\sigma^{(+)}$

We define $S^{(+)}$ to be the subset of S, consisting of those configurations, s, for which all but a finite number of components s_n take the value $+1$. Thus, $S^{(+)}$ is denumerable,[7] and consists of local modifications of the configuration, $s^{(+)}$,

---

[7] By contrast, S is a Cantor set.

whose components are all equal to $+1$. We define $\mathcal{H}^{(+)}$ to be the Hilbert space of square-summable functions on $S^{(+)}$, that is,

$$\left\{ f : S^{(+)} \to \mathbf{C} \mid \sum_{s \in S^{(+)}} |f(s)|^2 < \infty \right\},$$

with inner product

$$(f, g)^{(+)} = \sum_{s \in S^{(+)}} \bar{f}(s) g(s). \tag{2.3.5}$$

A complete orthonormal basis for this space is provided by the vectors $\{\phi_s^{(+)} \mid s \in S^{(+)}\}$, as defined by the formula

$$\phi_s^{(+)}(s') = \delta_{ss'} \quad \forall s, s' \in S^{(+)}. \tag{2.3.6}$$

Evidently, the correspondence between the vectors, $\phi_s^{(+)}$, and the configurations, s, is one-to-one.

We define the operators $\{\sigma_{n,u}^{(+)} \mid n \in \mathbf{Z}; u = x, y, z\}$ in $\mathcal{H}^{(+)}$ in such a way that the action of $\sigma_{n,u}^{(+)}$ on $\phi_s^{(+)}$ is the canonical analogue of that given by Eq. (2.2.10) for an isolated Pauli spin. Thus, defining $\sigma_n^{(+)} = (\sigma_{n,x}^{(+)}, \sigma_{n,y}^{(+)}, \sigma_{n,z}^{(+)})$,

$$\sigma_n^{(+)} \phi_s = \left(s_n \phi_{\theta_n s}^{(+)}, i s_n \phi_{\theta_n s}^{(+)}, s_n \phi_s^{(+)} \right) \quad \forall n \in \mathbf{Z}, \ s \in S^{(+)}. \tag{2.3.7}$$

It now follows easily from this formula and Eq. (2.3.4) that the algebraic relations (2.3.1) are valid on the basis vectors $\phi_s^{(+)}$, and that therefore the operators $\sigma_{n,u}^{(+)}$ provide a representation of these relations in $\mathcal{H}^{(+)}$. Its irreducibility follows from the fact that the passage between any two of the basis vectors, ϕ_s and $\phi_{s'}$, can be effected by a finite number of spin reversals, implemented by the action of a monomial in the operators $\sigma_{n,x}^{(+)}$.

We now obtain a simple global property of this representation in terms of the polarisation observable

$$m_N^{(+)} = \frac{1}{2N + 1} \sum_{n=-N}^{N} \sigma_n^{(+)}. \tag{2.3.8}$$

We see from Eqs. (2.3.7) and (2.3.8) that

$$\left(\phi_s^{(+)}, m_N^{(+)} \phi_s^{(+)} \right) = \left(0, 0, \frac{1}{2N + 1} \sum_{n=-N}^{N} s_n \right)$$

and therefore, since all but a fixed finite number of the s_n are $+1$, the rest being -1,

$$\lim_{N \to \infty} \left(\phi_s^{(+)}, m_N^{(+)} \phi_s^{(+)} \right) = k \quad \forall s \in S^{(+)}, \tag{2.3.9}$$

where k is the unit vector along Oz. Similarly, it also follows from Eqs. (2.3.7) and (2.3.8) that

$$\lim_{N\to\infty} \left(\phi_s^{(+)}, m_N^{(+)} \phi_{s'}^{(+)} \right) = 0 \quad \text{for } s \neq s'. \tag{2.3.10}$$

In order to extend these results to arbitrary vectors in $\mathcal{H}^{(+)}$, we note that, by Eq. (2.3.7), the norms of the operators $\sigma_{n,u}$ are all equal to unity and therefore, by Eq. (2.3.8), $m_N^{(+)}$ is uniformly bounded. Hence, as $\{\phi_s^{(+)}\}$ is a basis in $\mathcal{H}^{(+)}$, it follows from Eqs. (2.3.9) and (2.3.10) that, for all unit vectors $f^{(+)}$ in $\mathcal{H}^{(+)}$,

$$\lim_{N\to\infty} \left(f^{(+)}, m_N^{(+)} f^{(+)} \right) = k. \tag{2.3.11}$$

This result represents a global property of the representations $\sigma^{(+)}$, which stems from the fact that the states that it carries are local modifications of the one where all the spins are aligned parallel to Oz.

2.3.2 The Representation $\sigma^{(-)}$

We may similarly construct another representation of the relations (2.3.1), based this time on the subset $S^{(-)}$ of S, consisting of configurations s, which take the value -1 except at a finite set of sites of $\mathbf{Z}$. Thus, we define the Hilbert space $\mathcal{H}^{(-)}$, the basis vectors $\phi_s^{(-)}$, and the operators $\sigma^{(-)}$ and $m_N^{(-)}$ by the condition that they have the same relationship to $S^{(-)}$ as $\mathcal{H}^{(+)}$, $\phi_s^{(+)}$, $\sigma^{(+)}$ and $m_N^{(+)}$, respectively, have to $S^{(+)}$. In this way, we obtain an irreducible representation, $\sigma^{(-)}$, of the relations (2.3.1); and since it is based on the $S^{(-)}$ configurations, it carries the global property obtained by replacing the superscripts $(+)$ by $(-)$ and k by $-k$ in Eq. (2.3.11). Thus,

$$\lim_{N\to\infty} \left(f^{(-)}, m_N^{(-)} f^{(-)} \right) = -k \tag{2.3.12}$$

for any unit vector $f^{(-)}$ in $\mathcal{H}^{(-)}$.

2.3.3 Inequivalence of $\sigma^{(\pm)}$

We see immediately from Eqs. (2.3.11) and (2.3.12) that the representations $\sigma^{(\pm)}$ of the spin algebra are globally different, in that the states that they carry have polarisations $\pm k$, respectively. This implies that they cannot be unitarily equivalent, as the following argument shows.

If the representations $\sigma^{(\pm)}$ were unitarily equivalent, there would be a unitary mapping, W, of $\mathcal{H}^{(+)}$ onto $\mathcal{H}^{(-)}$ such that $W\sigma_n^{(+)} W^{-1} = \sigma_n^{(-)}$ for all n, which would imply, by Eq. (2.3.8), that

$$Wm_N^{(+)} W^{-1} = m_N^{(-)}.$$

This, in turn, would imply that, if $f^{(\pm)}$ were unit vectors in $\mathcal{H}^{(\pm)}$ such that $f^{(+)} = W^{-1} f^{(-)}$, then the following equation would hold:

$$\left(f^{(+)}, m_N^{(+)} f^{(+)}\right) = \left(f^{(-)}, m_N^{(-)} f^{(-)}\right).$$

This cannot be valid, however, since, by Eqs. (2.3.11) and (2.3.12), its left- and right-hand sides tend to $\pm k$, respectively, as $N \to \infty$. We conclude, therefore, that the assumption of the unitary equivalence of the representations $\sigma^{(\pm)}$ is untenable.

2.3.4 Other Inequivalent Representations

The above argument shows that the inequivalence of the representations $\sigma^{(\pm)}$ stems from the fact that they are based on globally different sets $S^{(\pm)}$ of configurations of the z-components of the spins. With this in mind, it is now a simple matter to construct various other inequivalent representations of the spin algebraic relations (2.3.1). For example, one could employ the above procedure to construct an irreducible representation based on finite modifications of the configuration of alternating $+1$'s and -1's. One could likewise construct representations based on configurations in which the spins were aligned either "ferromagnetically" along an arbitrary direction u or "antiferromagnetically" along the directions $\pm u$. Thus, by contemplating these and other similarly based constructions, we see that there is an infinity of globally different, and hence unitarily inequivalent, irreducible representations of the algebraic relations (2.3.1).

2.4 OPERATOR ALGEBRAIC INTERLUDE

Since the observables of infinite quantum systems admit inequivalent representations, we adopt the view of Segal [Seg] and Haag and Kastler [HK] that theories of such systems should be based on the algebraic structure of their observables, rather than on particular representations thereof. This requires a formulation of quantum mechanics within an extended version of the framework described in Section 2.2.3 for finite systems. We devote this section to a simple and concise presentation of the operator algebraic theory required for the construction of this extended framework. For comprehensive treatments, see the books of Dixmier [Dix1, Dix2] and Sakai [Sa].

2.4.1 Algebras: Basic Definitions and Properties

(1) An *algebra* over the field of complex numbers, $\mathbf{C}$, is defined to be a set, $\mathcal{A}$, of elements, $(A, B, \ldots)$, that satisfies the following conditions.

    (a) $\mathcal{A}$ is closed with respect to a commutative, associative operation of binary addition, and is a group with respect to this operation. The identity element of this group is termed the zero element, 0, of $\mathcal{A}$, that is, $A = A + 0$.

(b) $\mathcal{A}$ is closed with respect to binary multiplication, which is associative and distributive with respect to addition, that is, $A(B + C) = AB + AC$ and $(B + C)A = BA + CA$. It is not necessarily commutative.

(c) $\mathcal{A}$ is closed with respect to multiplication by complex numbers: this operation satisfies the conditions that, for $A, A_1, A_2 \in \mathcal{A}$ and $\lambda, \lambda_1, \lambda_2 \in \mathbf{C}$,

$$\lambda(A_1 + A_2) = \lambda A_1 + \lambda A_2, \quad (\lambda_1 + \lambda_2)A = \lambda_1 A + \lambda_2 A,$$

$$\lambda_1(\lambda_2 A) = (\lambda_1 \lambda_2)A, \quad (\lambda A_1)A_2 = A_1(\lambda A_2) = \lambda(A_1 A_2).$$

The algebra is termed *abelian* if it is commutative with respect to multiplication. It is termed *unital* if it has an identity element, I, characterised by the condition that $AI = IA = A$, $\forall A \subset \mathcal{A}$.

(2) An algebra $\mathcal{A}$ is termed a *-*algebra* if it is closed with respect to an operation $A \to A^{\star}$, such that $(A^{\star})^{\star} \equiv A$, $(AB)^{\star} \equiv B^{\star}A^{\star}$ and $(\lambda A)^{\star} \equiv \bar{\lambda}A^{\star}$. Such an operation is termed an *involution*.

(3) A *-algebra $\mathcal{A}$ is said to be *normed* if it is also equipped with a mapping $A \to \|A\|$ of $\mathcal{A}$ into the nonnegative real numbers, such that, for $A, B \in \mathcal{A}$ and $\lambda \in \mathbf{C}$,

$$\|A^{\star}\| = \|A\|, \quad \|\lambda A\| = |\lambda| \, \|A\|,$$

$$\|A + B\| \leq \|A\| + \|B\|, \quad \|AB\| \leq \|A\|\|B\|$$

and $\|A\| = 0$ if and only if $A = 0$. The mapping $\|\cdot\|$ is termed a norm. Norm convergence of A_n to A is then defined to be equivalent to convergence of $\|A_n - A\|$ to zero. Likewise, normwise Cauchy sequences $\{A_n\}$ in $\mathcal{A}$ are characterised by the property that, for any positive ϵ, there is a finite number N_ϵ, such that $\|A_m - A_n\| < \epsilon$ if m and n both exceed N_ϵ.

(4) A normed *-algebra $\mathcal{A}$ is said to be *norm-complete* if it contains the limits of its normwise Cauchy sequences. Otherwise, the norm completion of this algebra is obtained by the addition of the limits of these sequences.

(5) A $C^{\star}$-*algebra* is defined to be a norm-complete *-algebra possessing the property that $\|A^{\star}A\| \equiv \|A\|^2$.

We define the *unit ball*, $\mathcal{A}_1$, of a $C^{\star}$- algebra, $\mathcal{A}$, to be the set of its elements whose norms do not exceed unity. Examples of $C^{\star}$-algebras are the following.

Example 1 The field of complex numbers, **C**, with the usual definitions of addition and multiplication, and the norm defined as the modulus, that is, $\|z\| = |z|$.

Example 2 The set, $\mathcal{B}(\mathcal{H})$, of all bounded linear operators in a Hilbert space $\mathcal{H}$, with addition, multiplication and norm defined in the usual way for operators.

Example 3 The set, $C(X)$, of all complex-valued, bounded, continuous functions, f, on a closed, bounded region, X, of a Euclidean space (or, more generally, on a compact space, X), with addition and multiplication defined in the usual way, and norm defined by the formula $\|f\| = \sup_{x \in X} |f(x)|$.[8]

Note The algebras of Examples 1 and 3 are abelian.

(6) If $\mathcal{A}$ is a *-algebra (not necessarily $C^{\star}$) of bounded linear transformations of a Hilbert space, $\mathcal{H}$, then the *commutant*, $\mathcal{A}'$, of $\mathcal{A}$ is defined to be the set of bounded operators in $\mathcal{H}$ that commute with all elements of $\mathcal{A}$. Thus, $\mathcal{A}'$ is also a $\star$-algebra, and hence so too is its commutant, $\mathcal{A}''$, which is termed the *bicommutant* of $\mathcal{A}$. Moreover, it follows easily from these definitions that $\mathcal{A} \subset \mathcal{A}'' = (\mathcal{A}'')''$. In other words, $\mathcal{A}''$ contains $\mathcal{A}$, and is closed with respect to bicommutation.

(7) A $W^{\star}$-*algebra*, or *Von Neumann algebra*, is defined to be a *-algebra of bounded operators in a Hilbert space, $\mathcal{H}$, that is closed with respect to bicommutation. Remarkably, this latter condition is equivalent here to those of both weak and strong closure [Dix1]. Correspondingly, if $\mathcal{A}$ is an arbitrary *-algebra of bounded operators in a Hilbert space, $\mathcal{H}$, then the $W^{\star}$-algebra $\mathcal{A}''$ is identical to both the strong and the weak closures of $\mathcal{A}$.

Note Since norm convergence is stronger than weak convergence, it follows that a $W^{\star}$-algebra is norm complete. Hence, since operators in a Hilbert space enjoy the properties that $\|AB\| \leq \|A\|\|B\|$ and $\|A^{\star}A\| = \|A\|^2$, $W^{\star}$-*algebras are also $C^{\star}$-algebras.*

Example of a $W^{\star}$-algebra The set of all bounded operators in a Hilbert space, that is, the $C^{\star}$ Example 2 above.

Example of a $C^{\star}$-algebra that is not $W^{\star}$ This is the abelian algebra, $C(X)$, of the $C^{\star}$ Example 3 above, acting multiplicatively on the Hilbert space $L^2(X)$

---

[8] The term "sup" means supremum, that is, least upper bound.

according to the formula

$$(Af)(x) = A(x)f(x) \quad \forall A \in C(X),\ f \in L^2(X).$$

One sees that this algebra is not $W^\star$ from the fact that among its weak limit points are discontinuous functions, for example step functions, and these do not belong to $C(X)$.

(8) The $W^\star$-algebra generated by a *-algebra, $\mathcal{A}$, of operators in a Hilbert space, $\mathcal{H}$, is defined to be $\mathcal{A}''$ or, equivalently, the weak or the strong closure of $\mathcal{A}$. More generally, the $W^\star$-algebra generated by a set, $\mathcal{B}$, of bounded operators in $\mathcal{H}$ is the bicommutant, or equivalently the weak or strong closure, of the *-algebra of polynomials in the elements of $\mathcal{B}$ and their adjoints.

(9) The *centre*, $Z(\mathcal{A})$, of a $W^\star$-algebra, $\mathcal{A}$, is defined to be $\mathcal{A} \cap \mathcal{A}'$. Thus, $Z(\mathcal{A})$ is an abelian $W^\star$-algebra, consisting of the elements of $\mathcal{A}$ that commute with every element of this algebra.

(10) A $W^\star$-algebra, $\mathcal{A}$, is termed *primary*, or a factor, if its centre consists of scalar multiples of the identity, that is, if $Z(\mathcal{A}) = \{\lambda I \mid \lambda \in \mathbf{C}\}$. Otherwise, $Z(\mathcal{A})$ may be resolved into primaries, $\mathcal{A}_\alpha$, according to the *central decomposition*, which takes the form

$$\mathcal{H} = \int^\oplus d\mu(\alpha)\mathcal{H}_\alpha, \quad \mathcal{A} = \int^\oplus d\mu(\alpha)\mathcal{A}_\alpha, \quad \mathcal{A}_\alpha \subset \mathcal{B}(\mathcal{H}_\alpha),$$

where μ is a probability measure. This decomposition corresponds to a resolution of all central projectors into one-dimensional ones [Dix1].

Note Since every algebra we employ for physical purposes has an identity element, we henceforth take the term "algebra" to mean "unital algebra", as defined at the end of item (1).

2.4.2 States and Representations

(11) A *state* on a $C^\star$-algebra, $\mathcal{A}$, is defined to be a positive, normalised, linear functional, ρ, on $\mathcal{A}$, that is, a mapping, ρ, of $\mathcal{A}$ into $\mathbf{C}$, such that

$$\rho(\lambda_1 A_1 + \lambda_2 A_2) = \lambda_1 \rho(A_1) + \lambda_2 \rho(A_2),$$

$$\overline{\rho(A)} = \rho(A^\star), \quad \rho(A^\star A) \geq 0, \quad \rho(I) = 1,$$

for all $\lambda_1 \lambda_2 \in \mathbf{C}$ and $A, A_1, A_2 \in A$. This mathematical definition of states stems from its physical connotation (cf. Section 2.2.1). We sometimes denote $\rho(A)$ by $\langle \rho; A \rangle$.

(12) We define $S(\mathcal{A})$ to be the set of all states on $\mathcal{A}$. Thus, $S(\mathcal{A})$ is *convex*, that is, if ρ_1, ρ_2 are elements of this set, then so too is $\lambda \rho_1 + (1 - \lambda)\rho_2$, for all $\lambda \in (0, 1)$. The extremal elements of $S(\mathcal{A})$, that is, those that cannot be expressed as weighted sums, or mixtures, of different states, are termed the *pure states* on $\mathcal{A}$.

We define the norm, $\|\rho - \rho'\|$, of the difference between states ρ and ρ' by the formula

$$\|\rho - \rho'\| = \sup_{A \in \mathcal{A}} \frac{|\rho(A) - \rho'(A)|}{\|A\|}.$$

(13) There are two topologies, that is, definitions of convergence or continuity, for $S(\mathcal{A})$. These are the norm and weak*, or $w^\star$, topologies, which are defined as follows.

$$\text{norm:} \lim_{\alpha} \rho_\alpha = \rho \text{ signifies that } \lim_{\alpha} \|\rho_\alpha - \rho\| = 0.$$

$$w^\star: \lim_{\alpha} \rho_\alpha = \rho \text{ signifies that } \lim_{\alpha} \rho_\alpha(A) = \rho(A) \quad \forall A \in \mathcal{A}.$$

Norm convergence is therefore stronger than $w^\star$ convergence. Further, the state space $S(\mathcal{A})$ is compact with respect to the $w^\star$-topology [BR].

(14) A state, ρ, is termed *faithful* if $\rho(A^\star A) = 0$ only if $A = 0$.

(15) A state, ρ, on a $W^\star$-algebra, $\mathcal{A}$, is termed *normal* if it corresponds to a density matrix, $\hat{\rho}$, in the Hilbert space, $\mathcal{H}$, in which $\mathcal{A}$ acts, according to the formula

$$\rho(A) = \text{Tr}(\hat{\rho}A).$$

Equivalently [Dix1], normality may be defined by the so-called *ultraweak* continuity condition that it conserves weak continuity in the unit ball, that is, if A is the weak limit of a sequence $\{A_n\}$ in $\mathcal{A}_1$, then $\lim_{n \to \infty} \rho(A_n) = \rho(A)$. We denote by $\mathcal{N}(\mathcal{A})$ the set of all normal states on $\mathcal{A}$, and remark that it is norm complete, that is, it contains the limits of its normwise Cauchy sequences.

(16) A *representation* of a $C^\star$-algebra, $\mathcal{A}$, in a Hilbert space, $\mathcal{H}$, is a mapping, π, of $\mathcal{A}$ into the bounded operators in $\mathcal{H}$ that preserves its *-algebraic structure, that is,

$$\pi(\lambda_1 A_1 + \lambda_2 A_2) = \lambda_1 \pi(A_1) + \lambda_2 \pi(A_2),$$

$$\pi(A_1 A_2) = \pi(A_1)\pi(A_2), \quad \pi(A)^\star = \pi(A^\star)$$

for all $\lambda_1 \lambda_2 \in C$ and $A, A_1, A_2 \in A$. The representation is termed *faithful* if π is invertible, or, equivalently, if $\pi(A) = 0$ implies that $A = 0$. In fact, every $C^\star$- algebra has at least one faithful representation [Dix2].

Note

(i) $\pi(\mathcal{A})$ is itself a $C^\star$-algebra with respect to the operator norm in $\mathcal{H}$. In general, $\|\pi(A)\| \leq \|A\|$, equality occurring, for all elements A of $\mathcal{A}$, if and only if π is faithful.

(ii) The density matrices, $\hat{\rho}$, in the Hilbert space of a representation, π, of $\mathcal{A}$ correspond to states, ρ, on that algebra according to the formula $\rho(A) = \mathrm{Tr}(\hat{\rho}\pi(A))$.

(iii) If $\mathcal{A}$ is a $C^\star$-algebra of operators in a Hilbert space, $\mathcal{H}$, it may still be represented in another Hilbert space, $\mathcal{K}$. For example, if $\mathcal{K}$ is a tensor product $\mathcal{H} \otimes \mathcal{H}'$ and π is the mapping of $\mathcal{A}$ into $\mathcal{B}(\mathcal{K})$ defined by $\pi(A) = A \otimes I$, then π is a representation of $\mathcal{A}$ in $\mathcal{K}$.

(17) The representation, π, is termed *irreducible* if it cannot be expressed as a direct sum of other representations. It is termed *primary* if the $W^\star$-algebra $\pi(\mathcal{A})''$ is primary. It is termed *cyclic* if there is a vector, Φ, in the representation space, $\mathcal{H}$, such that $\pi(\mathcal{A})\Phi \; (= \{\pi(A)\Phi \mid A \in \mathcal{A}\})$ is dense in $\mathcal{H}$, in which case Φ is termed a *cyclic vector* for π. In the case where $\mathcal{A}$ is a $W^\star$-algebra, π is termed *normal* if it conserves weak convergence in the unit ball, that is, if A is the weak limit of a sequence $\{A_n\}$ in $\mathcal{A}_1$, then $w : \lim_{n \to \infty} \pi(A_n) = \pi(A)$.

(18) A state, ρ, on a $C^\star$-algebra, $\mathcal{A}$, induces a representation, π_ρ, of $\mathcal{A}$ in a Hilbert space, $\mathcal{H}_\rho$, with cyclic vector Φ_ρ, such that

$$\rho(A) = (\Phi_\rho, \pi_\rho(A)\Phi_\rho) \quad \forall A \in \mathcal{A}.$$

This is the *Gelfand–Naimark–Segal (GNS) representation*, denoted by $(\mathcal{H}_\rho, \pi_\rho, \Phi_\rho)$, and is unique, up to unitary equivalence.

Note The GNS construction is based on the fact that ρ induces an inner product on $\mathcal{A}$ according to the formula $(A, B) = \rho(A^\star B)$.

(19) The state ρ is pure, in the sense defined in item (12), if and only if its GNS representation is irreducible.

(20) The state ρ is termed *primary*, or a factor, if the GNS representation π_ρ is primary. The resolution of algebras into primaries according to the central decomposition of item (10) carries a corresponding resolution of states into primaries, that is,

$$\rho = \int d\mu_\alpha \rho_\alpha.$$

(21) We define the modification, ρ_B, of an arbitrary state, ρ, due to the action of an element, B, of $\mathcal{A}$, by the formula

$$\rho_B(A) = \frac{\rho(B^\star AB)}{\rho(B^\star B)}.$$

Thus, it follows from item (18) that ρ_B is the state corresponding to the vector

$$\Phi_\rho^{(B)} = \frac{\pi_\rho(B)\Phi_\rho}{\|idem\|}.$$

(22) A set, $\mathcal{F}$, of states on a $C^\star$-algebra, $\mathcal{A}$, is termed a *folium* if

(i) it is closed with respect to the modifications $\rho \rightarrow \rho_B$, as defined in item (21);

(ii) it is convex; and

(iii) it is norm complete.

(23) In particular, the *normal folium*, $\mathcal{F}_\rho$, of a state ρ is defined to be the norm completion of the convex set of states of the form

$$\sum_n w_n \rho_{B_n}, \quad \text{with } \sum_n w_n = 1.$$

Thus, in view of the last remark of item (21), $\mathcal{F}_\rho$ is the set of states on $\mathcal{A}$ corresponding to the density matrices in the GNS space $\mathcal{H}_\rho$. Evidently, it corresponds to the set of normal states on the $W^\star$-algebra $\pi_\rho(\mathcal{A})''$. In particular, the state $\tilde{\rho}$ on this latter algebra, defined by

$$\tilde{\rho}(B) = (\Phi_\rho, B\Phi_\rho) \quad \forall B \in \pi(\mathcal{A})''$$

is termed the *canonical extension* of ρ to $\pi(\mathcal{A})''$.

2.4.3 Automorphisms and Antiautomorphisms

(24) An *automorphism* of a $C^\star$-algebra, $\mathcal{A}$, is an invertible, norm preserving, linear transformation, α, of $\mathcal{A}$, such that

$$\alpha(AB) = (\alpha A)(\alpha B) \quad \text{and} \quad (\alpha A)^\star = \alpha A^\star \quad \forall A, B \in \mathcal{A}.$$

We denote by $\text{Aut}(\mathcal{A})$ the set of all automorphisms of $\mathcal{A}$. Evidently, $\text{Aut}(\mathcal{A})$ is a group with respect to multiplication.

(25) An *antiautomorphism* of a $C^\star$-algebra, $\mathcal{A}$, is an invertible, norm preserving, linear transformation, β, of $\mathcal{A}$, such that

$$\beta(AB) = (\beta B)(\beta A) \quad \text{and} \quad (\beta A)^\star = \beta A^\star \quad \forall A, B \in \mathcal{A}.$$

We denote by $\text{Ant}(\mathcal{A})$ the set of all antiautomorphisms of $\mathcal{A}$.

Note. $\text{Ant}(\mathcal{A})$, unlike $\text{Aut}(\mathcal{A})$, is not a group, since it follows from the above definition that, if $\beta_1, \beta_2 \in \text{Ant}(\mathcal{A})$, then $\beta_1\beta_2 \in \text{Aut}(\mathcal{A})$.

Example of an Antiautomorphism Let $\mathcal{A}$ be the $C^\star$-algebra of 2×2 matrices. Thus, $\mathcal{A}$ is the linear span of the Pauli matrices, defined by Eq. (2.2.8), and the identity, and its algebraic structure is governed by the relations

$$\sigma_x^2 = I, \text{ etc.,} \quad \sigma_x\sigma_y = i\sigma_z, \text{ etc.}$$

We define β to be the linear transformation of $\mathcal{A}$ given by

$$\beta(\sigma_u) = -\sigma_u \quad \text{for } u = x, y, z, \ \beta(I) = I.$$

It follows immediately from these specifications that

$$\beta(A)\beta(B) = \beta(BA) \quad \text{and} \quad (\beta A)^\star = \beta A^\star \quad \forall A, B \in \mathcal{A},$$

and hence that β is an antiautomorphism of $\mathcal{A}$.

(26) An automorphism or antiautomorphism, α, of a $W^\star$-algebra, $\mathcal{A}$, is termed *normal* if it is ultraweakly continuous, that is, if its preserves weak convergence in the unit ball.

(27) If $\alpha \in \text{Aut}(\mathcal{A})$ or $\text{Ant}(\mathcal{A})$, then its *dual* is the transformation, $\alpha^\star$, of $S(\mathcal{A})$, defined by the formula

$$\langle \alpha^\wedge \rho; A \rangle = \langle \rho; \alpha A \rangle \quad \forall \rho \in S(\mathcal{A}), \ A \in \mathcal{A}.$$

Note In the case where $\mathcal{A}$ is a $W^\star$ algebra and α is normal, the set, $\mathcal{N}(\mathcal{A})$, of its normal states is stable under $\alpha^\star$.

(28) A mapping, α, of a group, G, into $\text{Aut}(\mathcal{A})$ is termed a *representation of G in* $\text{Aut}(\mathcal{A})$ if it preserves the algebraic structure of the group, that is, if $\alpha(g_1g_2) = \alpha(g_1)\alpha(g_2)$ for all $g_1, g_2 \in G$.

(29) An automorphism, α, of a $C^\star$-algebra, $\mathcal{A}$, is said to be *unitarily implementable* in the Hilbert space, $\mathcal{H}$, of a representation, π, of this algebra if there is a unitary transformation, U, of $\mathcal{H}$, such that $\pi(\alpha A) = U\pi(A)U^{-1}$.

Likewise, an antiautomorphism β of $\mathcal{A}$ is said to be *antiunitarily implementable* in $\mathcal{H}$ if there is an antiunitary transformation, V, of $\mathcal{H}$, such that $\pi(\beta A) = V\pi(A^\star)V^{-1}$.

In the case where $\mathcal{A} = \mathcal{B}(\mathcal{H})$ and $\mathcal{H}$ is finite dimensional, then all automorphisms (respectively antiautomorphisms) of $\mathcal{H}$ are unitarily (respectively antiunitarily) implemented in this space.

(30) If $\mathcal{H}$ is a Hilbert space, and α and π are representations of a group, G, and a $C^\star$-algebra, $\mathcal{A}$, in $\mathrm{Aut}(\mathcal{A})$ and $\mathcal{H}$, respectively, then α is said to be implemented by a unitary representation, U, of G in $\mathcal{H}$ if $\pi(\alpha(g)A)) = U(g)\pi(A)U(g)^{-1}$. In the particular case where $G = \mathbf{R}$, the additive reals, and the representation U of $\mathbf{R}$ is strongly continuous, it follows from Stone's theorem that $U(\mathbf{R})$ has an infinitesimal generator, $iH/\hbar$, where H is self-adjoint. Thus, $U(t) = \exp(iHt/\hbar)$, and so H is a Hamiltonian, in the standard sense of quantum mechanics (cf. Section 2.2). This means that, in the algebraic picture, *the concept of a Hamiltonian is derived from the more basic one of the automorphism group $\alpha(\mathbf{R})$*.

(31) If $\alpha \in \mathrm{Aut}(\mathcal{A})$ and ρ is an α-invariant state, that is, if $\rho(\alpha A) \equiv \rho(A)$, then α is implemented in the GNS space, $\mathcal{H}_\rho$, of ρ by the unitary operator, U_ρ, defined by the formula

$$U_\rho \pi_\rho(A)\Phi_\rho = \pi(\alpha A)\Phi_\rho$$

where π_ρ, Φ_ρ are as specified in item (18).

Similarly, if ρ is invariant with respect to a one-parameter group, $\alpha(\mathbf{R})$, of automorphisms of $\mathcal{A}$, then $\alpha(\mathbf{R})$ is implemented by a unitary representation, U_ρ, of $\mathbf{R}$, defined by the formula

$$U_\rho(t)\pi_\rho(A)\Phi_\rho = \pi_\rho(\alpha(t)A)\Phi_\rho.$$

This formula, together with the specifications of the GNS representation in item (18), implies that

$$\rho(B^\star A) = (\pi_\rho(B)\Phi_\rho, U_\rho(t)\pi_\rho(A)\Phi_\rho).$$

Hence, if $\rho(B^\star \alpha(t)A)$ is continuous in t, for all A, B in $\mathcal{A}$, then, in view of the cyclicity of Φ_ρ, U_ρ is continuous and therefore is generated by $i/\hbar$ times a Hamiltonian operator, H_ρ.

2.4.4 Tensor Products

Tensor products of $C^\star$ and $W^\star$ algebras are basic to the theory of coupled and open systems, and are defined as follows.

(32) Let $\mathcal{A}_1$ and $\mathcal{A}_2$ be $C^\star$-algebras with faithful representations, π_1 and π_2, in Hilbert spaces $\mathcal{H}_1$ and $\mathcal{H}_2$, respectively. Then the $C^\star$ tensor product, $\mathcal{A}_1 \otimes \mathcal{A}_2$, of these algebras is defined to be the $C^\star$- algebra, whose faithful representation in $\mathcal{H}_1 \otimes \mathcal{H}_2$ is the norm completion of the linear span of the operators $\{\pi_1(A_1) \otimes \pi_2(A_2) \mid A_1 \in \mathcal{A}_1, A_2 \in \mathcal{A}_2\}$.

(33) If $\mathcal{A}_1$ and $\mathcal{A}_2$ are $W^\star$-algebras of operators in Hilbert spaces $\mathcal{H}_1$ and $\mathcal{H}_2$, respectively, then the $W^\star$ tensor product $\mathcal{A}_1 \otimes \mathcal{A}_2$ is defined to be the weak closure of the linear span of the operators $\{A_1 \otimes A_2 \mid A_1 \in \mathcal{A}_1, A_2 \in \mathcal{A}_2\}$.

2.4.5 Quantum Dynamical Systems

(34) A quantum dynamical system, Σ, is a triple $(\mathcal{A}, S, \alpha)$, where $\mathcal{A}$ is a $C^\star$-algebra, $\alpha(\mathbf{R})$ is a one-parameter group of automorphisms of $\mathcal{A}$, and S is a folium of states on $\mathcal{A}$ that is stable under $\alpha^\star(\mathbf{R})$. In particular, Σ is termed a $C^\star$-*dynamical system* if S is the set of all states on $\mathcal{A}$ and $\alpha(t)A$ is norm continuous in t: it is termed a $W^\star$-*dynamical system* if $\mathcal{A}$ is a $W^\star$-algebra, S is the set of normal states on $\mathcal{A}$ and the automorphisms $\alpha(\mathbf{R})$ are normal, with $\alpha(t)A$ weakly continuous in t. In fact, the quantum dynamical systems that we consider are all either $C^\star$ or $W^\star$.

Note

(i) The model of Section 2.2.3 is evidently a $W^\star$-dynamical system.

(ii) If ρ is an $\alpha(\mathbf{R})$-invariant state of a $C^\star$-system $\Sigma = (\mathcal{A}, S, \alpha)$, then it follows from the specifications of items (23) and (30) that the restriction of Σ to the normal folium of ρ is a $W^\star$-system, $(\mathcal{A}_\rho, S_\rho, \alpha_\rho)$, where $\mathcal{A}_\rho = \pi_\rho(\mathcal{A})''$, S_ρ is the normal folium of states on $\mathcal{A}_\rho$ and $\alpha_\rho(\mathbf{R})$ is the one-parameter group of automorphisms of this algebra defined by the formula[9]

$$\alpha_\rho(t)B = U_\rho(t)BU_\rho(t)^{-1} \quad \forall B \in \pi_\rho(\mathcal{A})''.$$

(35) A *classical* $C^\star$ or $W^\star$-dynamical system, Σ_{cl}, is one for which the algebra of observables, $\mathcal{A}$ is abelian. This is equivalent to the standard definition of a classical system in terms of a one-parameter group, $T(\mathbf{R})$, of transformations of a phase space, Ω, since, by the Gelfand isomorphism [Dix1,2], any abelian $C^\star$ (respectively $W^\star$) algebra is isomorphic with the algebra of bounded continuous (respectively essentially bounded[10]) functions on the space, Ω, of pure states on $\mathcal{A}$, equipped with the $w^\star$ topology. This isomorphism maps the observables, A, and the states, ρ, into functions $\hat{A}$, and probability measures $\hat{\rho}$, respectively,

---

[9] In fact, by item (28), this formula specifies $\alpha_\rho(t)$ as a group of automorphisms of $\pi(\mathcal{A})$, and this may be extended by continuity to the weak closure of this algebra, since weak convergence is preserved by unitary transformations.

[10] An essentially bounded function is one that is uniformly bounded, almost everywhere with repect to a certain probability measure, in this case determined by the $W^\star$-algebra $\mathcal{A}$.

on Ω, according to the formulae

$$\hat{A}(\omega) = \omega(A), \quad \int_{\Omega} \hat{A} d\hat{\rho} = \rho(A),$$

the pure states being given by the Dirac measures. Thus, defining the transformation, $T(t)$, of Ω to be the restriction of $\alpha^{\star}(t)$ to the pure states, these formulae serve to define the phase space description in terms of the algebraic one. The converse, which follows from the above specifications, is that the time-dependent expectation values of the observables of Σ_{cl} are given by the flow $T(\mathbf{R})$ over Ω, according to the formula

$$\langle \rho; \alpha(t)A \rangle = \int_{\Omega} \hat{A}(T(t)\omega) \, d\hat{\rho}(\omega).$$

2.4.6 Derivations of *-Algebras and Generators of Dynamical Groups [BR]

Stone's theorem for the generators of one-parameter unitary groups of transformations of a Hilbert space has its counterpart, which we now describe, in the theory of generators of one-parameter groups of automorphisms of $C^{\star}$ and $W^{\star}$ algebras.

(36) A derivation of a $C^{\star}$ (respectively $W^{\star}$) algebra, $\mathcal{A}$, is a linear mapping, δ, of a norm dense (respectively weakly dense) *-subalgebra, $\mathcal{D}(\delta)$, of $\mathcal{A}$ into $\mathcal{A}$, such that

(i) $\delta(A^{\star}) = \delta(A)^{\star}$ for all $A \in \mathcal{D}(\delta)$; and

(ii) δ satisfies the Leibnitz condition that

$$\delta(AB) = A\delta B + \delta(A)B \quad \forall A, B \in \mathcal{D}(\delta).$$

Thus, for example, if $\mathcal{A}$ is a *-algebra of operators in a Hilbert space, $\mathcal{H}$, then any transformation of $\mathcal{A}$ of the form $i[K, \cdot]^{11}$, with K self-adjoint, is a derivation of this algebra.

(37) A one-parameter group, $\alpha(\mathbf{R})$, of automorphisms of a $C^{\star}$ (respectively $W^{\star}$) algebra is termed continuous if $\alpha(t)A$ converges normwise (respectively weakly) to A, as $t \to 0$, for all $A \in \mathcal{A}$. In this case, the group has an infinitesimal generator, δ, which is a derivation of $\mathcal{A}$ and is defined by the formula

$$\frac{d}{dt} \alpha(t)A = \alpha(t)\delta A = \delta\alpha(t)A \quad \forall A \in \mathcal{D}(\delta), t \in \mathbf{R}$$

the left-hand side being the norm or weak limit of $h^{-1}(\alpha(t+h)A-$

---

[11]   The dot in the square brackets represents the argument of the transformation, which thus sends A to $i[K, A]$.

$\alpha(t)A)$, as $h \to 0$, according to whether $\alpha(t)$ is normwise or weakly continuous.

(38) Conversely, and subject to certain technical conditions,[12] a derivation of a $C^\star$ or $W^\star$ algebra is the infinitesimal generator of a one-parameter group $\alpha(\mathbf{R})$ of its automorphisms.

2.5 ALGEBRAIC FORMULATION OF INFINITE SYSTEMS

We come now to the general operator algebraic formulation of an infinitely extended system of particles, Σ, occupying a space X, which we take to be either a Euclidean continuum or a simple cubic lattice. Thus, in either case, X is a group with respect to addition. The model of Σ is constructed as a quantum dynamical system according to the following strategy. The observables of Σ in each bounded spatial region, Λ, are assumed to be just those of a finite system, $\Sigma(\Lambda)$, of particles of the given species occupying Λ. The states of Σ are taken to be the expectation functionals on these observables that reduce in each such Λ to states of the finite system $\Sigma(\Lambda)$, according to the prescription of Section 2.2. The dynamics of Σ is assumed to be given by a "natural" limiting form of that of $\Sigma(\Lambda)$ as Λ increases to the full space X. The nature of this limit, and correspondingly the question of whether the resultant dynamical system is $C^\star$ or $W^\star$, depends on the interactions in Σ.

2.5.1 The General Scheme

We now describe the general structure of the above scheme, leaving the construction of its explicit form for lattice and continuous systems to Sections 2.5.2 and 2.5.3, respectively.

We define L to be the set of bounded open regions, Λ, of the space X. For each such region, Λ, we take as given the model of the finite quantum system, $\Sigma(\Lambda)$, consisting of particles of the given species occupying Λ. Thus, in view of our observation at the end of Section 2.4, we assume that this is a $W^\star$-dynamical system, $\Sigma_\Lambda = (\mathcal{A}_\Lambda, S_\Lambda, \alpha_\Lambda)$, where $\mathcal{A}_\Lambda$ is its algebra of observables, which operate in a Hilbert space $\mathcal{H}_\Lambda$, S_Λ is the set of its (normal) states, and $\alpha_\Lambda(\mathbf{R})$ is the one-parameter groups of automorphisms of $\mathcal{A}_\Lambda$, representing its dynamics.

We assume that the local algebras $\mathcal{A}_\Lambda$ possess the following two natural properties.

(i) $\mathcal{A}_\Lambda \subset \mathcal{A}_{\Lambda'}$ for $\Lambda \subset \Lambda'$. This is the property of *isotony*.

(ii) The elements of $\mathcal{A}_\Lambda$ commute with those of $\mathcal{A}_{\Lambda'}$ if Λ and Λ' are

---

[12] These conditions are specified in [BR], Vol. 1, Section 3.2.4.

disjoint. This is the property of *local commutativity*, which signifies that observables located in mutually disjoint regions are compatible.

In view of the isotony of $\mathcal{A}_\Lambda$ with respect to Λ, the union, $\mathcal{A}_L = \bigcup_{\Lambda \in L} \mathcal{A}_\Lambda$, is well defined. Furthermore, it is a normed *-algebra, with norm inherited in a consistent way from the $\mathcal{A}_\Lambda$ and thus satisfying the $C^\star$ condition $\|A^\star A\| \equiv \|A\|^2$. We term $\mathcal{A}_L$ *the algebra of bounded local observables* of Σ. We define $\mathcal{A}$ to be its norm completion, and term $\mathcal{A}$ the $C^\star$-*algebra of quasi-local bounded observables* of the system.

We assume that there is a representation, ξ, of the additive group X in Aut($\mathcal{A}$), corresponding to space translations, such that the action of the automorphisms $\xi(x)$ on the local algebras, $\mathcal{A}_\Lambda$, satisfies the following covariance conditions.

(C.1) $\xi(x)\mathcal{A}_\Lambda = \mathcal{A}_{\Lambda+x}$, where $\Lambda + x$ is the space translate of Λ by x.

(C.2) $\xi(x)\alpha_\Lambda(t)\xi(-x) = \alpha_{\Lambda+x}(t)$, which signifies that the interactions are translationally invariant.

We demand that the physical states are positive linear normalised functionals, ρ, on $\mathcal{A}$, that reduce to normal functionals on the local algebras $\mathcal{A}_\Lambda$, as required for finite systems.[13] Thus, defining L to be the set of states, ρ, on $\mathcal{A}$, whose restrictions, ρ_Λ, to the algebras $\mathcal{A}_\Lambda$ are normal, we stipulate that the physical states of the system comprise L or possibly a subset thereof: this last qualification is designed to accommodate the possibility, discussed below, that some elements of L might not support a dynamics corresponding to a natural limit of that for $\Sigma(\Lambda)$. We term L the set of *locally normal* states on $\mathcal{A}$.

Note

(i) If ρ is locally normal, then so too are the states of its normal folium [Se3].

(ii) The GNS representation spaces of the locally normal states are separable [Ru2]. This stems essentially from the fact that the GNS spaces of the normal states on the local algebras, $\mathcal{A}_\Lambda$, which serve to generate $\mathcal{A}$, are separable.

(iii) In the case of lattice systems, such as the Heisenberg ferromagnet, for which the local algebras $\mathcal{A}_\Lambda$ are finite dimensional, all states on $\mathcal{A}$ are locally normal.

We also term a representation, π, of $\mathcal{A}$ locally normal if its restrictions to the local algebras $\mathcal{A}_\Lambda$ are all normal. Thus, in view of the correspondence between states and representations, the local normality of π implies that of all

---

[13]   In fact, as we shall see in Section 2.5.3, this local normality condition is essential, since its violation would imply a catastrophic collapse of an infinity of particles in some bounded region.

states corresponding to density matrices in its Hilbert space; and conversely, the local normality of a state, ρ, implies that of its GNS representation.

The dynamics of Σ is assumed to correspond to a limiting form of that given by the automorphisms $\alpha_\Lambda(\mathbf{R})$ of the local algebras $\mathcal{A}_\Lambda$. Here, we have the following two possibilities.

(a) For each $t \in \mathbf{R}$ and $A \in \mathcal{A}_L$, $\alpha_\Lambda(t)A$ converges in norm to some limit $\alpha(t)A$. In this case, which is realised by lattice systems with finite dimensional local algebras and short range interactions [St, Ro1], $\{\alpha(t) \mid t \in \mathbf{R}\}$ extends by continuity[14] to a one-parameter group of automorphisms of $\mathcal{A}$.

(b) $\alpha_\Lambda(t)A$ does not converge in norm as Λ increases to X, but for certain privileged locally normal representations, π, of $\mathcal{A}$, $\pi(\alpha_\Lambda(t)A)$ is strongly convergent. This situation prevails under rather general conditions for continuous systems [Se4], and implies that the limit dynamics in the representation π corresponds to a one-parameter group, $\alpha_\pi(\mathbf{R})$, of automorphisms of $\pi(\mathcal{A})''$. Thus, denoting by Π the maximal locally normal representation for which the strong convergence condition is fulfilled and defining $\mathcal{A}_\Pi$ to be the $W^\star$-algebra $\Pi(\mathcal{A})''$, we see that the dynamics corresponds to the representation α_Π of $\mathbf{R}$ in $\mathrm{Aut}(\mathcal{A}_\Pi)$.

To summarise, the model of Σ is a $C^\star$ or $W^\star$ dynamical system according to whether $\alpha(t)A$ converges in norm, or merely strongly in dynamically determined representations, as Λ increases to X. In the former case, where (a) is valid, the model is given by $(\mathcal{A}, S, \alpha)$, where S is the set of all states on $\mathcal{A}$. In the latter case, where (b) holds, it is $(\mathcal{A}_\Pi, S_\Pi, \alpha_\Pi)$, where S_Π is the set of normal states, ρ_Π, on $\mathcal{A}_\Pi$. These correspond to locally normal states, ρ, on $\mathcal{A}$ according to the formula $\rho(A) = \rho_\Pi(\Pi(A))$.

It is a straightforward matter to extend the $C^\star$ and $W^\star$ models obtained in this way so as to include space translations. In the $C^\star$ case (a), the extended model for Σ is $(\mathcal{A}, S, \alpha, \xi)$; and, by the covariance conditions (C.1) and (C.2), $\alpha(\mathbf{R})$ and $\xi(X)$ intercommute.

In the $W^\star$ case (b), we define the automorphisms $\xi_\Pi(x)$ of $\Pi(\mathcal{A})$ by the formula $\xi_\Pi(x)\Pi(A) = \Pi(\xi(x)A)$, and employ the covariance conditions (C.1) and (C.2) to extend this definition of $\xi_\Pi(x)$, by strong continuity, to $\mathcal{A}_\Pi$. Thus, the extended $W^\star$ model of Σ is $(\mathcal{A}_\Pi, S_\Pi, \alpha_\Pi, \xi_\Pi)$, and by (C.1) and (C.2), the automorphisms $\alpha_\Pi(\mathbf{R})$ and $\xi_\Pi(X)$ intercommute.

Note   Even in the $C^\star$ case, one may formulate the model as a $W^\star$-system of the above form, with Π the maximal, or universal [Dix2], representation of $\mathcal{A}$. Hence the $W^\star$ model subsumes the $C^\star$ one.

---

[14]   Thus, if $\{A_n\}$ is a sequence in $\mathcal{A}_L$ that converges in norm to $A(\in\mathcal{A})$, then $\alpha(t)A = \mathrm{norm} : \lim_{n\to\infty} \alpha(t)A_n$.

Note on the KMS Equilibrium Conditions As we saw in Section 2.2, the equilibrium states of finite systems are those that satisfy the KMS conditions, represented formally by Eq. (2.2.25), that is,

$$\rho(A(t)B) = \rho(BA(t + i\hbar\beta)) \; \forall A, B \in \mathcal{A},$$

where $A(t) = \alpha(t)A$. These conditions are also well defined for infinite systems, whereas the Gibbs density matrix $\exp(-\beta H)/\text{Tr}(idem)$ is not, for the following two reasons. Firstly, the normalisation factor, $\text{Tr}(\exp(-\beta H))$, is equal to $\exp(-\beta F)$, where F is the free energy, and so would be infinite for an infinite system; and, secondly, neither the Hamiltonian nor its representation space is given *a priori* for such a system.

The KMS conditions are of great significance for both physics and mathematics. Their key physical import is that

(a) they imply local normality [TW], and

(b) as we shall see in Chapter 5, they are equivalent to dynamical and thermodynamical stability conditions that characterise thermal equilibrium states of infinite, as well as finite, systems. For these reasons, they are generally accepted as the conditions for thermal equilibrium. On the mathematical level, they are at the centre of a major development in the theory of operator algebras [Ta].

Note on Unbounded Local Observables (cf. [Se3]) The model of an infinite system may be extended so as to incorporate its unbounded observables in the following way. We define the unbounded observables for the region $\Lambda(\in L)$ to be the unbounded self-adjoint operators, Q_Λ, in $\mathcal{H}_\Lambda$ whose spectral projectors belong to $\mathcal{A}_\Lambda$. This definition then permits canonical extensions of both the GNS representation and the KMS conditions for locally normal states, ρ, to those unbounded local observables, Q_Λ, for which $\text{Tr}(Q_\Lambda^\star \hat{\rho}_\Lambda Q_\Lambda)$ is finite. This extension also carries through for all closed, densely defined, unbounded operators in $\mathcal{H}_\Lambda$ that satisfy this condition and commute with the commutant of $\mathcal{A}_\Lambda$ in $\mathcal{B}(\mathcal{H}_\Lambda)$.

2.5.2 Construction of the Lattice Model (cf. [St, Ro1])

We assume here that Σ is a system of identical atoms, or spins, located on the sites of the d-dimensional lattice $\mathbf{Z}^d$. To build the model of this system, we start with that of a single atom of the species, when isolated. Thus, we assume that the algebra of observables, $\mathcal{A}_{at}$, of the atom consists of the operators[15] in a finite-dimensional Hilbert space, $\mathcal{H}_{at}$. We then assign to each site, x, of X copies $\mathcal{H}_x$ and $\mathcal{A}_x$ of $\mathcal{H}_{at}$ and $\mathcal{A}_{at}$, respectively. $\mathcal{A}_x(= \mathcal{B}(\mathcal{H}_x))$ is designated as the algebra of observables of the atom at x.

---

[15] The finite dimensionality of $\mathcal{H}_{at}$ ensures that all these operators are bounded.

To pass from the individual atoms to the whole system, Σ, we first note that the bounded open regions, Λ, of the lattice X are simply the finite point subsets thereof. We define the Hilbert space, $\mathcal{H}_\Lambda$, and the algebra of observables, $\mathcal{A}_\Lambda$, for the region $\Lambda(\in L)$ to be $\otimes_{x\in\Lambda} \mathcal{H}_x$ and $\otimes_{x\in\Lambda} \mathcal{A}_x$, respectively. Thus, $\mathcal{A}_\Lambda = \mathcal{B}(\mathcal{H}_\Lambda)$. It follows immediately from these definitions that if $\Lambda \subset \Lambda'$ then $\mathcal{A}_{\Lambda'} = \mathcal{A}_\Lambda \otimes \mathcal{A}_{\Lambda'\backslash\Lambda}$. Accordingly, we identify each element A of $\mathcal{A}_\Lambda$ with $A_\Lambda \otimes I_{\Lambda'\backslash\Lambda}$. Under this identification, the local $W^\star$- algebras $\{\mathcal{A}_\Lambda\}$ become endowed with the properties of isotony and local commutativity, specified in Section 2.5.1. We define $\mathcal{A}_L$ to be $\bigcup_{\Lambda\in L}\mathcal{A}_\Lambda$ and $\mathcal{A}$ to be the norm completion of $\mathcal{A}_L$. Thus, $\mathcal{A}$ is the norm completion of the algebra of polynomials in $\{a_x \in \mathcal{A}_x \mid x \in X\}$, where a_x is the copy in $\mathcal{A}_x$ of an element a of $\mathcal{A}_{at}$.

We define the representation ξ of the additive group X in $\text{Aut}(\mathcal{A})$, corresponding to space translations, by the formula

$$\xi(x)a_y = a_{x+y} \quad \forall x, y \in X. \tag{2.5.1}$$

To formulate the dynamics of the system, we start by introducing the Hamiltonian operator, H_Λ, to be the self-adjoint element[16] of $\mathcal{A}_\Lambda$ that represents the sum of the kinetic and potential energies of the particles in Λ, due to their mutual interactions. Thus, the local dynamical automorphisms, $\alpha_\Lambda(t)$, that it engenders are given by

$$\alpha_\Lambda(t)A = \exp(iH_\Lambda t/\hbar)A \exp(-iH_\Lambda t/\hbar) \quad \forall A \in \mathcal{A}_\Lambda. \tag{2.5.2}$$

We assume that the interactions of the system are translationally invariant, that is, that $\xi(x)H_\Lambda = H_{\Lambda+x}$, which ensures that the covariance conditions (C.1) and (C.2) are fulfilled.

Now it has been proved to follow from these specifications that the norm convergence condition (a) of Section 2.5.1 is fulfilled if the interactions are of finite range [St] or even if they fall off sufficiently fast at large distances [Ro1]. Thus, for a lattice system with short range, translationally invariant interactions, the dynamics is given by a one-parameter semigroup, $\alpha(\mathbf{R})$, of automorphisms of $\mathcal{A}$, defined by the formula

$$\alpha(t)A = \text{norm} : \lim_{\Lambda\uparrow X} \alpha_\Lambda(t)A \quad \forall A \in \mathcal{A}_L, \, t \in \mathbf{R}. \tag{2.5.3}$$

These automorphisms commute with the space translational ones, $\xi(X)$, in view of the covariance properties (C.1) and (C.2).

We take the states of the system to comprise the full set, $S(\mathcal{A})$, of positive linear normalised functionals on $\mathcal{A}$, since the normality of their restrictions to the local algebras $\mathcal{A}_\Lambda$ is guaranteed by the finite dimensionality of these algebras.

---

[16] Note that H_Λ, like all operators in $\mathcal{H}_\Lambda$, must be bounded, because of the finite dimensionality of $\mathcal{H}_\Lambda$.

Thus, the model of Σ is given by the $C^\star$-system $(\mathcal{A}, S, \alpha, \xi)$, where $S = S(\mathcal{A})$.

2.5.3 Construction of the Continuum Model (cf. [HHW, DS, Se4])

We assume here that Σ is a system of particles of one species of bosons or fermions in a d-dimensional Euclidean space, $X = \mathbf{R}^d$. We formulate the model of Σ in second quantisation, representing the system of particles by a quantum field, ψ. In the case where the particles are bosons, ψ is a scalar field operator, quantised according to the canonical commutation relations (CCR),

$$[\psi(x), \psi^\star(x')] = \delta(x - x'), \quad [\psi(x), \psi(x')] = 0. \qquad (2.5.4a)$$

In the case where the particles are fermions, ψ is a two-component spinor field operator, $\left(\begin{smallmatrix} \psi_1 \\ \psi_{-1} \end{smallmatrix} \right) \equiv \left(\begin{smallmatrix} \psi_\uparrow \\ \psi_\downarrow \end{smallmatrix} \right)$, quantised according to the canonical anti-commutation relations (CAR),

$$[\psi_s(x), \psi_{s'}^\star(x')]_+ = \delta(x - x')\delta_{ss'}, \quad [\psi_s(x), \psi_{s'}(x')]_+ = 0 \qquad (2.5.4b)$$

where $[A, B]_+ = AB + BA$. Evidently, these equations for the CCR and CAR are merely formal, not only because no Hilbert space has been specified, but also because $\delta(x - x')$ is infinite when the points x, x' coincide and so cannot be an operator in any such space. Thus, in order to provide a precise mathematical definition of the quantum field ψ, we need to specify the Hilbert space in which it acts and tame the δ-function singularity by integrating ψ against suitable test functions, f. In other words, we need to formulate the field in terms of operators $\psi(f) = \int dx \psi(x) f(x)$ in a suitable Hilbert space, $\mathcal{H}$. In fact, $\psi(f)$, rather than $\psi(x)$, is the basic object of the description, the latter being a merely formal quantity that may heuristically be considered to generate the operators $\psi(f)$ by integration against test functions.

Accordingly, we define the field ψ and the Hilbert space, $\mathcal{H}$, in which it acts, by the following specifications.

(1) ψ, in the case of bosons, and ψ_s, in the case of fermions, is a map from $L^2(X)$ into the closed, densely defined operators in $\mathcal{H}$.

(2) In the bosonic case, ψ satisfies the CCR,

$$[\psi(f), \psi(g)^\star] = (g, f)_{L^2(X)}, \quad [\psi(f), \psi(g)] = 0. \qquad (2.5.5a)$$

In the fermionic case, ψ satisfies the CAR,

$$[\psi_s(f), \psi_{s'}(g)^\star]_+ = (g, f)_{L^2(X)}\delta_{ss'}, \quad [\psi_s(f), \psi_{s'}(g)]_+ = 0. \qquad (2.5.5b)$$

(3) $\mathcal{H}$ contains a vector, Φ, such that

   (a) $\psi(f)\Phi = 0$ for all $f \in L^2(X)$; and

(b) the set of vectors obtained by the action on Φ of the polynomials in $\{\psi(f) \mid f \in L^2(X)\}$ (for bosons) or $\{\psi_s(f) \mid f \in L^2(X), s = \pm 1\}$ (for fermions) is dense in $\mathcal{H}$. Thus $\mathcal{H}$ is generated by the action of these polynomials on Φ.

Note (cf. [Ru1])

(i) $\mathcal{H}$ is termed the *Fock space*, and is the direct sum, $\oplus_{n=0}^{\infty}\mathcal{H}_n$, of the irreducible representation spaces for systems of n particles of the given species in the space X.

(ii) $\psi(f)^{\star}$ and $\psi(f)$ are creation and annihilation operators, in that their actions lead from $\mathcal{H}_n$ to $\mathcal{H}_{n+1}$ and $\mathcal{H}_{n-1}$, respectively.

(iii) Φ is the vacuum vector: in fact, $\mathcal{H}_0$, the particle-free subspace of $\mathcal{H}$, consists of the scalar multiples of Φ.

(iv) In the case of fermions, the operators $\psi_s(f)$ are bounded. In the case of bosons, the operators $\psi(f)$ are unbounded, and so it is convenient to express these in terms of the Weyl operators

$$W(f) = \exp(i(\psi(f) + \psi(f)^{\star})), \qquad (2.5.6)$$

noting that the CCR may be expressed in the form

$$W(f)W(g) = W(f + g)\exp\left(\frac{i}{2}\mathrm{Im}(f, g)\right). \qquad (2.5.7)$$

This implies that the polynomials in the $W(f)$ reduce to linear combinations of these operators.

(v) The correspondence between the first and second quantisation schemes may be formally expressed as follows. If A is an operator in $\mathcal{H}_1$, corresponding to an observable for a single particle of the species, then the n-particle projection of $\int dx \psi^{\star} A \psi(x)$ is equal to the canonical sum, $\sum_{j=1}^{n} A_j$, of copies, A_j, of A for the n particles.

(vi) The observables of the n-particle system are generated by functions of such permutation-invariant sums, and so are invariant under the global[17] gauge transformation $\psi \to \psi \exp(i\theta)$, for arbitrary constant θ. Thus, the observables of $\sum$ must always be *gauge invariant* operators in $\mathcal{H}$.

Note   It is sometimes both harmless and notationally economical to formulate properties of the continuum model in terms of the formal object $\psi(x)$, rather than the bona fide operator $\psi(f)$; and on some such occasions, we take the liberty of doing so.

---

[17] The local gauge transformations are of similar form, but with θ position-dependent.

Local Structure

As previously, we define L to be the family of bounded open regions, Λ, of X. For each Λ in L, we define $\mathcal{H}_\Lambda$ to be the subspace of $\mathcal{H}$ generated by the action on Φ of the polynomials in $\{\psi(f)^\star \text{ (or } \psi_s(f)^\star) \mid f \in L^2(\Lambda), s = \pm 1\}$. Thus, $\mathcal{H}_\Lambda$ is the Fock space for a system of particles of the species confined to Λ. We define $\mathcal{A}_\Lambda^{(F)}$ to be the $W^\star$-algebra, $\mathcal{B}(\mathcal{H}_\Lambda)$, of all bounded operators in $\mathcal{H}_\Lambda$. This is the $W^\star$-algebra of operators in $\mathcal{H}_\Lambda$ generated, in the bosonic case, by the Weyl operators, $W(f)$, and, in the fermionic case, by the field operators $\psi_s(f), \psi_s(f)^\star$, with $f \in L^2(\Lambda)$[18]. Thus, $\mathcal{A}_\Lambda^{(F)}$ may be canonically identified with the weak closure of the *-algebra of operators in $\mathcal{H}$ generated by the Weyl operators, $W(f)$, or the field operators $\psi_s(f)$, for which f vanishes outside Λ. Under this identification, $\mathcal{A}_\Lambda^{(F)}$ is isotonic with respect to Λ.

Observable and Field Algebras

We now define $\mathcal{A}_L^{(F)}$ to be $\bigcup_{\Lambda \in L} \mathcal{A}_\Lambda^{(F)}$ and $\mathcal{A}^{(F)}$ to be its completion with respect to the $\mathcal{H}$-operator norm. Thus, $\mathcal{A}^{(F)}$ is a $C^\star$-algebra, which we term the *field algebra*. It is equipped with groups of automorphisms $\gamma(\mathbf{R})$ and $\xi(X)$, representing gauge transformations and space translations, respectively, as defined by the formulae

$$\gamma(\lambda)\psi(f) = \psi(f)\exp(i\lambda) \tag{2.5.8}$$

and

$$\xi(x)\psi(f) = \psi(f_x) \quad \text{with } f_x(y) = f(y - x). \tag{2.5.9}$$

Note that this last equation may be written formally as $\xi(x)\psi(y) = \psi(x + y)$.

It follows immediately from these definitions that $\mathcal{A}^{(F)}$ contains elements, such as $W(f)$ or $\psi_s(f)$, that are not gauge invariant. Consequently, we do *not* designate it to be the algebra of observables of the system, since, as noted above, observables should be gauge invariant. We therefore define the *algebras of observables* for the region Λ and for the whole system to be the gauge invariant subalgebras, $\mathcal{A}_\Lambda$ and $\mathcal{A}$, of $\mathcal{A}_\Lambda^{(F)}$ and $\mathcal{A}^{(F)}$, respectively. Thus, $\mathcal{A}_\Lambda$ (respectively $\mathcal{A}$) = $\{A \in \mathcal{A}_\Lambda^{(F)} \text{ (respectively } \mathcal{A}^{(F)}) \mid \gamma(\lambda)A \equiv A\}$, from which it follows that $\mathcal{A}_\Lambda$ is isotonic with respect to Λ and that $\mathcal{A}$ is the norm completion of their union, $\mathcal{A}_L$. Furthermore, the subalgebras $\mathcal{A}_\Lambda$ satisfy the requirement of local commutativity, *even in the fermionic case*. There, $\mathcal{A}_\Lambda$ is generated by the operators $\{\psi_s(f)^\star \psi_{s'}(f') \mid f, f' \in L^2(\Lambda); s, s' = \pm 1)\}$ and the identity, and it follows from the CAR that these commute with similar

---

[18] This may be seen from the fact that the commutant of the $W^\star$-algebra so generated must be that of the scalar multiples of the identity, since these are the only operators in $\mathcal{H}_\Lambda$ that commute with the field operators $\{\psi(f), \psi(f)^\star \mid f \in L^2(\Lambda)\}$.

operators for a disjoint region Λ'. By contrast, the fermionic field algebras $\mathcal{A}_\Lambda^{(F)}$ do not satisfy local commutativity, since $\psi_s(f)$ and $\psi_s(g)$ anticommute, but do not commute, when the supports of f and g are disjoint. Thus, the restriction of the observable algebra to the gauge invariant elements of $\mathcal{A}^{(F)}$ brings a bonus of local commutativity.

It follows from the above definitions that both the observable and field algebras are covariant with respect to space translations, that is,

$$\xi(x)\mathcal{A}_\Lambda = \mathcal{A}_{\Lambda+x} \quad \text{and} \quad \xi(x)\mathcal{A}_\Lambda^{(F)} = \mathcal{A}_{\Lambda+x}^{(F)}. \qquad (2.5.10)$$

Note Any state ρ on $\mathcal{A}$ may be canonically identified with a gauge invariant state $\rho^{(F)}$ on $\mathcal{A}^{(F)}$ according to the following prescription. For each $B^{(F)} \in \mathcal{A}^{(F)}$, we define

$$B = \frac{1}{2\pi} \int_0^{2\pi} d\lambda\, \gamma(\lambda) B^{(F)},$$

which is the average of $B^{(F)}$ over gauge transformations, and is manifestly gauge invariant. Correspondingly, for each state ρ on $\mathcal{A}$, we define $\rho^{(F)}$, the canonical gauge invariant extension of ρ to $\mathcal{A}^{(F)}$ by the formula

$$\rho^{(F)}(B) = \rho(B^{(F)}).$$

Locally Normal States

An important theorem by Dell'Antonio, Doplicher and Ruelle [DDR] has established that the locally normal states of a system of bosons or fermions in a continuum are precisely those for which the probability of finding an infinite number of particles in a bounded region is zero. In other words, *local normality is equivalent to local finiteness*. Hence, the demand that the physical states be locally normal serves to exclude catastrophies in which an infinite number of particles collapse into a finite region.

Physical States and Dynamics

We assume that the Hamiltonian of the finite version, Σ_Λ, of Σ is a self-adjoint operator, H_Λ, in $\mathcal{H}_\Lambda$, defined in the usual way as the sum of the kinetic and potential energies of this system. Thus, if the particles interact via translationally invariant two-body forces, this Hamiltonian is of the form

$$H_\Lambda =$$

$$\frac{\hbar^2}{2m} \int_\Lambda dx \nabla \psi^\star(x).\nabla \psi(x) + \frac{1}{2} \int_\Lambda dx \int_\Lambda dx' \psi^\star(x)\psi^\star(x')V(x-x')\psi(x')\psi(x),$$

$$(2.5.11)$$

where, in the case of fermions, ψ is a spinor. The dynamics of Σ_Λ is given by the one-parameter group, $\alpha_\Lambda(\mathbf{R})$, of automorphisms of $\mathcal{A}_\Lambda$, defined by the formula

$$\alpha_\Lambda(t)A = \exp(iH_\Lambda t/\hbar)A \exp(-iH_\Lambda t/\hbar). \qquad (2.5.12)$$

The question now arises as to whether the dynamics of the infinite system Σ can be constructed as a limit, as Λ increases to X, of that governed by local automorphisms $\alpha_\Lambda(\mathbf{R})$. Here, there arises a serious problem, which stems from the fact that there is no upper limit to the speeds at which nonrelativistic particles can move through a continuum. This implies that Σ cannot support a dynamics corresponding to automorphisms of $\mathcal{A}$ according to scheme (a) of Section 2.5.1, for the following two reasons.

(1) Such a putative dynamics would delocalise initially local observables and thus remove them from $\mathcal{A}$ [DS].

(2) It would lead to catastrophies wherein an infinite number of particles travel from infinity to a bounded region in a finite time, thus destroying the local normality of the states and rendering the resultant forces on the particles indeterminate [Ra1].

In view of this situation, we are left with scheme (b) of Section 2.5.1, in which the dynamics corresponds to a one-parameter group, $\alpha_\Pi(\mathbf{R})$, of automorphisms of $\Pi(\mathcal{A})''$, where Π is the maximal locally normal representation in which the $\alpha_\Lambda(t)A$ converges strongly as Λ increases to X (cf. [Se4]). Thus,

$$s : \lim_{\Lambda \uparrow X} \Pi(\alpha_\Lambda(t)A) = \alpha_\Pi(t)\Pi(A) \ \forall A \in \mathcal{A}_L, t \in \mathbf{R}. \qquad (2.5.13)$$

The $W^\star$ Model

We conclude from the above that Σ is the $W^\star$-dynamical system $(\mathcal{A}_\Pi, S_\Pi, \alpha_\Pi, \xi_\Pi)$, where $\mathcal{A}_\Pi = \Pi(\mathcal{A})''$, S_Π is the set of normal states on this algebra, and ξ_Π is the representation of the space translation group, X, in $\mathrm{Aut}(\mathcal{A}_\Pi)$ defined by

$$\xi_\Pi(x)\Pi(A) = \Pi(\xi(x)A). \qquad (2.5.14)$$

We note that the automorphisms $\alpha_\Pi(\mathbf{R})$ and $\xi_\Pi(X)$ intercommute if the interactions of the system are translationally invariant.

2.6 THE PHYSICAL PICTURE

In this section, we discuss the physical relevance of some of the structures in the generic model of an infinite system, as formulated in Section 2.5. For simplicity, we confine our explicit considerations here to the $C^\star$-model; the $W^\star$-model of Section 2.5 may be treated analogously. Since we do not distinguish here between lattice and continuous systems, we employ a notation in which $\int_\Lambda dx$ is taken to mean $\sum_{x\in\Lambda}$ in the case where X is a lattice. Also, we define $|\Lambda|$ to be $\int_\Lambda dx$, that is, the volume of Λ or the number of sites in this region, according to whether X is a continuum or a lattice.

2.6.1 Normal Folia as Local Modifications of Single States

Suppose that ρ is a state of a $C^\star$-system $\Sigma = (\mathcal{A}, S, \alpha)$ and that B is an element of the local algebra of observables $\mathcal{A}_L$. Then it follows from the specifications of items (21) and (23) of Section 2.4 that the normal folium, $\mathcal{F}_\rho$, of ρ is generated by the local modifications $\{\rho \to \rho_B \mid B \in \mathcal{A}_L\}$ of the single state ρ.

This has particularly interesting implications if $\mathcal{F}_\rho$ is stable under the dynamical group $\alpha^\star(\mathbf{R})$, as is the case if, for example, ρ is time-translationally invariant[19]. The stability of $\mathcal{F}_\rho$ under $\alpha^\star(\mathbf{R})$ ensures that neither the natural evolution of the system nor the application of local operations can lead to an escape from this folium. In this case, we term $\mathcal{F}_\rho$ an *island*.

2.6.2 Space-translationally Invariant States

We define S_X to be the set of states on $\mathcal{A}$ that are invariant under the space translation group $\xi^\star(X)$, dual to $\xi(X)$. Evidently, S_X is convex. We denote by $\mathcal{E}_X$ the set of its extremal elements, that is, those that cannot be expressed in the form $w_1\rho_1 + w_2\rho_2$, where ρ_1, ρ_2 are different translationally invariant states and w_1, w_2 are positive numbers whose sum is unity. The $\mathcal{E}_X$-class states are important for the following reasons (cf. [Ru1, Se5]).

(1) $\mathcal{E}_X$ consists of precisely those elements, ρ, of S_X for which the GNS representation, π_ρ, of the space average of any local observables, A, is equal to $\rho(A)$, that is,

$$w : \lim_{\Lambda \uparrow X} |\Lambda|^{-1} \int_\Lambda dx\, \pi_\rho(\xi(x)A) = \rho(A)I \quad \forall\, A \in \mathcal{A}. \qquad (2.6.1)$$

In view of this property, the $\mathcal{E}_X$-class states are termed the *spatially ergodic* ones.

---

[19] This follows from the fact that, if $\alpha^\star(t)\rho = \rho$, then, in the notation of item (21), $\alpha^\star(t)\rho_B = \rho_{\alpha(-t)B}$.

(2) Any space-translationally invariant state, ρ, has a *unique* decomposition into extremals,[20] as given by the formula

$$\rho = \int_{\mathcal{E}_X} \sigma d\mu_\rho(\sigma), \qquad (2.6.2)$$

where μ_ρ is a probability measure on $\mathcal{E}_X$ that is completely determined by ρ. This decomposition is termed the spatially ergodic one.

2.6.3 Primary States have Short Range Correlations

This result has been established by Ruelle, in the following sense [Ru2].[21] If a state, ρ, is primary, then for any $A \in \mathcal{A}_L$ and $\epsilon > 0$, there is a bounded region, Λ, such that for any observable, C, localised outside Λ,

$$\mid \rho(AC) - \rho(A)\rho(C) \mid < \epsilon \|C\|. \qquad (2.6.3)$$

Since this clustering condition is uniform with respect to the observables outside Λ, it is stronger than that given by

$$\lim_{|x| \to \infty} \left(\rho(AC(x)) - \rho(A)\rho(C(x)) \right) = 0 \quad \forall A, C \in \mathcal{A}, \qquad (2.6.4)$$

where

$$C(x) = \xi(x)C \quad \forall C \in \mathcal{A}, \ x \in X. \qquad (2.6.5)$$

Moreover, if ρ is translationally invariant, the condition (2.6.3) implies that [LaRu]

$$\lim_{\min|x_j - x_k| \to \infty} \rho\left(A_1(x_1) \cdots A_n(x_n)\right) = \rho(A_1) \cdots \rho(A_n) \quad \forall A_1, ..., A_n \in \mathcal{A}. \quad (2.6.6)$$

Note
(1) This last condition implies that given by Eq. (2.6.1) for spatial ergodicity. Hence the translationally invariant primary states are spatially ergodic.

(2) The uniform clustering property (2.6.3) extends to all states in the normal folium of ρ for the following reasons. By local commutativity,

---

[20] The existence of such a decomposition follows from the Krein-Millman theorem and the fact that S_X is $w^\star$-compact. Its uniqueness depends on the quasi-local $C^\star$-algebraic structure of $\mathcal{A}$ and the local normality of the state space.

[21] The key to this result [Ru2] is that, if $\{B_n\}$ is a sequence of uniformly bounded observables, localised in regions $\{\Lambda_n\}$, respectively, such that the minimum distance, d_n, of Λ_n from the origin tends to infinity with n, then it follows from local commutativity that any weak limit of $\pi_\rho(B_n)$ belongs to the centre of $\pi_\rho(\mathcal{A})''$ and so is a scalar if ρ is primary.

any fixed B of $\mathcal{A}_L$ commutes with the elements, C, of this algebra that are localised outside a sufficiently large region Λ. Consequently, since $\mathcal{F}_\rho$ is generated by the local modifications ρ_B of ρ, as defined in item (21) of Section 2.4, it follows from Eq. (2.6.3) that all states in this folium satisfy that equation.

(3) Any mixture of primary translationally invariant states cannot possess the clustering property (2.6.6) for the following reason. If $\rho = \lambda\rho_1 + (1 - \lambda)\rho_2$, with $\rho_1 \neq \rho_2$ and $0 < \lambda < 1$, then, by Eq. (2.6.6),

$$\lim_{x\to\infty}\left[\rho(A^\star(x)A) - \rho(A^\star)\rho(A)\right] = \lambda(1 - \lambda) \mid \rho_1(A) - \rho_2(A) \mid^2 .$$

Since $\rho_1 \neq \rho_2$, the right-hand side of this equation is nonzero for some $A \in \mathcal{A}$, which implies that ρ does not satisfy the clustering condition (2.6.6), even for $n = 2$.

2.6.4 Decay of Time Correlations and Irreversibility

The statistical mechanics of irreversible processes depends heavily on the decay properties of time correlation functions [Ku]. Indeed such decay represents the loss of "memory effects" and so corresponds, in some measure, to the stochasticity at the root of irreversibility.

For finite systems, any putative mathematical theory of decaying correlations, and thus of irreversibility, is thwarted by Poincaré recurrences, which arise from the discreteness of the spectra of the Hamiltonians of these systems.[22] On the other hand, these arguments are not applicable to infinite systems, and therefore it is reasonable to enquire whether their time correlations can have the decay properties required for a theory of irreversibility.

For this purpose, we start by asking whether the decay of space correlations in primary states, as given by Eqs. (2.6.4)–(2.6.6), are paralleled by corresponding decay properties of time correlations. The answer is not straightforward, since the derivation of the results cited in Section 2.6.3 was based on the spatial local commutativity property of the observables, and there is no temporal counterpart to this, as observables for disjoint time intervals do not in general intercommute.[23] However, there is a realisable, weakened version of temporal local commutativity that does lead to a decay of time

---

[22] The proof of this spectral property is that, for finite systems with realistic interactions, the partition function, $\mathrm{Tr}(\exp(-\beta H))$, has been shown to be finite [Ru1].

[23] For example, in the case of an ideal Bose gas, it follows from the construction of Section 2.5.3 (cf. [DS]) that the time translate of the Weyl operator $W(f)$ is $W(f_t)$, where f_t is the evolute of f according to the single particle Schrödinger equation; and by Eq. (2.5.7), $W(f)$ and $W(f_t)$ do not intercommute.

correlations in primary states. This weaker property is given by the condition that $\mathcal{A}$ is *asymptotically abelian* with respect to time translations, that is, that $[\alpha(t)A, B]$ tends to zero, whether in norm or strongly, in the relevant representation space, as $t \to \infty$, for all $A, B \in \mathcal{A}$. Moreover, this condition is fulfilled in various tractable models [Ro2], and it implies the temporal counterpart of Eqs. (2.6.4). Thus, if ρ is primary and $A(t) = \alpha(t)A$, then

$$\lim_{t \to \infty} \big(\rho(A\alpha(t)B) - \rho(A)\rho(\alpha(t)B)\big) = 0 \quad \forall A, B \in \mathcal{A}, \qquad (2.6.7)$$

and, further, if ρ is time-translationally invariant,

$$\lim_{\min|t_j - t_k| \to \infty} \rho\big(A_1(t_1)\cdots A_n(t_n)\big) = \rho(A_1)\cdots\rho(A_n) \quad \forall A_1, ..., A_n \in \mathcal{A}. \quad (2.6.8)$$

To throw more light on this result, we note that, for $n = 2$, the temporal clustering condition (2.6.8) reduces to

$$\lim_{t \to \infty} \rho(A(t)B) = \rho(A)\rho(B) \quad \forall A, B \in \mathcal{A},$$

and, by items (18) and (29) of Section 2.4, this is equivalent to the following formula for the unitary group, $U_\rho(\mathbf{R})$, that implements time translations in the GNS representation space of ρ.

$$w : \lim_{t \to \infty} U_\rho(t) = P(\Phi_\rho). \qquad (2.6.9)$$

Most importantly, this result implies that the spectrum of the Hamiltonian, H_ρ, is continuous on the orthogonal complement of the cyclic vector Φ_ρ. By contrast, as noted above, the Hamiltonians of finite systems have discrete spectra.

Note By the same argument that was used in the third note of Section 2.6.3, mixtures of stationary states cannot satisfy the temporal clustering condition (2.6.8).

2.6.5 Global Macroscopic Observables (cf. [He])

These are global intensive observables of infinitely extended systems, and generally are functionals on the state space, rather than elements of the quasi-local algebra $\mathcal{A}$. The simplest ones are infinite volume limits of observables, $\hat{A}_\Lambda$, given by space averages of local observables, $A \ (\in \mathcal{A}_L)$, over bounded spatial regions Λ, that is,

$$\hat{A}_\Lambda = \frac{1}{|\Lambda|} \int_\Lambda dx A(x). \qquad (2.6.10)$$

It follows from this definition that $\hat{A}_\Lambda \in \mathcal{A}$ for each Λ in L. We define $\mathcal{D}(\hat{A})$ to be the family of states, ρ, for which $\rho(\hat{A}_\Lambda)$ converges to a limit, which we denote by $\hat{A}(\rho)$, as $\Lambda\uparrow X$. Thus,

$$\hat{A}(\rho) = \lim_{\Lambda\uparrow X} \rho(\hat{A}_\Lambda) \quad \forall \rho \in \mathcal{D}(\hat{A}), \tag{2.6.11}$$

which signifies that $\hat{A}$ is a functional on the state space with domain $\mathcal{D}(\hat{A})$. We term it a *global macroscopic observable*, since, by Eqs. (2.6.10) and (2.6.11), it corresponds to a global space average of $A(x)$.

We note here that it follows from our definition of $\mathcal{D}(\hat{A})$ that this domain contains all the translationally invariant states, since their expectation values of A and $\hat{A}_\Lambda$ are identical, by Eq. (2.6.10). Moreover, it follows from Section 2.6.2 that the extremals, $\mathcal{E}_X$, are the only translationally invariant states that render the observables $\hat{A}_\Lambda$ dispersionless in the limit where Λ increases to X, that is, $\mathcal{E}_X$ comprises just those elements, ρ, of S_X that satisfy the condition

$$\lim_{\Lambda\uparrow X}\left[\rho(\hat{A}_\Lambda^2) - \rho(\hat{A}_\Lambda)^2\right] = 0 \quad \forall A \in \mathcal{A}.$$

A more general definition of macroscopic observables, due to Hepp [He], represents these as arithmetic means of uniformly bounded sequences $\{A_n \in \mathcal{A}_{\Lambda_n}\}$ of local observables, such that the distance of Λ_n from the origin tends to infinity with n. Thus, the macroscopic observable $\hat{A}$ is defined as the functional on the state space given by

$$\hat{A}(\rho) = \lim_{N\to\infty} \rho(\hat{A}_N), \tag{2.6.12}$$

where

$$\hat{A}_N = N^{-1} \sum_{n=1}^{N} A_n, \tag{2.6.13}$$

the domain, $\mathcal{D}(\hat{A})$, of $\hat{A}$ consisting of the states for which the limit in Eq. (2.6.12) is well defined. It is easily seen that this definition of $\hat{A}$ reduces to that of Eq. (2.6.10) in the special case where $A_n = \int_{C_n} dx A(x)$, $A \in \mathcal{A}_L$ and $\{C_n\}$ is a sequence of unit cubes that form a partition of X and are arranged in such a way that, for any $L \in N$, the region $\bigcup_{n=1}^{L^d} C_n$, with d the dimensionality of X, is a cube of side L.

The macroscopic observables $\hat{A}_N$ have the following key properties [He], which stem from the definition (2.6.12), together with the local commutativity of $\mathcal{A}$ and the short range correlations of primary states, as given by Eq. (2.6.3).

(M.1) If ρ is primary, the $\hat{A}$ takes the same value on all states in the normal folium, $\mathcal{F}_\rho$, of ρ.

(M.2) If ρ and ρ' are disjoint primary states, that is, if their normal folia have no states in common, then there is some macroscopic observable $\hat{A}$ such that $\hat{A}(\rho) \neq \hat{A}(\rho')$.

In other words, *the states of the normal folium of a primary are macroscopically equivalent, whereas the folia of disjoint primaries are macroscopically different from one another.*

GNS Representation of Macroscopic Observables.

Suppose now that ρ is a primary state that lies in the domain of a macroscopic observable $\hat{A}$. Then it follows from (M.1) that, if ψ is an arbitrary vector in the GNS space $\mathcal{H}_\rho$ and ρ_ψ is the corresponding state on $\mathcal{A}$, then

$$\lim_{N\to\infty} (\psi, \pi_\rho(\hat{A}_N)\psi) = \hat{A}(\rho_\psi) = \hat{A}(\rho),$$

which signifies that $\pi_\rho(\hat{A}_N)$ converges weakly to $\hat{A}(\rho)I_\rho$ as $N \to \infty$. Accordingly, we define this limit to be the GNS representation, $\hat{A}_\rho$, of $\hat{A}$ in $\mathcal{H}_\rho$, that is,

$$\hat{A}_\rho = w : \lim_{N\to\infty} \pi_\rho(\hat{A}_N) = \hat{A}(\rho)I_\rho. \qquad (2.6.14)$$

Furthermore, in certain cases, for example, when $\hat{A}$ is the space average of a local observable, as in Eq. (2.6.11), the weak limit of Eq. (2.6.14) is also a strong one.[24] In such cases,

$$w : \lim_{N\to\infty} \pi_\rho(\hat{A}_N^2) = \left(\hat{A}(\rho)\right)^2 I_\rho, \qquad (2.6.15)$$

which implies that $\hat{A}_N$ is dispersion-free for any state in the folium $\mathcal{F}_\rho$. We take this to signify that the macroscopic observable $\hat{A}$ is dispersion-free in that folium.

2.6.6 Consideration of Pure Phases

From a phenomenological point of view, it is natural to characterise a pure phase of a system as a state, or family of states, that satisfies the following conditions.

(1) The macroscopic observables are all sharply defined, by contrast with mixed phases, where they may fluctuate.[25]

(2) The values of the macroscopic observables are unaffected by local modifications of the state, that is, they are uniform over its folium.

---

[24] This follows from the local commutatativity of $\mathcal{A}$ and the cyclicity of the GNS representation.

[25] In the case of a statistical mixture of two states of different density, for example, states of H_2O corresponding to ice and water, the dispersion of the global density would be nonzero.

It follows from Section 2.6.4 that the states that meet these conditions are just the primaries. Accordingly, we designate them to be the pure phases.

2.6.7 Fluctuations and Mesoscopic Observables (cf. [GVV])

The classical theory of large-scale fluctuations in complex systems is dominated by the central limit theorem, which, in its simplest form, may be expressed as follows [PR]. If $x_1, \ldots, x_N$ are independent, equivalently distributed random variables, each with mean $\bar{x}$ and dispersion Δ, then, in the limit $N \to \infty$ and under certain mild technical conditions, the variable $X_N = N^{-1/2} \sum_{n=1}^{N} (x_n - \bar{x})$ has a Gaussian distribution, whose mean and dispersion are 0 and Δ, respectively. Evidently, X_N represents the fluctuations of the sum $(x_1 + \cdots + x_N)$ about its mean, $N\bar{x}$, as normalised by the crucial factor $N^{1/2}$. Most importantly, this theorem may be generalised to statistically dependent random variables, x_n, whose correlations are "sufficiently weak" [PR].

One might well expect a similar result for the observables of the infinite quantum system, Σ, of the form

$$\tilde{A}_\Lambda = |\Lambda|^{-1/2} \int_\Lambda (A(x) - \rho(A)I), \qquad (2.6.16)$$

where ρ is a translationally invariant state. Evidently, $\tilde{A}_\Lambda$ represents the fluctuations of the extensive observable $\int_\Lambda dx A(x)$ about its mean value for the state ρ, and the normalisation factor $|\Lambda|^{1/2}$ corresponds to the $N^{1/2}$ of the above classical case. Since this normalisation is the square root of the natural one for the macroscopic observables $\hat{A}_\Lambda$ (cf. Eq. (2.6.10)), the observables $\tilde{A}_\Lambda$ are sometimes termed *mesoscopic*, that is, intermediate between microscopic and macroscopic.[26]

Goderis, Verbeure and Vets (GVV) have established a quantum central limit theorem for the mesoscopic observables $\tilde{A}_\Lambda$ [GVV]. Specifically, they have proved that these observables have Gaussian distributions, in the limit $\Lambda \uparrow X$, subject to certain conditions on the decay properties of the spatial correlations of the local observables, A, for the state ρ. Thus, under those conditions,

$$\lim_{\Lambda \uparrow X} \langle \rho; \exp(i\lambda \tilde{A}_\Lambda) \rangle = \exp\left(-\frac{1}{2} C(A, A)\right), \qquad (2.6.17)$$

where C is the bilinear form on the local observables given by

$$C(A, B) = \frac{1}{2} \int_X dx \big(\rho(AB(x)) - \rho(A)\rho(B)\big). \qquad (2.6.18)$$

---

[26] The term "mesoscopic" appears to have been introduced into the physics literature by Van Kampen [VK1].

Fluctuation Observables

Although $\pi_\rho(\tilde{A}_\Lambda)$ does not converge, even weakly, as Λ increases to X, GVV have been able to provide a precise representation of the mesoscopic observables in this limit. Specifically, they have shown that the statistics of the observables $\tilde{A}_\Lambda$ in the state ρ reduce, in the limit $\Lambda \uparrow X$, to those of corresponding elements, $\tilde{A}_\rho$, of an associated $W^\star$- algebra, $\tilde{\mathcal{A}}_\rho$, in a state, ω_ρ. To be precise, $\tilde{\mathcal{A}}_\rho$, ω_ρ and the mapping $A \to \tilde{A}_\rho$ of the local observables into $\tilde{\mathcal{A}}_\rho$ are completely specified by the following prescription.

(a)

$$\omega_\rho\Big(\exp(i\lambda\tilde{A}_\rho)\Big) = \lim_{\Lambda\uparrow X} \langle\rho; \exp(i\lambda\tilde{A}_\Lambda)\rangle. \qquad (2.6.19)$$

(b) The algebra $\tilde{\mathcal{A}}_\rho$ is generated by $\{\exp(i\tilde{A}_\rho) \mid A = A^\star \in \mathcal{A}_L\}$.

(c) The structure of this algebra is governed by the commutation relations

$$[\tilde{A}_\rho, \tilde{B}_\rho] = \int_X dx\rho([A(x), B])I, \qquad (2.6.20)$$

which, in view of (b), implies that $\tilde{\mathcal{A}}_\rho$ is the weak closure of the linear combinations of the Weyl-type operators $\exp(i\lambda\tilde{A}_\rho)$, with $A = A^\star \in \mathcal{A}_L$ and $\lambda \in \mathbf{R}$.

Thus, the $\tilde{A}_\rho$ correspond to mesoscopic observables, in an infinite volume limit, and these serve to generate the "fluctuation algebra", $\tilde{\mathcal{A}}_\rho$, associated with the algebra $\mathcal{A}$ in the state ρ. Further, by Eqs. (2.6.17) and (2.6.19), the state ω_ρ on this algebra has the Gaussian property

$$\omega_\rho\Big(\exp(i\lambda\tilde{A}_\rho)\Big) = \exp\Big(-\frac{1}{2}C(A,A)\Big). \qquad (2.6.21)$$

2.7 OPEN SYSTEMS

An open system, Σ, is one that is coupled to a second system, Σ_R, in such a way that the composite, Σ_C, of Σ and Σ_R is conservative. We restrict our considerations to the situation where Σ is a finite system and Σ_R an infinitely extended one that plays the role of a reservoir. Further, we confine our formulation here to the case where both of these are $W^\star$- dynamical systems, and where, correspondingly, so too is $\Sigma_C = (\mathcal{A}_C, S_C, \alpha_C)$. This latter system is then specified by the following prescription (cf. [PSSW]).

(a) $\mathcal{A}_C$ is the tensor product, $\mathcal{A} \otimes \mathcal{A}_R$, of the algebras of observables of Σ and Σ_R, the former consisting of the bounded operators in a separable Hilbert space $\mathcal{H}$. We identify elements A of $\mathcal{A}$ with $A \otimes I_R$ ($\in \mathcal{A}_C$).

(b) S_C is the set of all normal states on $\mathcal{A}_C$.

(c) $\alpha_C(\mathbf{R})$ is the one-parameter group of automorphisms of $\mathcal{A}_C$. A typical form for its generator, δ_C is given by the formula

$$\delta_C = \delta \otimes I_R + I \otimes \delta_R + \frac{i}{\hbar}[V, \cdot], \qquad (2.7.1)$$

where V is an self-adjoint element of $\mathcal{A}_C$, representing the energy of interaction between Σ and Σ_R, and δ and δ_R are the generators of the autmorphism groups governing the dynamics of those systems when uncoupled.

We now formulate the dynamics of Σ in the situation where this system and Σ_R have been prepared independently of one another in states ρ and ρ_R, respectively, and then coupled together at time $t = 0$. The time-dependent expectation value of an arbitrary observable, A, of Σ in the subsequent evolution of this open system is then

$$\langle A \rangle_t = \langle \rho \otimes \rho_R; \alpha(t)[A \otimes I_R] \rangle \quad \forall t \in \mathbf{R}_+. \qquad (2.7.2)$$

In order to express this quantity in terms of an evolution of the observables of Σ, we define the ultraweakly continuous linear transformations $P : \mathcal{A}_C \rightarrow \mathcal{A}$ and $\theta(t) : \mathcal{A} \rightarrow \mathcal{A}$ by the formulae

$$P(A \otimes A_R) = \rho_R(A_R)A \quad \forall A \in \mathcal{A}, \ A_R \in \mathcal{A}_R \qquad (2.7.3)$$

and

$$\theta(t) = P\alpha(t)P. \qquad (2.7.4)$$

It follows now from Eqs. (2.7.2–2.7.4) that

$$\langle A \rangle_t = \langle \rho; \theta(t)A \rangle \quad \forall A \in \mathcal{A}, \ \rho \in S, \ t \in \mathbf{R}_+, \qquad (2.7.5)$$

which signifies that the dynamics of the open system Σ is governed by the transformations $\theta(t)$ of $\mathcal{A}$. As we shall see in Chapter 4, this dynamics can correspond to an irreversible evolution, whereas that of the full conservative system Σ_C is reversible. Moreover, the irreversibilty of the Σ-dynamics becomes particularly transparent in a certain weak coupling limit where, as proved by Davies [Da], the transformations $\{\theta(t) \mid t \in \mathbf{R}_+\}$ reduce to a one-parameter Markovian semigroup, that is, $\theta(t)\theta(s) \equiv \theta(t + s)$, subject to certain viable conditions on the decay of time correlation functions of the isolated system Σ_R in the state ρ_R.

2.8 CONCLUDING REMARKS

The operator algebraic form of quantum mechanics provides a framework for the theory of both conservative and open systems which possesses physically relevant structures not carried by the traditional theory of finite systems. In particular, it provides (a) classifications of states and phases in terms of inequivalent representations of the algebras of observables (b) sharp characterisations of different levels of macroscopicality, and (c) decay properties of time correlation functions, that are needed for a theory of irreversibility.

APPENDIX A: HILBERT SPACES

Hilbert spaces are the natural generalisations of Euclidean spaces to arbitrary, possibly infinite, dimensionality. Here we provide a rudimentary treatment of these spaces, which suffices for the purposes of this book. Comprehensive treatments are provided in the books by Riesz and Nagy [RN] on functional analysis and Von Neumann on the foundations of quantum mechanics [VN1].

(1) *A Hilbert space,* $\mathcal{H}$, is a set of elements $(f, g, h, ...)$, possessing the following properties.

   (i) $\mathcal{H}$ is a complex vector space, that is, it is closed with respect to (a) binary addition, which is commutative, associative and invertible, and (b) multiplication by complex numbers, possessing the distributive properties that

$$\lambda(f_1 + f_2) = \lambda f_1 + \lambda f_2,$$

$$(\lambda_1 + \lambda_2)f = \lambda_1 f + \lambda_2 f,$$

$$\lambda_1(\lambda_2 f) = (\lambda_1 \lambda_2)f,$$

   for all f, f_1, f_2 in $\mathcal{H}$ and $\lambda, \lambda_1, \lambda_2$ in $\mathbf{C}$.

   (ii) $\mathcal{H}$ is equipped with a scalar product according to the following specifications. For each $f, g \in \mathcal{H}$, there is a complex number (f, g), termed the scalar product of f with g, possessing the properties that

$$(f, \lambda g + \mu h) = \lambda(f, g) + \mu(f, h) \quad \forall \lambda, \mu \in \mathbf{C},$$

$$(f, g) = \overline{(g, f)}$$

   and

$$(f, f) \geq 0,$$

   equality occurring if and only if $f = 0$. Thus, $\mathcal{H}$ is equipped with a norm, $\|\cdot\|$, defined by

$$\|f\| = (f, f)^{1/2}.$$

   It follows from these specifications that

$$| (f,g) | \leq \|f\| \|g\| \quad \text{(Schwartz inequality)},$$

$$\|\lambda f\| = | \lambda | \, \|f\| \quad \text{and} \quad \|f + g\| \leq \|f\| + \|g\|.$$

Thus, the norm serves as a metric for $\mathcal{H}$.

(iii) $\mathcal{H}$ is complete with respect to the norm, that is, if a sequence, $\{f_n\}$, of vectors in $\mathcal{H}$ possesses the Cauchy property that $\|f_m - f_n\|$ tends to zero as m, n tend independently to infinity, then there is a (unique) element f of $\mathcal{H}$ such that $\lim_{n \to \infty} \|f_n - f\| = 0$. In this case, f is termed the strong limit of the sequence $\{f_n\}$.

The following are typical examples of Hilbert spaces.

Example 1 $\mathcal{H} = \mathbf{C}^N$, the N-dimensional space whose elements are the sequences $f = (f_1, ..., f_N)$, of N complex numbers (coordinates), with scalar product

$$(f, g) = \sum_{n=1}^{N} \bar{f}_n g_n.$$

Example 2 $\mathcal{H} = l^2(\mathbf{N})$, which is an infinite dimensional version of Example 1. Thus, its elements are infinite sequences $f = (f_1, f_2, ..., f_n, ...)$ of complex numbers, such that $\sum_{n=1}^{\infty} | f_n |^2$ is finite, and its scalar product is given by the formula

$$(f, g) = \sum_{n=1}^{\infty} \bar{f}_n g_n.$$

Example 3 $\mathcal{H} = L^2(X)$, the Hilbert space of complex-valued, square integrable functions on the Euclidean space X, that is, $\{f : X \to \mathbf{C} \, | \, \int_X | f(x) |^2 \, dx < \infty\}$, with scalar product

$$(f, g) = \int_X \overline{f(x)} g(x) dx.$$

The symbol $\mathcal{H}$ will henceforth always denote a Hilbert space.

(2) *Uniformly Bounded Subsets.* A subset, Δ, of $\mathcal{H}$ will be termed uniformly bounded if the norms of its elements are uniformly bounded.

(3) *Dense Subsets.* A subset, Δ, of vectors in $\mathcal{H}$ is termed dense if, for any $f \in \mathcal{H}$ and $\epsilon > 0$, there is an element f' of Δ such that $\|f - f'\| < \epsilon$.

(4) *Separability.* A Hilbert space is termed separable if it has a dense denumerable subset of vectors. The examples given at the end of item (1) are all separable. Furthermore, all the Hilbert spaces that arise in the physical contexts with which we are concerned are separable, and therefore *we henceforth assume the separability of the Hilbert spaces treated here.*

(5) *Strong and Weak Convergence.* These are defined as follows. A sequence $\{f_n\}$ of vectors in $\mathcal{H}$ converges weakly to f ($\in \mathcal{H}$) if $\lim_{n\to\infty}(g, f_n - f) = 0$ for all vectors g in $\mathcal{H}$: it converges strongly to f if $\lim_{n\to\infty} \|f_n - f\| = 0$. We denote weak and strong limits by $w : \lim$ and $s : \lim$, respectively.

Note

(i) The relation between strong and weak convergence may be summarised as follows. $s : \lim_{n\to\infty} f_n = f$ if and only if (a) $w : \lim_{n\to\infty} f_n = f$ and (b) $\lim_{n\to\infty} \|f_n\| = \|f\|$.

(ii) Every uniformly bounded sequence of vectors in $\mathcal{H}$, that is, every sequence of vectors whose norms are uniformly bounded, has a weakly convergent subsequence.

(iii) If a linear mapping L of $\mathcal{H}$ into $\mathbf{C}$ is strongly continuous, that is, if the (strong) convergence of f_n to f implies that of $L(f_n)$ to $L(f)$, then the action of L is induced by a unique vector, g, in $\mathcal{H}$ according to the formula

$$L(f) = (g, f) \ \forall f \in \mathcal{H}.$$

This is Riesz's representation theorem.

(6) *Orthonormality.* A vector f in $\mathcal{H}$ is termed *normalised* if $\|f\| = 1$. Vectors f, g are termed orthogonal if $(f, g) = 0$. A sequence of vectors, $\{\phi_n\}$, in $\mathcal{H}$ is termed orthonormal if the ϕ_n are normalised and mutually orthogonal. It is termed a basis if, further, its linear combinations span the space $\mathcal{H}$, that is, if any vector in $\mathcal{H}$ can be expressed in the form

$$f = \sum_{n=1}^{\infty} c_n \phi_n \equiv s : \lim_{N\to\infty} \sum_{n=1}^{N} c_n \phi_n,$$

where the c_n are complex numbers. The assumed separability of $\mathcal{H}$ guarantees, and is indeed equivalent to, the existence of a (nonunique) denumerable basis for this space.

(7) *Subspaces.* A subspace, $\mathcal{K}$, of $\mathcal{H}$ is defined to be a subset of its vectors that themselves constitute a Hilbert space. The orthogonal complement, $\mathcal{K}^{\perp}$, of $\mathcal{K}$ in $\mathcal{H}$ is then defined to be the subspace of $\mathcal{H}$ comprising the vectors that are orthogonal to all those of $\mathcal{K}$. Further, any vector, f, of $\mathcal{H}$ has a unique decomposition into components $f_{\mathcal{K}}$ and $f_{\mathcal{K}^{\perp}}$ that lie in $\mathcal{K}$ and $\mathcal{K}^{\perp}$, respectively. Here, $f_{\mathcal{K}}$ is the vector f' in $\mathcal{K}$ for which $\|f - f'\|$ takes its minimum value.

(8) *Linear Manifolds.* A linear manifold in $\mathcal{H}$ is a subset, $\mathcal{M}$, of its vectors that is closed with respect to finite linear combinations, that is, if

$f_1, ..., f_n \in \mathcal{M}$ and $\lambda_1, ..., \lambda_n \in \mathbf{C}$, then $\sum_{r=1}^{n} \lambda_r f_r \in \mathcal{M}$. The closure, $\overline{\mathcal{D}}(A)$, of $\mathcal{D}(A)$, obtained by adding to this manifold the limits of its Cauchy sequences, is then a subspace of $\mathcal{H}$.

(9) *Operators in* $\mathcal{H}$. A linear (respectively antilinear) operator A in $\mathcal{H}$ is a mapping of a linear manifold $\mathcal{D}(A)$ ($\subset \mathcal{H}$) into $\mathcal{H}$, such that

$$A(c_1 f_1 + c_2 f_2) = c_1 A f_1 + c_2 A f_2 \text{ (respectively } \overline{c}_1 A f_1 + \overline{c}_2 A f_2)$$

$$\forall f_1, f_2 \in \mathcal{H}, \ c_1, c_2 \in \mathbf{C}.$$

$\mathcal{D}(A)$ is termed the *domain* of A. It may be assumed, without loss of generality, that it is dense in $\mathcal{H}$, since it may always be extended to be so by the simple expedient of defining A to be zero on the orthogonal complement of the closure of $\mathcal{D}(A)$.

We henceforth use the term "operator" to signify "linear operator" unless otherwise stated.

(10) *Bounded Operators.* An operator A is termed *bounded* if $\|Af\|$ is uniformly bounded as f runs through the normalised vectors in $\mathcal{D}(A)$. In this case, we may extend the domain of A to the whole space $\mathcal{H}$ by the following procedure. We note that, by (9), the domain of A may be assumed to be dense in $\mathcal{H}$, and therefore that any vector in this space is the strong limit of a sequence $\{f_n\}$ in $\mathcal{D}(A)$. Further, by the above definition, the boundedness of A implies that the sequence $\{Af_n\}$ is Cauchy and therefore converges to a limit, g, in $\mathcal{H}$. We define $Af = g$, thereby extending the domain of A by continuity to the whole of $\mathcal{H}$.

We denote the set of all the bounded linear operators in $\mathcal{H}$ by $\mathcal{B}(\mathcal{H})$. This is itself is a vector space, with norm defined by

$$\|A\| = \sup_{f \in \mathcal{H}} \|Af\| / \|f\|,$$

where "sup" means supremum, that is, least upper bound. It follows from this definition that the operator norm possesses the following properties.

$$\|A + B\| \le \|A\| + \|B\|, \quad \|AB\| \le \|A\| \|B\|, \quad \text{and} \quad \|\lambda A\| = |\lambda| \|A\|,$$

for all A, B in $\mathcal{B}(\mathcal{H})$ and λ in $\mathbf{C}$. We term a subset, $\mathcal{B}$, of $\mathcal{B}(\mathcal{H})$ *uniformly bounded* if the norms of its elements are uniformly bounded.

(11) *Convergence in* $\mathcal{B}(\mathcal{H})$. A sequence of bounded operators $\{A_n\}$ in $\mathcal{H}$ is said to converge weakly (respectively strongly) to $A (\in \mathcal{B}(\mathcal{H}))$ if $A_n f$ converges weakly (respectively strongly) to Af for all $f \in \mathcal{H}$. The sequence is said to converge uniformly, or in norm, to A if $\|A_n - A\| \to 0$ as $n \to \infty$. Thus, norm convergence implies strong convergence, which in turn implies weak convergence: the converse statements are

valid only if $\mathcal{H}$ is finite dimensional. We denote norm, strong and weak operator limits by norm : lim, s : lim and w : lim, respectively.

(12) *Closed Unbounded Operators.* We define an unbounded linear operator, A, in $\mathcal{H}$ to be closed if $\mathcal{D}(A)$ contains the limit, f, of any of its strongly convergent sequences $\{f_n\}$, for which $\{Af_n\}$ also converges, in which case $Af = s : \lim_{n \to \infty} Af_n$.

Note   The domain of an unbounded operator, A, cannot be extended by continuity to the whole of $\mathcal{H}$, since the convergence of an *arbitrary* sequence $\{f_n \in \mathcal{D}(A)\}$ does not imply that of $\{Af_n\}$.

(13) *Adjoint of an Operator.* Let A be an operator whose domain $\mathcal{D}(A)$ is dense in $\mathcal{H}$. The adjoint, $A^\star$, of A is defined as follows.

   We define the domain of $A^\star$ to consist of those vectors, g, in $\mathcal{H}$ for which the mapping $f \to (Af, g)$ of $\mathcal{D}(A)$ into $\mathbf{C}$ is continuous. Thus, by the Riesz representation theorem, specified in Note (iii) of item (9), there is a vector, g_A, in $\mathcal{H}$, such that

$$(Af, g) = (f, g_A) \quad \forall f \in \mathcal{D}(A).$$

We define $A^\star$ to be the (manifestly linear) operator that sends g to g_A, that is,

$$A^\star g = g_A.$$

Thus,

$$(f, A^\star g) = (Af, g) \quad \forall f \in \mathcal{D}(A), \ g \in \mathcal{D}(A^\star).$$

Note   In the case where A is bounded, so too is $A^\star$, and further, it follows from the specifications of items (10) and (13) that

$$(A^\star)^\star = A, \quad \|A^\star\| = \|A\| \quad \text{and} \quad \|A^\star A\| = \|A\|^2.$$

(14) *Self-adjoint Operators.* An operator, A, in $\mathcal{H}$ is termed self-adjoint if $A = A^\star$. Note that this implies that $\mathcal{D}(A) = \mathcal{D}(A^\star)$. A self-adjoint operator A is termed *positive* if $(f, Af) \geq 0$ for all f in $\mathcal{D}(A)$.

(15) *Projection Operators.* If $\mathcal{K}$ is a subspace of $\mathcal{H}$, then (cf. item (7)) any vector f in $\mathcal{H}$ has a unique resolution into components $f_{\mathcal{K}}$ and $f_{\mathcal{K}^\perp}$, belonging to $\mathcal{K}$ and $\mathcal{K}^\perp$, respectively. The projection operator, P, from $\mathcal{H}$ to $\mathcal{K}$ (sometimes simply termed the projection operator of $\mathcal{K}$) is defined by the formula

$$Pf = f_{\mathcal{K}} \quad \forall f \in \mathcal{H}.$$

Note that it follows from this definition, together with that of a subspace in item (7), that

$$P^2 = P^\star = P.$$

In fact, these conditions characterise a projection operator, since they imply that the manifold $\{Pf \mid f \in \mathcal{H}\}$ is a subspace of $\mathcal{H}$.

In the particular case where $\mathcal{K}$ is one-dimensional, that is, when it consists of the scalar multiples of a normalised vector ϕ, we denote $P_{\mathcal{K}}$ by $P(\phi)$. It follows from the above definitions that this projector is given by the formula

$$P(\phi)f = (\phi, f)\phi \quad \forall f \in \mathcal{H}.$$

(16) *Unitary and Antiunitary Transformations.* A unitary operator, U, in $\mathcal{H}$ is an invertible element of $\mathcal{B}(\mathcal{H})$ that conserves the values of the inner products between vectors in this space, that is, that satisfies the condition that

$$(Uf, Ug) = (f, g) \; \forall f, g \in \mathcal{H}.$$

Since this latter condition is equivalent to $(f, U^\star Ug) \equiv (f, g)$, that is, $U^\star U = I$, and since U is invertible, it follows that the unitarity of U is characterised by the relation

$$U^\star = U^{-1}.$$

We denote the set of all unitary operators in $\mathcal{H}$ by $\mathcal{U}(\mathcal{H})$.

A mapping, U, of $\mathcal{H}$ onto another Hilbert space, $\mathcal{K}$, is termed unitary if (a) it is invertible, and (b) it preserves the value of inner products, that is,

$$(Uf, Ug)_{\mathcal{K}} = (f, g)_{\mathcal{H}}.$$

Thus, defining $f_U = Uf$ and, for $A \in \mathcal{B}(\mathcal{H})$, $A_U = UAU^{-1} (\in \mathcal{B}(\mathcal{K}))$,

$$(f_U, A_U g_U) \equiv (f, Ug),$$

which signifies that the operator U effects a one-to-one correspondence between the operators, as well as the vectors, in the spaces $\mathcal{H}$ and $\mathcal{K}$.

An *antiunitary* operator, V, in $\mathcal{H}$ is an invertible antilinear transformation of this space, such that

$$(Vf, Vg) = (g, f) \; \forall f, g \in \mathcal{H}.$$

(17) *Unitary Representations of Groups.* A unitary representation of a group, G, in $\mathcal{H}$ is a mapping, U, of G into $\mathcal{U}(\mathcal{H})$ that preserves its group structure, that is,

$$U(g_1)U(g_2) = U(g_1 g_2) \quad \forall g_1, g_2 \in G.$$

In particular, $U(g)^\star = U(g)^{-1} = U(g^{-1})$.

(18) *Stone's Theorem.* Let U be a unitary representation of the additive group of real numbers, $\mathbf{R}$, in $\mathcal{H}$, such that $U(t)$ is strongly continuous with respect to the real variable t, that is, $s : \lim_{t \to t_0} U(t) = U(t_0)$ for all $t_0 \in \mathbf{R}$. Then Stone's theorem asserts that there is a (unique) self-adjoint operator, K, in $\mathcal{H}$, such that

$$s : \frac{d}{dt} U(t)f = iU(t)Kf = iKU(t)f \quad \forall f \in \mathcal{D}(K), \ t \in \mathbf{R}.$$

The operator iK is termed the infinitesimal generator of the group $U(\mathbf{R})$.

Conversely, the above equation for dU/dt, together with the initial condition that $U(0) = I$, ensures that U is a strongly continuous representation of $\mathbf{R}$ in $\mathcal{H}$.

(19) *Density Matrices and Traces.* A density matrix in $\mathcal{H}$ is an operator of the form

$$\hat{\rho} = \sum_n w_n P(\phi_n),$$

where $\{\phi_n\}$ is an orthonormal sequence of vectors in $\mathcal{H}$ and $\{w_n\}$ is a sequence of positive numbers, whose sum is unity. Thus, $\hat{\rho}$ is a weighted sum of orthogonal projectors.

The Trace of a positive operator, B, is defined as

$$\mathrm{Tr}(B) = \sum_n (\psi_n, B\psi_n),$$

where $\{\psi_n\}$ is an orthonormal basis. In fact, the value of $\mathrm{Tr}(B)$ here, which might be infinite, is independent of the choice of basis.

Note

(i) It follows from these definitions that density matrices are positive operators of unit Trace. In fact, their above definition is equivalent to this description.

(ii) The Trace has the *cyclic property* that $\mathrm{Tr}(ABC) \equiv \mathrm{Tr}(CAB)$.

(20) *Spectral Analysis.* If A is a self-adjoint operator in $\mathcal{H}$ and ϕ is a vector such that $A\phi = \lambda\phi$, then ϕ is said to be an eigenvector of A, with eigenvalue λ, which is necessarily real. In the case where A may be expressed in the form $\sum_n \lambda_n P(\phi_n)$, A is said to have a discrete spectrum. In this case, by arranging the eigenvalues so that λ_n increases with n, we may express A in the form

$$A = \sum_n \lambda_n (E_n - E_{n-1}), \qquad (*)$$

where E_n is the projector given by the sum of the (orthogonal) eigenprojectors $P(\phi_m)$, corresponding to eigenvalues less than λ_n.

The generalisation of the formula (*) to all self-adjoint operators, regardless of whether their spectra are discrete, has been obtained in the following form [VN1].

$$A = \int_{-\infty}^{\infty} \lambda dE(\lambda), \qquad (**)$$

where $\int$ is a Stieltjes integral and $\{E(\lambda) \mid \lambda \in \mathbf{R}\}$ is a unique family of projectors, termed the *spectral projectors*, with the following properties.

$$E(\lambda) \leq E(\lambda') \quad \text{for } \lambda < \lambda', \quad s: \lim_{\lambda' \downarrow \lambda} E(\lambda') = E(\lambda),$$

$$E(-\infty) = 0, \quad E(\infty) = I.$$

The spectral resolution of A given by the above Stieltjes integral may be understood to signify that

$$(f, Ag) = \int_{-\infty}^{\infty} \lambda d(f, E(\lambda)g).$$

(21) *Direct Sums of Hilbert Spaces.* The direct sum of two Hilbert spaces, $\mathcal{H}_1$ and $\mathcal{H}_2$, is the Hilbert space, $\mathcal{H}_1 \oplus \mathcal{H}_2$, defined by the following conditions.

(i) For each pair of vectors $f_1(\in \mathcal{H}_1)$ and $f_2(\in \mathcal{H}_2)$, there is a vector denoted by $f_1 \oplus f_2$ in $\mathcal{H}_1 \oplus \mathcal{H}_2$, such that

$$\lambda(f_1 \oplus f_2) = \lambda f_1 \oplus \lambda f_2$$

and

$$(f_1 \oplus f_2, g_1 \oplus g_2)_{\mathcal{H}_1 \oplus \mathcal{H}_2} = (f_1, g_1)_{\mathcal{H}_1} + (f_2, g_2)_{\mathcal{H}_2}$$

for all $\lambda \in \mathbf{C}$, $f_1, g_1 \in \mathcal{H}_1$ and $f_2, g_2 \in \mathcal{H}_2$.

(ii) $\mathcal{H}_1 \oplus \mathcal{H}_2$ is the strong closure, with respect to the norm corresponding to the above inner product, of the finite linear combinations of vectors $f_1 \oplus f_2$.

Likewise, if A_1, A_2 are operators in $\mathcal{H}_1$ and $\mathcal{H}_2$, respectively, then $A_1 \oplus A_2$ is the operator in $\mathcal{H}_1 \oplus \mathcal{H}_2$ defined on the domain $\mathcal{D}(A_1) \oplus \mathcal{D}(A_2)$ by the formula

$$(A_1 \oplus A_2)(f_1 \oplus f_2) = A_1 f_1 \oplus A_2 f_2.$$

These definitions have obvious generalisations to direct sums and integrals of arbitrarily many Hilbert spaces and operators therein. In general, we denote the direct sum of Hilbert spaces $\{\mathcal{H}_n \mid n \in K\}$, where K is some index set, by $\oplus_{n \in K} \mathcal{H}_n$. Direct integrals, denoted by $\int^{\oplus} \mathcal{H}_\alpha d\mu(\alpha)$, with μ a probability measure, are defined analogously.

(22) *Tensor Products.* The tensor product, $\mathcal{H}_1 \otimes \mathcal{H}_2$, of two Hilbert spaces $\mathcal{H}_1$ and $\mathcal{H}_2$ is the Hilbert space defined by the following conditions.

   (i) For each $f_1 \in \mathcal{H}_1$ and $f_2 \in \mathcal{H}_2$, there is a vector $f_1 \otimes f_2$ in $\mathcal{H}_1 \otimes \mathcal{H}_2$, such that

$$\lambda(f_1 \otimes f_2) = (\lambda f_1) \otimes f_2 = f_1 \otimes (\lambda f_2)$$

and

$$(f_1 \otimes g_1, \ f_2 \otimes g_2)_{\mathcal{H}_1 \otimes \mathcal{H}_2} = (f_1, g_1)_{\mathcal{H}_1} (f_2, g_2)_{\mathcal{H}_2}$$

for all $\lambda \in \mathbf{C}$, $f_1, g_1 \in \mathcal{H}_1$ and $f_2, g_2 \in \mathcal{H}_2$.

   (ii) $\mathcal{H}_1 \otimes \mathcal{H}_2$ is the completion of the space of finite sums $\sum_{j=1}^{n} f_{j,1} \otimes f_{j,2}$ with respect to the norm corresponding to this scalar product.

   The tensor product $A_1 \otimes A_2$ of operators A_1, A_2 in $\mathcal{H}_1$, $\mathcal{H}_2$, respectively, is defined on the domain $\mathcal{D}(A_1) \otimes \mathcal{D}(A_2)$ by the formula

$$(A_1 \otimes A_2)(f_1 \otimes f_2) = A_1 f_1 \otimes A_2 f_2.$$

   These definitions of tensor products have their obvious generalisations to finite and even denumerably infinite numbers of Hilbert spaces and their operators. In general, the tensor product of spaces $\{\mathcal{H}_n \mid n \in K\}$, where K is a denumerable index set, is denoted by $\otimes_{n \in K} \mathcal{H}_n$.

(23) *Linear Transformations of one Hilbert Space onto Another.* The formulation of these is similar to that of operators in a Hilbert space. Thus, if $\mathcal{H}$ and $\mathcal{K}$ are Hilbert spaces, then a bounded linear transformation, V, of $\mathcal{H}$ onto $\mathcal{K}$ is a one for which the ratio of $\|Vf\|$ to $\|f\|$ is uniformly bounded as f runs through $\mathcal{H}$. The adjoint of V, is then defined as the bounded linear transformation, $V^\star$, of $\mathcal{K}$ into $\mathcal{H}$ given by the formula

$$(V^\star g, f)_{\mathcal{H}} = (g, Vf)_{\mathcal{K}} \quad \forall f \in \mathcal{H}, \ g \in \mathcal{K}.$$

Further, V is termed unitary if

$$V^\star V = I_{\mathcal{H}} \quad \text{and} \quad VV^\star = I_{\mathcal{K}}.$$

Chapter 3

On symmetry, entropy and order

The concepts of symmetry, entropy and order play cardinal, interrelated roles in the theory of complex systems, and may be qualitatively described as follows. Symmetry consists of the invariance of physical properties of a system under transformations of its observables. The entropy of a state is an information theoretic measure of its degree of impurity and thus, in a particular sense, of its disorder. Order, on the other hand, constitutes the organisation of the components of a system into structured patterns. Most importantly, order and disorder are *not necessarily* characterised by low and high entropy, respectively, since, on the one hand, certain pure states are spatially chaotic, while, on the other hand, highly organised structures can carry large entropy, just as highly coherent signals may be accompanied by intense noise.

We devote this chapter to a formulation of the concepts of symmetry, entropy, order and disorder within the operator algebraic framework of Chapter 2, our generic model being the quantum dynamical system $(\mathcal{A}, S, \alpha)$, as specified for finite systems in Section 2.2.3 and for infinite systems in Section 2.5.1.

We start in Section 3.1 by representing symmetry groups as automorphisms of $\mathcal{A}$. In Section 3.2, we provide a concise formulation of entropy and its basic mathematical properties, followed by a discussion of its relationship with disorder. In Section 3.3, we formulate the concepts of order and coherence, as arising from symmetry breakdown, and there we demonstrate that order is not necessarily characterised by low entropy. In Section 3.4, we summarise our conclusions and indicate the need for more far-reaching conceptions of both order and disorder than those available at present.

3.1 SYMMETRY GROUPS

A symmetry group of Σ is defined as a group, G, with a faithful representation, θ, in Aut($\mathcal{A}$). It is termed a *dynamical symmetry* if $\theta(G)$ commutes with the dynamical automorphisms $\alpha(\mathbf{R})$. Thus, in general, the action of a symmetry group conserves the kinematics, as represented by the structure of its algebra of observables, while that of a dynamical symmetry group also conserves the dynamics.

We term a symmetry group G *internal* if, for each g *in* G, $\theta(g)$ commutes with the space translations $\xi(x)$ and leaves each of the local algebras $\mathcal{A}_\Lambda$ invariant; and we term it *spatial* if its elements, g, are volume conserving transformations of the configuration space X and if, correspondingly, $\theta(g)$ maps $\mathcal{A}_\Lambda$ onto $\mathcal{A}_{g\Lambda}$. More specifically, we assume that, in the notation of Sections 2.5.2 and 2.5.3, the action of a spatial symmetry group on $\mathcal{A}$ is given by the formulae

$$\theta(g)a_x = a_{gx} \quad \text{for lattice systems}$$

and, formally,

$$\theta(g)\psi(x) = \psi(gx) \quad \text{for continuous systems.}$$

We henceforth restrict our considerations to symmetry groups that are either internal or spatial, or possibly a product of the two.

In the case of a finite system, we assume that $\theta(G)$ is implemented by a unitary representation, U, of G in the irreducible representation space of the observables, that is, that

$$\theta(g)A = U(g)AU(g^{-1}).$$

In the case of an infinitely extended system, we assume a local version of this formula, namely that the restriction of $\theta(G)$ to each local algebra, $\mathcal{A}_\Lambda$, takes the form

$$\alpha(g)A = U_\Lambda(g)AU_\Lambda(g^{-1}) \quad \forall A \in \mathcal{A}_\Lambda,$$

where $U_\Lambda(g)$ is a unitary mapping of $\mathcal{H}_\Lambda$ onto either itself or $\mathcal{H}_{g\Lambda}$, according to whether the group G is internal or spatial.

3.2 ENTROPY

3.2.1 Classical Preliminaries [SW, Kh1]

Suppose that E is an experiment with n possible outcomes $(e_1, ..., e_n)$. A probability measure p on E comprises N nonnegative numbers $(p_1, ..., p_n)$, whose sum is unity, with p_j the probability that the experiment yields the result e_j. The set $\mathcal{P}$ of all such probability measures is then the state space of E. Evidently, $\mathcal{P}$ is a convex set, and its extremal elements are the pure states, π, as characterised by the condition that π_j is unity for one value of j and zero for the others.

The entropy of a state p is defined to be

$$S(p) = -\sum_{j=1}^{n} p_j \ln p_j, \tag{3.2.1}$$

and the relative entropy of p with respect to another state, q, is defined as

$$S_{\text{rel}}(q \mid p) = \sum_{j=1}^{n} (p_j \ln p_j - p_j \ln q_j). \tag{3.2.2}$$

The following simple argument demonstrates that $S(p)$ is a measure of the degree of impurity, or stochastic disorder, of p, and that $S(q \mid p)$ represents the relative inaccessibility of this state from q.

Suppose that the experiment E described above is repeated N times. Then the number of ways of obtaining the outcome that e_j occurs N_j times, for $j = 1, ..., n$, is

$$\Omega(N_1, ..., N_n \mid N) = \frac{N!}{N_1! .... N_n!}, \tag{3.2.3}$$

while the probability of obtaining this result when the system is in the state q is

$$P_q(N_1, ..., N_n \mid N) = \Omega(N_1, ..., N_n \mid N) q_1^{N_1} .... q_n^{N_n}. \tag{3.2.4}$$

Thus, if N_j/N converges to p_j as $N \to \infty$, for $j = 1, ..., n$, then, for large N, $\Omega(N_1, ..., N_n \mid N)$ and $P_q(N_1, ..., N_n \mid N)$ provide measures of the degree of impurity of p and its relative accessibility from q, respectively, in this limit. Furthermore, it follows from Eqs. (3.2.1)–(3.2.4) that, by Sterling's theorem,

$$\lim_{N \to \infty} N^{-1} \ln \Omega(N_1, ..., N_n \mid N) = S(p), \tag{3.2.5}$$

and

$$- \lim_{N \to \infty} N^{-1} \ln P_q(N_1, ..., N_n \mid N) = S_{\text{rel}}(q \mid p). \tag{3.2.6}$$

Hence, $S(p)$ and $S_{\text{rel}}(p \mid q)$, like Ω and P_q, serve to quantify the degree of impurity of p and its inaccessibility from q, respectively.

We note here that S and S_{rel} are extensive quantities, in the following sense. Suppose that $E = (e_1, ..., e_n)$ and $E' = (e'_1, ..., e'_m)$ are two experiments, whose outcomes are statistically independent. Then the states on the composite $E \times E'$ are the probability measures $p \otimes p'$, that is, those for which the joint probability of obtaining the results e_j and e'_k is of the form $p_j p'_k$. It follows then from Eqs. (3.2.1) and (3.2.2) that S and S_{rel} satisfy the extensivity conditions

$$S(p \otimes p') = S(p) + S(p') \tag{3.2.7}$$

and

$$S_{\text{rel}}(q \otimes q' \mid p \otimes p') = S_{\text{rel}}(q \mid p) + S_{\text{rel}}(q' \mid p'). \tag{3.2.8}$$

3.2.2 Finite Quantum Systems

We turn now to the generic finite quantum system of Section 2.2.3, in which the states, ρ, are faithfully represented by the density matrices, $\hat{\rho}$, in the irreducible representation space, $\mathcal{H}$, of the observables. For this system, the entropy, $\hat{S}$, and the relative entropy, $\hat{S}_{\text{rel}}$, are defined by the quantum analogues of the formulae (3.2.1) and (3.2.2), namely [VN1]

$$\hat{S}(\rho) = -\text{Tr}(\hat{\rho}\ln\hat{\rho}) \tag{3.2.9}$$

and [Um, Li1]

$$\hat{S}_{\text{rel}}(\sigma \mid \rho) = \text{Tr}(\hat{\rho}\ln\hat{\rho} - \hat{\rho}\ln\hat{\sigma}). \tag{3.2.10}$$

Equivalently, $\hat{S}_{\text{rel}}$ may be expressed in terms of $\hat{S}$ by the following formula [Li2]:

$$\hat{S}_{\text{rel}}(\sigma \mid \rho) = \lim_{\lambda \downarrow 0} \lambda^{-1}\Big(\hat{S}(\lambda\rho + (1-\lambda)\sigma) - \lambda\hat{S}(\rho) - (1-\lambda)\hat{S}(\sigma)\Big). \tag{3.2.10$'$}$$

Note The definition of $\hat{S}$ given by Eq. (3.2.9) is the standard one of statistical physics in units where Boltzmann's constant is equal to unity. We use such units throughout this book.

As in the classical case, $\hat{S}$ provides a measure of the impurity of the states, as may be seen on expressing the density matrix $\hat{\rho}$ in the canonical form

$$\hat{\rho} = \sum_{n=0}^{\infty} p_n P(\psi_n), \tag{3.2.11}$$

where p_n and $P(\psi_n)$ are the occupation probability and the projection operator, respectively, of its eigenstate ψ_n. It follows then from Eqs. (3.2.9) and (3.2.11) that

$$\hat{S}(\rho) = -\sum_n p_n \ln p_n, \tag{3.2.12}$$

which is precisely the entropy of the classical probability, $p = \{p_n\}$, governing the admixture of the pure states $\{\psi_n\}$.

The relative entropy, on the other hand, is not so simply related to its classical counterpart, but is nevertheless a very powerful tool for both statistical mechanics and the theory of quantum dynamical systems [Li2, Li3, Ar1, CNT, AF].

Among the key mathematical properties of $\hat{S}$ and $\hat{S}_{\text{rel}}$ are the following [Weh, Th].

(S.1) (a) In view of our assumption of Section 3.1 that the symmetry groups of finite systems are unitarily implemented, it follows from Eqs. (3.2.9) and (3.2.10) that their actions on the states leave both $\hat{S}$ and $\hat{S}_{\text{rel}}$ invariant, that is,

$$\hat{S}(\theta^{\star}(g)\rho) \equiv \hat{S}(\rho) \quad \text{and} \quad \hat{S}_{\text{rel}}(\theta^{\star}(g)\sigma \mid \theta^{\star}(g)\rho) \equiv \hat{S}_{\text{rel}}(\sigma \mid \rho).$$

(b) Likewise, $\hat{S}$ and $\hat{S}_{\text{rel}}$ are invariant under antiunitarily implemented antiautomorphisms of $\mathcal{A}$.

(S.2) As in the classical case, both $\hat{S}$ and $\hat{S}_{\text{rel}}$ are extensive, that is, if ρ, σ are states of Σ and ρ', σ' are states of another finite system Σ', then

$$\hat{S}(\rho \otimes \rho') = \hat{S}(\rho) + \hat{S}(\rho') \qquad (3.2.13)$$

and

$$\hat{S}_{\mathrm{rel}}(\sigma \otimes \sigma' \mid \rho \otimes \rho') = \hat{S}_{\mathrm{rel}}(\sigma \mid \rho) + \hat{S}_{\mathrm{rel}}(\sigma' \mid \rho'). \qquad (3.2.14)$$

(S.3) (a) $\hat{S}(\rho)$ is zero if ρ is pure, and positive if it is mixed. Thus $\hat{S}$ is minimised by the pure states.

(b) For given σ, $\hat{S}_{\mathrm{rel}}(\sigma \mid \rho)$ is zero if $\rho = \sigma$ and is otherwise positive.

(S.4) The functional $\hat{S}$ is concave, and $\hat{S}_{\mathrm{rel}}$ is jointly convex in its two arguments, that is,

$$\hat{S}(\lambda\rho_1 + (1-\lambda)\rho_2) \geq \lambda\hat{S}(\rho_1) + (1-\lambda)\hat{S}(\rho_2) \qquad (3.2.15)$$

and

$$\hat{S}_{\mathrm{rel}}\big(\lambda\sigma_1 + (1-\lambda)\sigma_2 \mid \lambda\rho_1 + (1-\lambda)\rho_2\big)$$

$$\leq \lambda\hat{S}_{\mathrm{rel}}(\sigma_1 \mid \rho_1) + (1-\lambda)\hat{S}_{\mathrm{rel}}(\sigma_2 \mid \rho_2), \qquad (3.2.16)$$

for $0 < \lambda < 1$.

(S.5) (a) If $\mathcal{H}$ is finite dimensional, then both $\hat{S}$ and $\hat{S}_{\mathrm{rel}}$ are continuous with respect to the $w^\star$ topology of the state space, that is, if ρ and σ are the $w^\star$ limits of sequences $\{\rho_n\}$ and $\{\sigma_n\}$, respectively, then $\hat{S}(\rho_n)$ converges to $\hat{S}(\rho)$ and $\hat{S}_{\mathrm{rel}}(\sigma_n \mid \rho_n)$ to $\hat{S}_{\mathrm{rel}}(\sigma \mid \rho)$ as $n \to \infty$.

(b) If $\mathcal{H}$ is infinite dimensional, then $\hat{S}_{\mathrm{rel}}$ is upper semi-continuous[1] with respect to the $w^\star$ topology of the state space, that is, if ρ_n, σ_n and $\hat{S}_{\mathrm{rel}}(\sigma_n \mid \rho_n)$ converge to ρ, σ and $\overline{S}_{\mathrm{rel}}$, respectively, as $n \to \infty$, then $\hat{S}_{\mathrm{rel}}(\sigma \mid \rho) \leq \overline{S}_{\mathrm{rel}}$.

(S.6) If σ is the canonical equilibrium state for the inverse temperature β, as represented by the density matrix $\exp(-\beta H)/\mathrm{Tr}(idem)$, then

$$\hat{S}_{\mathrm{rel}}(\sigma \mid \rho) = \beta\big(\hat{F}_\beta(\rho) - \Phi(\beta)\big), \qquad (3.2.17)$$

where $\hat{F}_\beta(\rho)$ is the free energy of the state ρ and $\Phi(\beta)$ is its equilibrium value at inverse temperature β, that is,

$$\hat{F}_\beta(\rho) = \rho(H) - \beta^{-1}\hat{S}(\rho), \qquad (3.2.18)$$

and

---

[1] For a general definition of lower semicontinuity, see [Ch, p. 29].

$$\Phi(\beta) = -\beta^{-1} \ln \mathrm{Tr}(\exp(-\beta H)). \qquad (3.2.19)$$

Hence, by (S.3b) and Eq. (3.2.17), the free energy $\hat{F}_\beta$ attains its minimum value, $\Phi(\beta)$, at the canonical equilibrium state only.

(S.7) (a) Entropy is *subadditive* [Ru1], that is, if ρ is a normal state on $\mathcal{A} = \mathcal{B}(\mathcal{H})$, where $\mathcal{H} = \mathcal{H}_1 \otimes \mathcal{H}_2$, and if ρ_1, ρ_2 are the restrictions of ρ to $\mathcal{B}(\mathcal{H}_1)$, $\mathcal{B}(\mathcal{H}_2)$, respectively, then

$$\hat{S}(\rho) \leq \hat{S}(\rho_1) + \hat{S}(\rho_2). \qquad (3.2.20)$$

In other words, the entropy of a system is less than the sum of the entropies of its component parts.

(b) Moreover, as proved by Lieb and Ruskai [LiRus], entropy enjoys the even stronger, and technically important,[2] property of *strong subadditivity* [LiRus], which may be expressed as follows. If ρ is a normal state on $\mathcal{A} = \mathcal{B}(\mathcal{H})$, where $\mathcal{H} = \mathcal{H}_1 \otimes \mathcal{H}_2 \otimes \mathcal{H}_3$, and ρ_j, ρ_{jk} are the restrictions of ρ to $\mathcal{B}(\mathcal{H}_j)$, $\mathcal{B}(\mathcal{H}_j \otimes \mathcal{H}_k)$, respectively for $j, k = 1, 2, 3$, then

$$\hat{S}(\rho) + \hat{S}(\rho_3) \leq \hat{S}(\rho_{13}) + \hat{S}(\rho_{23}). \qquad (3.2.21)$$

3.2.3 Infinite Systems

We describe an infinitely extended system, Σ, according to the scheme of Section 2.5.1, in which the state space, S, is a folium of locally normal states on the algebra, $\mathcal{A}$, of quasi-local observables, and is stable under the space translational automorphisms $\xi(X)$. We denote by S_X the set of translationally invariant elements of S: thus $S_X = \{\rho \in S \mid \xi^\star(x)\rho = \rho \; \forall x \in X\}$.

By the local normality of the state space, S, the restrictions, ρ_Λ and σ_Λ, of states ρ and σ to a bounded spatial region Λ are normal states of a finite system Σ_Λ (cf. Section 2.5.1). We define the local entropy and relative entropy functionals, $\hat{S}_\Lambda$ and $\hat{S}_{\mathrm{rel},\Lambda}$, on S and $S \times S$, respectively, by the formulae

$$\hat{S}_\Lambda(\rho) = \hat{S}(\rho_\Lambda) \qquad (3.2.22)$$

and

$$\hat{S}_{\mathrm{rel},\Lambda}(\sigma \mid \rho) = \hat{S}(\sigma_\Lambda \mid \rho_\Lambda), \qquad (3.2.23)$$

where the right-hand sides of these equations are the entropy and relative entropy of a finite system, namely Σ_Λ, and therefore are specified by the formalism of Section 3.2.2.

It follows from the local structure of the algebra $\mathcal{A}$ and the strong subad-

---

[2] Compare Eq. (3.2.24) below.

ditivity of entropy, as given by Eq. (3.2.21), that, for any disjoint spatial regions, Λ_1, Λ_2 and Λ_3,

$$\hat{S}_{\Lambda_1 \cup \Lambda_2 \cup \Lambda_3}(\rho) + \hat{S}_{\Lambda_3}(\rho) \leq \hat{S}_{\Lambda_1 \cup \Lambda_3}(\rho) + \hat{S}_{\Lambda_2 \cup \Lambda_3}(\rho). \qquad (3.2.24)$$

The Global Entropy Density

An important consequence of the strong subadditivity of entropy is that, for any translationally invariant state, ρ, the local entropy density, $\hat{S}(\rho_\Lambda)/|\Lambda|$, converges to a definite limit, $\hat{s}(\rho)$, as Λ increases to X over a suitably regular sequence of bounded regions [Ru1]. Thus,

$$\hat{s}(\rho) = \lim_{\Lambda \uparrow X} \frac{\hat{S}_\Lambda(\rho)}{|\Lambda|} \quad \forall \rho \in S_X. \qquad (3.2.25)$$

The following key properties of the entropy density functional have been established [Ru1].

(s.1) $\hat{s}$ is affine, that is,

$$\hat{s}(\lambda \rho_1 + (1-\lambda)\rho_2) = \lambda \hat{s}(\rho_1) + (1-\lambda)\hat{s}(\rho_2) \quad \forall \rho_1, \rho_2 \in S_X, \quad (3.2.26)$$

and its range is $[0, \infty]$.

(s.2) In the case of lattice systems, as formulated in Section 2.5.2, $\hat{s}$ is upper semi-continuous with respect to the $w^\star$ topology of the state space, that is, if ρ_n and $\hat{s}(\rho_n)$ converge to ρ and $\bar{s}$, respectively, then $\hat{s}(\rho) \leq \bar{s}$.

In addition, the following property of $\hat{s}$ is an immediate consequence of our assumption, in Section 3.1, of the local unitary implementability of a symmetry group, G.

(s.3) If the automophisms $\theta(g)$ commute with the space translations, $\xi(X)$, then they leave the functional $\hat{s}$ invariant, that is, $\hat{s}(\theta^\star(g)\rho) \equiv \hat{s}(\rho)$.

Note This picture of $\hat{s}$ may readily be extended to locally normal states that possess only the translational invariance of a normal subgroup of X, corresponding to a lattice.

Relative Entropy for Localised Modifications of States

The definition of relative entropy may be extended from finite to infinite systems [Ar1]. Here, we sketch a limited version of its formulation. Thus, we note that, for lattice systems, the local relative entropy $\hat{S}_{\text{rel},\Lambda}(\sigma \mid \rho)$ has been proved to converge to a definite limit (possibly infinite) for all states ρ and σ: no translational invariance is required here. Accordingly, we define

$$\hat{S}_{\text{rel}}(\sigma \mid \rho) = \lim_{\Lambda \uparrow X} \hat{S}_{\text{rel}}(\sigma_\Lambda \mid \rho_\Lambda). \qquad (3.2.27)$$

In fact, the condition for this quantity to be finite is essentially that ρ corresponds to a modification of σ by localised operations.

The relative entropy functional has the following key properties.

($S_{\text{rel}}.1$) $\hat{S}_{\text{rel}}$ is jointly convex in its two arguments, that is,

$$\hat{S}_{\text{rel}}\left(\lambda\sigma_1 + (1-\lambda)\sigma_2 \mid \lambda\rho_1 + (1-\lambda)\rho_2\right)$$

$$\leq \lambda\hat{S}_{\text{rel}}(\sigma_1 \mid \rho_1) + (1-\lambda)\hat{S}_{\text{rel}}(\sigma_2 \mid \rho_2) \quad \forall\rho, \sigma \in S, \ \lambda \in (0, 1), \quad (3.2.28)$$

and the range of $\hat{S}_{\text{rel}}$ is $[0, \infty]$.

($S_{\text{rel}}.2$) In view of the assumption of the local unitary implementability of the representation, θ, of a symmetry group, G, the relative entropy is invariant under the action of this group, that is,

$$\hat{S}_{\text{rel}}(\theta^\star(g)\sigma \mid \theta^\star(g)\rho) \equiv \hat{S}_{\text{rel}}(\sigma \mid \rho).$$

For continuous systems, the above definition and properties of $\hat{S}_{\text{rel}}$ are still applicable, when restricted to the pairs of states for which the limit on the right-hand side of Eq. (3.2.27) exists and is independent of the chosen sequence of Λ's.

3.2.4 On Entropy and Disorder

As defined above, the entropy of a state is a measure of its degree of impurity, and thus, in a probabilistic sense, of its disorder. However, it is clearly not the only measure of disorder, since pure states may have extremely chaotic structures, as the following simple example shows. Suppose that Σ is a system of Pauli spins, $\sigma(x) = (\sigma_1(x), \sigma_2(x), \sigma_3(x))$, located on the sites, x, of a lattice $X = \mathbf{R}^d$. Then a state, ρ, of Σ in which the spins are in uncorrelated pure states takes the form

$$\rho = \otimes_{x \in X}\, \rho_x, \tag{3.2.29}$$

where ρ_x is the state of the spin at the site x. We note that it follows from Eq. (3.2.29) and the purity of the ρ_x that the restriction of ρ to any finite point subset of X is pure, and hence, by Eqs. (3.2.22) and (3.2.25), that both the local entropy, $\hat{S}_\Lambda(\rho)$, and the global entropy density, $\hat{s}(\rho)$, vanish.

In order to examine the possible spatial structures of ρ, we note that, since the algebra of observables at the site x is just the linear span of the $\sigma(x)$ components and the identity, ρ_x is completely determined by the value of the vector

$$\langle \rho_x; \sigma(x) \rangle := u(x). \tag{3.2.30}$$

In view of the purity of ρ_x, $u(x)$ must be a unit vector, and consequently, by

Eqs. (3.2.29) and (3.2.30), the state ρ is given by the configuration of unit vectors $\{u(x) \mid x \in X\}$, representing the orientations of the spins. Since these may be chosen at random, that is, without any correlations between the values of $u(x)$ at different sites x, we see that the pure, entropy-free state ρ can carry a highly chaotic spatial structure.

In fact, such states occur naturally in disordered systems, and could be realised by the application of a spatially random magnetic field at low (ideally zero) temperature. The direction of $u(x)$ would then be that of the field at the point x.

3.3 ORDER AND COHERENCE

We take the concept of order in a complex system to connote organisation, in some sense, of the structures of its component parts. The particular formulation of this concept that we present here represents order as emanating from a loss of symmetry, and appears to cover the standard structures that arise in condensed matter physics (cf. [LL2, Se2]). Different kinds of order are clearly essential to other areas of the natural sciences, as we discuss briefly in Section 3.4.

3.3.1 Order and Symmetry

The connection between order, as defined in a rather natural way, and loss of symmetry may be qualitatively understood as follows. Suppose that ρ is a state of an infinitely extended system, Σ, that is not invariant under the action of a dynamical symmetry group, G. Then, defining

$$\rho_g = \theta^{\star}(g)\rho, \tag{3.3.1}$$

there is some local observable $\hat{F}$ for which the function $x \rightarrow F_g(x) := \rho_g(\xi(x)\hat{F})$ has a nontrivial g-dependence. Since G is a dynamical symmetry group, g is a *random* variable,[3] in that neither the algebraic structure of the observables nor the dynamics can discriminate between the different states in the set $\{\rho_g \mid g \in G\}$. Correspondingly, F_g is a random field, whose realisation is achieved by selection of g from G. Thus, if F_g is a function of position that is simple[4] in the sense of being specified by a finite number of parameters, then this realisation serves to interrelate the values of *a priori* random variables, $F_g(x)$, attached to the different points x. In this case, we term the state ρ

---

[3] Assuming that G is amenable, the probability measure governing this random variable can naturally be designated as the left Haar measure or, more generally, left invariant mean of the group.

[4] This qualification is designed to excludes chaotic fields, such as those discussed in the example of Section 3.2.4.

ordered. A canonical example of order, in this sense, is provided by the alignment of spins in a ferromagnet, corresponding to a loss of rotational symmetry: the observable $\hat{F}$ here is just the spin at the spatial origin.

   To formulate this concept of order precisely, we assume that the system Σ is equipped with a *G-field*, which we define to be an *n*-component real or complex field $\hat{\eta}(x) = (\hat{\eta}^{(1)}(x), ..., \hat{\eta}^{(n)}(x))$ ($\in \mathcal{A}^n$) with the following properties.

   (i) $\hat{\eta}$ is covariant with respect to space translations, that is, $\xi(x)\hat{\eta}(y) = \hat{\eta}(x + y)$.

   (ii) In the case where the symmetry group G is internal, the action of $\theta(G)$ on $\hat{\eta}$ is given by

$$\theta(g)\hat{\eta}(x) = [L(g)\hat{\eta}](x) \equiv \hat{\eta}_g(x), \qquad (3.3.2)$$

   where L is a unitary representation of G in $\mathbf{R}^n$ or $\mathbf{C}^n$ according to whether $\hat{\eta}$ is real or complex. Thus

$$\hat{\eta}_g^\star(x)\cdot\hat{\eta}_g(x') = \hat{\eta}^\star(x)\cdot\hat{\eta}(x') \quad \forall g \in G, \ x, x' \in X, \qquad (3.3.3)$$

   where

$$\hat{\eta}^\star(x)\cdot\hat{\eta}(x') \equiv \sum_{j=1}^{n} \hat{\eta}^{(j)\star}(x)\cdot\hat{\eta}^{(j)}(x').$$

   (iii) In the case where G is spatial, the action of $\theta(G)$ on $\hat{\eta}$ is implemented by that of G on X according to the formula

$$\theta(g)\hat{\eta}(x) = \hat{\eta}(gx). \qquad (3.3.4)$$

   Typical examples of G-fields are the following.

   (a) For a system of Pauli spins $\sigma(x)$ on the sites X of a lattice, X, with G the internal symmetry group of the three-dimensional rotations, $\hat{\eta}$ is simply σ.

   (b) For any infinitely extended system, as formulated in Section 2.5, with X the space translation group, $\hat{\eta}(x)$ may be chosen to be the space translate, $\xi(x)B$, of any local observable B.

   We now suppose that ρ is a primary state, corresponding to a pure phase of Σ. Thus, by Eq. (3.3.1), together with the specifications of Sections 2.5 and 3.1, the states ρ_g are also primary. We define

$$\eta(x) = \rho(\hat{\eta}(x)), \quad \eta_g(x) = \rho_g(\hat{\eta}(x)). \qquad (3.3.5)$$

Hence, by Eqs. (3.3.2) and (3.3.4),

$$\eta_g = L(g)\eta \quad \text{or} \quad \eta \circ g, \qquad (3.3.6)$$

according to whether G is internal or spatial.

We now recast the qualitative description of order, outlined above, in the following form. We term the state ρ *ordered* if the following conditions are fulfilled.

(O.1) The field $\eta_g(x)$ carries a nontrivial g-dependence, and, further, it does not vanish in the limit $x \to \infty$.

(O.2) This field is invariant under the space translation group, X, or some normal subgroup, Y, thereof, that corresponds to a lattice.

(O.3) The conditions (O.1, O.2) are stable under time translations, that is, they remain valid when ρ is replaced by $\alpha^\star(t)\rho$.

Comments

(1) These ordering conditions imply a *spontaneous* symmetry breakdown, in that ρ lacks the G-symmetry of the dynamics.

(2) If ρ satisfies these conditions, then so also do all the states $\{\rho_g \mid g \in G\}$.

(3) Condition (O.2) could certainly be weakened to a demand that the form of η_g was simple, in the sense discussed following Eq. (3.3.1).

(4) In some circumstances, the condition (O.3) should be reinforced by further conditions of stability, for example, against local perturbations of the dynamics. In fact, thermal equilibrium states do enjoy stability against these perturbations [Ar2], and, under certain technical conditions, are even characterised by it [HKT-P].

The following are typical examples of order, as formulated according to the above prescription.

Example 1   Ferromagnetic ordering. Here, we take Σ to be a system of Pauli spins, $\sigma(x)$, located on the sites, x, of a lattice X. Thus, its algebra of observables, $\mathcal{A}$, is generated by these spins. The three-dimensional rotation group, G, is then an internal symmetry group, with faithful representation, θ, in $\mathrm{Aut}(\mathcal{A})$ defined by the formula

$$\theta(g)\sigma(x) = L(g)\sigma(x),$$

where L is the canonical three-dimensional representation of G. Hence, if $\hat{\eta}(x) = \sigma(x)$ *and if ρ is* a translationally invariant equilibrium state of Σ that yields a nonzero expectation value, s, of each $\sigma(x)$, then this state satisfies the conditions (O.1)–(O.3) and so is ferromagnetically ordered.

Example 2   Antiferromagnetic ordering. Here, we take $\mathcal{A}$, X, , G, θ and $\hat{\eta}$ to be the same as in Example 1, but now we assume that the dynamics is such as to support an equilibrium state, ρ, that is invariant under the space translational groups of two sublattices, $Y_\pm$, whose union is X, and that satisfies the condition

$$\rho(\sigma(x)) = \pm s \; (\neq 0) \quad \text{for } x \in Y_\pm.$$

In this case, the conditions (O.1)–(O.3) are again fulfilled and the ordering is antiferromagnetic.

Example 3 Crystalline ordering. Here, Σ is a system with translationally invariant interactions, G is the space translation group X, and θ is the representation, ξ, of this group, specified in Section 2.5. We assume that the system supports a stationary state, ρ, that is not invariant under the full group X of space translations, but only under one of its normal subgroups, Y, that corresponds to a lattice. This signifies that there is some $\hat{F}$ in $\mathcal{A}_L$ such that $\rho(\xi(x)\hat{F}) = \rho(\hat{F})$ for $x \in Y$, but not for all $x \in X/Y$. Thus, defining $\hat{\eta}(x) = \xi(x)\hat{F}$, we see that the state ρ satisfies the conditions (O.1)–(O.3), and that it carries the crystalline order of invariance under the lattice group Y.

3.3.2 Coherence

The concept of *coherence*, due originally to Glauber [Gl], represents an extreme form of order, as exemplified by laser light. We may formulate it within the present framework as follows. We term a state, ρ, coherent, relative to the G-field $\hat{\eta}$, if its satisfies (O.1)–(O.3), together with the following.

$$\left\langle \rho; \hat{\eta}^{\#(j_1)}(x_1)\cdots\hat{\eta}^{\#(j_k)}(x_k) \right\rangle = \eta^{\#(j_1)}(x_1)\cdots\eta^{\#(j_k)}(x_k)$$

$$\forall x_1,\ldots,x_k \in X, \; j_1,\ldots,j_k \in (1,2,\ldots,k), \tag{3.3.7}$$

where $\hat{\eta}^\#(x_j)$ is either $\hat{\eta}(x_j)$ or $\hat{\eta}^\star(x_j)$, and correspondingly, $\eta^\#(x_j)$ is either $\eta(x_j)$ or $\overline{\eta}(x_j)$.

Comments

(1) The coherence condition (3.3.7) signifies that $\hat{\eta}$ behaves like a dispersionless classical field when the system is in the state ρ.

(2) Again, the condition (O.2) may be weakened to the demand that the function η be simple, in the sense of involving at most a finite number of parameters.

(3) A stronger version of the condition (3.3.7) is the corresponding one for the time-dependent G-field $\hat{\eta}(x,t) \equiv \alpha(t)\hat{\eta}(x)$, namely

$$\left\langle \rho; \hat{\eta}^{\#(j_1)}(x_1,t_1)\cdots\hat{\eta}^{\#(j_k)}(x_k,t_k)\right\rangle = \eta^{\#(j_1)}(x_1,t_1)\cdots\eta^{\#(j_k)}(x_k,t_k),$$

$$\forall x_1,...,x_k \in X, \; t_1,...,t_k \in \mathbf{R}, \tag{3.3.8}$$

where now η is a periodic, or at least suitably simple, function of position and time. This corresponds to Glauber's original characterisation of coherent radiation, of the type exhibited by lasers.

(4) The coherence conditions (3.3.7) and (3.3.8) can be generalised to unbounded G-fields, provided that the domains of definition of the monomials in $\hat{\eta}$ there are such that the left-hand sides of these formulae are well defined.

3.3.3 Long Range Correlations in G-invariant Mixtures of Ordered Phases

Returning to the picture of Section 3.3.1, wherein the state ρ is ordered by virtue of a G-symmetry breakdown, we now assume that G is equipped with a probability measure, μ, such that the μ-mean of the states ρ_g is G-invariant. Thus, defining

$$\rho_\mu = \int \rho_g d\mu(g), \tag{3.3.9}$$

$$\theta(g)\rho_\mu \equiv \rho_\mu. \tag{3.3.10}$$

Thus, in Examples 1 and 2, ρ_μ would be the average of ρ_g over all rotations[5] g; and in Example 3, it would be the mean of $\xi^\star(x)\rho$ ($\equiv \rho_g$), with respect to the Lebesgue measure dx, over the coset, or "cell", X/Y.

We may assume, without loss of generality, that

$$\rho_\mu(\hat{\eta}(x)) \equiv 0, \tag{3.3.11}$$

since this demand is satisfied when $\hat{\eta}(x)$ is modified by the subtraction of its expectation value for the state ρ_μ. Since the states ρ_g are primary, it follows from Eqs. (2.6.4) and (3.3.9) that

$$\lim_{|x'|\to\infty}\left[\left\langle\rho_\mu; \hat{\eta}^\star(x)\cdot\hat{\eta}(x+x')\right\rangle - \int d\mu(g)\eta_g(x)\cdot\eta_g(x+x')\right] = 0. \tag{3.3.12}$$

Consequently, by condition (O.2) and Eqs. (3.3.2)–(3.3.4), (3.3.6) and (3.3.12), the correlation function $\langle\rho_\mu; \hat{\eta}^\star(x)\cdot\hat{\eta}(x+y)\rangle$ converges to $|\eta(x)|^2$ if G is internal and to $\int d\mu(g)|\eta(gx)|^2$ if that group is spatial, as y tends to infinity over X or a lattice therein. Thus, regardless of whether G is internal or spatial, the state ρ_μ exhibits long range correlations in the field $\hat{\eta}$, in that

---

[5] Specifically, μ would be the left Haar measure [Hal] on the rotation group.

$\langle \rho_{\mu}; \hat{\eta}^{\star}(x).\hat{\eta}(x+y) \rangle$ tends to a positive limit as y tends to infinity over at least some lattice in X, while the expectation value of this field vanishes. On this basis, we say that ρ_{μ} has *long range order*.

3.3.4 Superfluidity and Off-diagonal Long Range Order

Order, as depicted above, entails symmetry breakdown in pure phases and long range correlations in mixed, symmetric ones. A nontrivial modification of this picture is required for the description of order in the superfluid phase, whether of liquid Helium or of superconductors.

In fact, the characterisation of the superfluid phase by long range correlations, first proposed by O. Penrose [Pe, PO] for Bose systems (He$_4$), may be expressed in the following form.

(Ia) There is a complex classical function, ϕ, on X, that does not tend to zero at infinity, such that

$$\lim_{|y|\to\infty}\left[\rho(\psi(x)\psi^{\star}(x+y)) - \phi(x)\overline{\phi}(x+y)\right] = 0. \qquad (3.3.13a)$$

This characterisation was subsequently extended by Yang [Ya] to systems of fermions by replacing the quantum and classical fields ψ and ϕ by corresponding two-point fields, Ψ and Φ, respectively, where

$$\Psi(x_1,x_2) = \psi_{\uparrow}(x_1)\psi_{\downarrow}(x_2). \qquad (3.3.14)$$

This field represents the pairing of electrons [Scha1,Co] and, most importantly, has bose-like properties. Thus, in the case of fermions, the proposed condition for superfluid order is the following.

(Ib) There is a classical two-point field, that is, a complex-valued function Φ on $X \times X$, such that $\Phi(x_1 + y, x_2 + y)$ does not tend to zero as y tends to infinity, and

$$\lim_{|y|\to\infty}\left[\rho(\Psi(x_1,x_2)\Psi^{\star}(x_1' + y, x_2' + y)) - \Phi(x_1,x_2)\overline{\Phi}(x_1' + y, x_2' + y)\right] = 0.$$

$$(3.3.13b)$$

States satisfying these conditions, whether for Bose or Fermi systems, are said to have *off-diagonal long range order*, or ODLRO [Ya]. Thus, ODLRO signifies long range correlations of the field ψ, for bosons, or Ψ, for fermions. Since these fields are gauge dependent, they are *not* observables, and therefore the order here is qualitatively different from that carried by the observable G-fields $\hat{\eta}$.

We reinforce the above ODLRO condition (Ia,b) by the following conditions, analogous to the order conditions (O.2) and (O.3).

(II) The field $\phi(x)$ or, for fixed u, $\Phi(x-u, x+u)$, is constant or periodic
      with respect to x (or, more generally, is a simple function of this vari-
      able); and

(III) the conditions (I) and (II) are preserved under time translations.

We shall now show that, although the ODLRO condition (I) pertains to long
range correlations of a nonobservable field, ψ or Ψ, its mathematical formulation can nevertheless be brought into line with that of Section 3.3.1 for the
ordered states by extending the algebraic description of the system from $\mathcal{A}$ to
the field algebra, $\mathcal{A}^{(F)}$, defined in Section 2.5.3. To this end, we note that the
gauge group, $G = S^{(1)}$ is a symmetry group of $\mathcal{A}^{(F)}$, with representation γ in
$\mathrm{Aut}(\mathcal{A}^{(F)})$ defined by Eq. (2.5.8), that is,

$$\gamma(\lambda)\psi(x) = \psi(x) \exp(i\lambda) \quad \text{for bosons}, \tag{3.3.15a}$$

and, by Eq. (3.3.14),

$$\gamma(\lambda)\Psi(x_1, x_2) = \Psi(x_1, x_2)\exp(2i\lambda) \quad \text{for fermions}, \tag{3.3.15b}$$

for $0 \le \lambda < 2\pi$. Hence, ψ, or Ψ, is a G-field with respect to $\mathcal{A}^{(F)}$. A state,
$\rho^{(F)}$, on $\mathcal{A}^{(F)}$ that is ordered with respect to this field enjoys the property (cf.
(O.1)) that

$$\rho^{(F)}(\psi(x)) = \phi(x) \quad \text{for bosons} \tag{3.3.16a}$$

or

$$\rho^{(F)}(\Psi(x_1, x_2)) = \Phi(x_1, x_2) \quad \text{for fermions}, \tag{3.3.16b}$$

where ϕ, Φ do not vanish at infinity. Further, defining

$$\rho_\lambda^{(F)} = \gamma^\star(\lambda)\rho^{(F)}, \tag{3.3.17}$$

it follows from Eqs. (3.3.15)–(3.3.17) that $\rho_\lambda^{(F)}$ enjoys the order property (O.1).

We now follow the line of argument of Section 3.3.3 by defining $\bar{\rho}_{\text{sym}}^{(F)}$ to be
the gauge invariant state on $\mathcal{A}^{(F)}$ given by the G-mean of the states $\rho_\lambda^{(F)}$, that
is,

$$\bar{\rho}_{\text{sym}}^{(F)} = \frac{1}{2\pi} \int_0^{2\pi} d\lambda \rho_\lambda^{(F)}. \tag{3.3.18}$$

Then, assuming that $\rho^{(F)}$, and hence $\rho_\lambda^{(F)}$, is primary, it follows easily from Eqs.
(2.6.4) and (3.3.16)–(3.3.18) that the state $\bar{\rho}_{\text{sym}}^{(F)}$ satisfies the ODLRO condition
(I); while the conditions (O.2) and (O.3) for $\rho_\lambda^{(F)}$ imply (II) and (III) for $\bar{\rho}_{\text{sym}}^{(F)}$.
Furthermore, since $\mathcal{A}$ is the gauge invariant subalgebra of $\mathcal{A}^{(F)}$, the states
$\rho_\lambda^{(F)}$, and hence also $\bar{\rho}_{\text{sym}}^{(F)}$, all reduce to the same state, ρ, on $\mathcal{A}$. Hence, *any
primary state, $\rho^{(F)}$, on $\mathcal{A}^{(F)}$ that is ordered relative to the $G(= S^1)$-field ψ, or
Ψ, induces an ODLRO state on the observable algebra $\mathcal{A}$.* Conversely [Se3,
Theorem 4.3], any translationally invariant state on $\mathcal{A}$ with ODLRO has a
canonical gauge invariant extension to a state, $\bar{\rho}_{\text{sym}}^{(F)}$, on $\mathcal{A}^{(F)}$ that decomposes

into states $\rho_\lambda^{(F)}$ in such a way that Eqs. (3.3.16)–(3.3.18) are satisfied. The states $\rho_\lambda^{(F)}$ are therefore ordered with respect to the $G(=S^{(1)})$-field ψ or Ψ.

3.3.5 On Entropy and Order

According to the definitions in Sections 3.2 and 3.3, entropy is a measure of impurity of a state, whereas order corresponds to structure realised by suitable "freezing" of the values of random variables comprising its symmetry group. Moreover, low entropy cannot be either necessary or sufficient for the advent of order since, on the one hand, an ideal Fermi gas at zero temperature carries neither entropy nor order, while, on the other hand, crystals with high melting points can possess both large entropy and crystallographic order at suitably high temperatures.

Indeed, a rather graphic illustration of the fact that order is not necessarily characterised by low entropy is provided by the comparison between an optical vacuum and a laser beam, which fufills coherence conditions of the form (3.3.8). The vacuum state carries neither order nor entropy, whereas the laser beam, although highly ordered, cannot have less entropy than the vacuum! Hence, low entropy alone does not characterise ordered optical states.

3.4 FURTHER DISCUSSION OF ORDER AND DISORDER

We conclude this chapter by arguing that there remains a need for more general characterisations of both disorder and order than we have at present. In this connection we first remark that these entities are not mutually exclusive, just as a highly coherent signal can coexist with intense background noise.

Now as we saw in Section 3.2.4, the entropy of a state is a measure of its degree of impurity, although not necessarily of its disorder, since some pure states have highly chaotic structures. This indicates that a comprehensive theory of disorder requires some characterisation of structural chaos that applies even to pure states. In fact, Kolmogorov's [Ko1] algorithmic entropy, which essentially represents the information required to specify a string of symbols of a given finite alphabet, does provide such a characterisation in a limited classical context. Thus, as Zurek [Zu] has suggested, it is tempting to conjecture that a general theory of disorder should entail a combination of the Von Neumann entropy, given by Eq. (3.2.9), and some generalised quantum version of algorithmic entropy.

Turning now to the description of order, we note that its formulation both here and elsewhere [LL2, Se2] in terms of symmetry groups stems from the fact that the group parameters, g, constitute a free source of random variables, whose realisations can evince structures with simple spatial ordering. A more general characterisation of order is surely needed for the vast variety of systems in physics and biology that lack the standard spatial and internal symmetries characteristic of simple physical systems. For example, Schrödinger [Sch] has pointed out that the latter structures bear the same kind of relationship to those of biological ones as a conventional wallpaper, with a simple repeated pattern, does to a Raphael tapestry.

At all events, we consider that any natural formulation of order should contain at least two of the ingredients of the above group theoretic one, namely (a) a set of random variables and (b) a specification of those of its realisations that are deemed to be ordered or organised, whether as a simple wallpaper design or as an artistic or natural masterpiece. The problem, evidently, is to characterise both the random variables and their organised realisations. To cite the concrete case of a DNA molecule, the random variables there presumably comprise its possible sequences of nucleotides, and the order lies in the pattern encoded in their realised configurations.

Chapter 4

Reversibility, irreversibilty and macroscopic causality

At the microscopic level, quantum mechanics does not distinguish between future and past. This may be seen form the fact that, if $f_t(x_1, ..., x_n)$ is the time-dependent wave function of a pure state of a conservative n-particle system, then its evolution is governed by the Schrödinger equation

$$i\hbar \frac{\partial f_t}{\partial t} = -\sum_{j=1}^{n} \frac{\hbar^2}{2m_j} \Delta f_t + V f_t,$$

where m_j is the mass of the jth particle and $V(x_1, ..., x_n)$ is the total potential energy of the system. Hence, by complex conjugation,

$$-i\hbar \frac{\partial \bar{f}_t}{\partial t} = -\sum_{j=1}^{n} \frac{\hbar^2}{2m_j} \Delta \bar{f}_t + V \bar{f}_t,$$

which means that $\bar{f}_t$ satisfies the Schrödinger equation obtained by replacing t by $-t$. In other words, the temporally forward evolution of f_t is precisely matched by the backward one of $\bar{f}_t$. Thus, quantum dynamics, like classical mechanics, carries no intrinsic distinction between past and future.

On the other hand, the second law of thermodynamics provides such a distinction through its decree that the entropy[1] of a closed system should always increase monotonically with time. Thus, we have the paradox that thermodynamics provides an "arrow of time" that does not stem from quantum mechanics alone, despite the fact that the macroscopic objects governed by the second law are, presumably, just quantum systems of particles.

This is the long-standing paradox of microscopic reversibility and macroscopic irreversibility. Its resolution lies in the fact that the former is a property of the *complete* dynamics of a many-particle system, whereas the latter pertains to the evolution of only a subset of its variables[2] (cf. [Uh, VK2, Zw]). Thus, theories of irreversibility must be centered on the dynamics induced in certain subsystems by the global time-development of many-particle quantum systems. In fact, as first appreciated by Boltzmann [Bol], irrever-

---

[1] This is the entropy of *classical thermodynamics*, as expressed in terms of a few macroscopic variables.

[2] In the thermodynamical context, this subset generally consists of very few variables.

sibility is generated by molecular chaos, or stochasticity. In the context of quantum many-body theory, this may now be interpreted as "fast" decay of correlations of the forces governing the evolution of the relevant macroscopic variables [Ku, Se6] .Here, the qualification "fast" signifies that, as viewed from the macroscopic standpoint, this decay is (almost) instantaneous and thus provides a continuous production of entropy or loss of "information".

The purpose of this chapter is to provide a general scheme, within the algebraic framework of Chapter 2, for the passage from the microscopically reversible dynamics of a conservative system to the induced irreversible evolution of a subsystem. We start in Section 4.1 with a formulation of the microscopic reversibilty principle for conservative quantum systems, whether finite or infinite. In Section 4.2, we move on to the quantum probabilistic concepts of complete positivity and conditional expectations that are needed for the formulation of the relationship between a system and its subsystems. In Section 4.3, we formulate the induced dynamics of a subsystem in terms of the conservative evolution of an entire system and the relevant conditional expectations. In Section 4.4, we provide a formulation of irreversible dynamics in the subsytem, as represented by the monotonic decrease of a Lyapunov functional of its time-dependent states. In Section 4.5, we extend this formulation to situations where the subdynamics corresponds to classical deterministic macroscopic laws. There are three appendices to the chapter, each devoted to an illustration of a general issue raised therein. Thus, Appendix A demonstrates the difference between positivity and complete positivity by means of a simple example; Appendices B and C provide treatments of simple models that realise the general schemes of Sections 4.4. and 4.5, respectively.

4.1 MICROSCOPIC REVERSIBILITY

4.1.1 Finite Systems

Let us first consider a single spinless particle in a Euclidean space X. Then its algebra of observables, $\mathcal{A}$, consists of the bounded operators in the Hilbert space $\mathcal{H} = L^2(X)$. The Wigner time reversal operator is defined to be the antiunitary transformation, T, of this space, given by the equation

$$(Tf)(x) = \bar{f}(x) \ \forall f \in L^2(X), \qquad (4.1.1)$$

and this operator implements an antiautomorphism, τ, of $\mathcal{A}$, according to the formula

$$\tau A = T A^\star T \quad \forall A \in \mathcal{A}. \qquad (4.1.2)$$

Since this is a normal antiautomorphism, it extends canonically to the unbounded observables. In particular, if $\hat{x}$ and $\hat{p}$ are the position and momen-

tum observables, as represented by the multiplicative and differential operators that send $f(x)$ to $xf(x)$ and $-i\hbar\nabla f(x)$, respectively, then

$$\tau\hat{p} = -\hat{p} \quad \text{and} \quad \tau\hat{x} = \hat{x}, \tag{4.1.3}$$

that is, τ reverses the particle's momentum and leaves its position unchanged.
Furthermore, assuming that the Hamiltonian of the particle has the standard
form

$$H = \frac{\hat{p}^2}{2m} + V(\hat{x}), \tag{4.1.4}$$

then, by Eq. (4.1.3),

$$\tau H = H. \tag{4.1.5}$$

Hence, as the time translational automorphisms, $\alpha(t)$, are defined by the
formula

$$\alpha(t)A = \exp(iHt/\hbar)A\exp(-iHt/\hbar), \tag{4.1.6}$$

it follows from Eqs. (4.1.5) and (4.1.6), together with the antiautomorphic
property of τ, that

$$\tau\alpha(t)\tau = \alpha(-t) \quad \forall t \in \mathbf{R}.$$

Since both $\alpha(-t)\alpha(t)$ and τ^2 reduce to the identity, this condition is equivalent
to

$$\tau\alpha(t)\tau\alpha(t) = I \quad \forall t \in \mathbf{R}_+. \tag{4.1.7}$$

This is the principle of *microscopic reversibility* (MR). It is easily extended to
a particle of spin 1/2, whose algebra of observables is the tensor product of
spatial and spin components and, correspondingly,

$$\tau = \tau_{\text{space}} \otimes \tau_{\text{spin}}, \tag{4.1.8}$$

where τ_{space} is the antiautomorphism denoted above by τ and τ_{spin} is defined, as
in the example in Section 2.4, item (27), by the equation

$$\tau_{\text{spin}}\sigma_u = -\sigma_u \quad \text{for } u = x, y, z. \tag{4.1.9}$$

Likewise, the MR formula (4.1.7) remains valid for many particle systems,
with τ defined as the canonical generalisation of the single particle time
reversal antiautomorphism.

Comments

(1) The MR condition has been effectively realised in "spin echo" experiments [Hah].

(2) In general, the essential content of the MR principle is that the obser-

vables of the system remain unchanged by the sequence of operations comprising free evolution over time t, followed by a reversal of the momenta and spins of all the particles, followed by a further free evolution over time t and then by another reversal of the particles' momenta and spins. Note that it is essential here that the momenta and spins of *all* the particles be reversed in the second and fourth operations of this sequence.

(3) As expressed by Eq. (4.1.7), the MR principle pertains exclusively to the evolution of a system over positive times and thus does not involve considerations of fictitious processes that "go backwards in time".

(4) Another way of looking at MR is to re-express Eq. (4.1.7) in terms of the action of its dual on the state space, that is,

$$\alpha^{\star}(t)\tau^{\star}\alpha^{\star}(t)\tau^{\star}\rho = \rho,$$

where ρ is an arbitrary state. Thus, replacing ρ by $\tau^{\star}\rho$ and noting that $(\tau^{\star})^2$, like τ^2, is equal to the identity, this last formula is equivalent to

$$\alpha^{\star}(t)\tau^{\star}\alpha^{\star}(t)\rho = \tau^{\star}\rho.$$

Hence, defining ρ^{τ} to be $\tau^{\star}\rho$, MR signifies that, *if $\rho_f = \alpha^{\star}(t)\rho_i$ is the evolute at time t of an initial state ρ_i, then the evolute of ρ_f^{τ} at time t is ρ_i^{τ}.*

4.1.2 Infinite Systems

We now extend the MR principle to infinite systems, starting with the model of Section 2.5.3 for a system, Σ, of particles of one species in a Euclidean continuum X.

We define the time reversal antiautomorphism, τ, of its quasi-local algebra, $\mathcal{A}$, in terms of the single particle Wigner operator T by the formula

$$\tau\psi(f) = \psi(Tf)^{\star} \equiv \psi(\bar{f})^{\star},$$

or, more precisely,

$$\tau W(f) = W(\bar{f}) \quad \text{for bosons,} \tag{4.1.10}$$

where $W(f)$ is the Weyl operator defined by Eq. (2.5.6), and

$$\tau\psi_s(f) = (\psi_{-s}(\bar{f}))^{\star} \quad \text{for fermions.} \tag{4.1.11}$$

It follows from these specifications that the action of τ preserves the CCR or CAR. Further, τ is formally related to the single particle time reversal anti-automorphism, which we now denote by τ_{part}, by the following equation.

$$\tau\left(\int \psi^{\star}(x)A\psi(x)dx\right) = \int \psi^{\star}(x)(\tau_{\text{part}}A)\psi(x),$$

where A is an arbitrary single particle observable and ψ is a scalar or a spinor according to whether the particles are bosons or fermions. Thus, the action of τ serves to reverse all the momenta and spins of the system, while leaving the particle positions unchanged.

In order to relate this antiautomorphism to the dynamics of an infinite system Σ, as formulated in Section 2.5, we note that, by Eq. (2.5.11) and the definition of τ, the local Hamiltonians H_Λ are invariant under the action of τ. Hence, by Eq. (2.5.12),

$$\tau\alpha_\Lambda(t)\tau = \alpha_\Lambda(-t).$$

On passing to the limit $\Lambda \uparrow X$, whether in norm or in the strong topology of an appropriate representation,[3] this yields the MR formula (4.1.7).

In a similar way, the MR formula for spin systems on a lattice may be derived within the framework of Section 2.5.2, the time reversal antiautomorphism τ acting so as to reverse the spins, as in Eq. (4.1.9).

4.2 FROM SYSTEMS TO SUBSYSTEMS: COMPLETELY POSITIVE MAPS, QUANTUM DYNAMICAL SEMIGROUPS AND CONDITIONAL EXPECTATIONS

4.2.1 Complete Positivity

The concept of completely positive (CP) operations, T, on a $C^\star$ algebra, $\mathcal{A}$, due originally to Stinespring [Sti], has been shown by Gorini and Kossakowski [GK] and by Lindblad [Li 2,3] to be crucial to the passage from the dynamics of quantum systems to that of their subsystems. These operations are characterised by the fact, that, if M is an arbitrary finite-dimensional matrix algebra, then not only T but also $T \otimes I_M$, in its action on $\mathcal{A} \otimes M$, preserves positivity. In other words, if Σ and S are independent systems, whose algebras of observables are $\mathcal{A}$ and M, respectively, then the positivity of the action of T on $\mathcal{A}$ extends to that of $T \otimes I_M$ on the observables of the composite, $\Sigma + S$. Remarkably, and unlike its classical counterpart, this is a strictly stronger property than that of the positivity of T; and moreover it is one that arises naturally in the dynamics of subsystems of quantum dynamical systems.

The precise definition of complete positivity is the following.

Definition 4.2.1

(1) If $\mathcal{A}$ and $\mathcal{B}$ are $C^\star$-algebras, than a linear mapping $f : \mathcal{A} \to \mathcal{B}$ is termed *positive* if $A > 0$ implies that $f(A) > 0$; and *completely positive*

---

[3] As pointed out in Section 2.5, Σ is a $C^\star$ or $W^\star$ dynamical system according to whether the first or second of these alternatives is realized.

[Sti], or CP, if, for any finite dimensional matrix algebra, M, the mapping, $f \otimes I$, of $\mathcal{A} \otimes M$ into $\mathcal{B} \otimes M$, is positive. In the case where $\mathcal{A}$ and $\mathcal{B}$ are a $W^{\star}$-algebras, we term f *normal* if it is ultraweakly continuous, that is, weakly continuous on the unit ball. In general we denote by $CP_1(\mathcal{A})$ the set of all linear, identity-preserving CP transformations of $\mathcal{A}$; and, in the case where this algebra is $W^{\star}$, we denote by $CP_1^{(N)}(\mathcal{A})$ the set of normal elements of $CP_1(\mathcal{A})$.

(2) The dual, $f^{\star}$, of a linear CP transformation, f, of $\mathcal{A}$ is defined by the formula

$$(f^{\star}\rho)(A) = \rho(fA). \qquad (4.2.1)$$

In the case where $\mathcal{A}$ is a $W^{\star}$ algebra and f is normal, this formula serves to define $f^{\star}$ as a transformation of $\mathcal{N}(\mathcal{A})$, the space of normal states on $\mathcal{A}$.

Comment   As mentioned above, the CP condition is strictly stronger than that of positivity [Sti], and is indeed very restrictive [GK, Li2, Li3]. The essential difference beween positivity and complete positivity may be interpreted, in physical terms, as being due to quantum intereference. We provide a simple example of a positive map that is not CP in Appendix A.

Examples of CP Maps

(1) The automorphisms of $C^{\star}$-algebras are CP.

(2) If $\mathcal{A}$ and $\mathcal{B}$ are the algebras of bounded operators in Hilbert spaces $\mathcal{H}$ and $\mathcal{K}$, respectively, where $\mathcal{K}$ is a subspace of $\mathcal{H}$ whose projection operator is P, then the mapping $f : \mathcal{A} \to \mathcal{B}$, given by $f(A) = PAP$ is CP, normal and identity preserving.

(3) In the generic model of an open system formulated in Section 2.7, the mapping P of $\mathcal{A}_C$ onto $\mathcal{A}$ and the transformation $\theta(t)$ of $\mathcal{A}$ are CP, normal and identity preserving.

Some General Properties of CP maps

The key theorem here is that of Stinespring [Sti], which asserts that, if $\mathcal{B}$ is identified with a faithful representation of this algebra in a Hilbert space, $\mathcal{H}$, then any linear CP map $f : \mathcal{A} \to \mathcal{B}$ takes the form

$$f(A) = V^{\star} \pi(A) V, \qquad (4.2.2)$$

where π is a representation of $\mathcal{A}$ in a second Hilbert space, $\mathcal{K}$, and V is a bounded linear mapping from $\mathcal{H}$ into $\mathcal{K}$. The following general properties of CP maps follow from this theorem.

(CP.1) A linear CP map, $f : \mathcal{A} \rightarrow \mathcal{B}$, is norm continuous, that is, if A_n converges in norm to A, then $f(A_n)$ converges to $f(A)$.

(CP.2) A linear, identity preserving CP map, f, does not neccessarily preserve algebraic structure, that is, it might violate the condition $f(A_1 A_2) = f(A_1)f(A_2)$. It does, however, satisfy the Schwartz-type inequality (cf. [Sto])

$$f(A^\star A) \geq f(A^\star)f(A). \tag{4.2.3}$$

(CP.3) [Kr] If $\mathcal{A}$ is the algebra of bounded operators in a separable Hilbert space $\mathcal{H}$, then any linear CP transformation, f, of $\mathcal{A}$ is of the form

$$f(A) = \sum_j V_j^\star A V_j, \tag{4.2.4}$$

where the V_j and $\sum_j V_j^\star V_j$ all belong to $\mathcal{A}$. It follows that f is normal, that is, its action on the unit ball $\mathcal{A}_1$ conserves weak continuity.

(CP.4) [Li2] If ρ and σ are normal states on the algebra $\mathcal{A} = \mathcal{B}(\mathcal{H})$ and $f \in CP_1^{(N)}(\mathcal{A})$, then

$$\hat{S}_{\text{rel}}(f^\star \sigma \mid f^\star \rho) \geq \hat{S}_{\text{rel}}(\sigma \mid \rho), \tag{4.2.5}$$

where $\hat{S}_{rel}$ is the relative entropy functional defined by Eq. (3.2.10).

4.2.2 Quantum Dynamical Semigroups

Definition 4.2.2 A quantum dynamical semi-group is a set $\{T(t) \mid t \in \mathbf{R}_+\}$ of elements of $CP_1(\mathcal{A})$ that possesses the semi-group properties

$$T(s)T(t) = T(s+t) \quad \forall s, t \in \mathbf{R}_+ \text{ and } T(0) = I.$$

Thus, the evolution of the open system of Section 2.7 is governed by a quantum dynamical semi-group in the weak coupling limit described there, and, as discussed below, the same is true in a wider context for the dynamics of subsystems of conservative quantum systems.

In the case where $T(t)A$ is normwise continuous in t for all $A \in \mathcal{A}$, the semi-group T has an infinitesimal generator, L, defined by the formula

$$\frac{d}{dt}T(t)A = LT(t)A = T(t)LA \quad \forall A \in \mathcal{A},$$

and Lindblad [Li3] has obtained the important result that, in the case where $\mathcal{A}$ is the algebra of bounded operators in a Hilbert space, L takes the form

$$LA = i[H, A] + \sum_j \left(V_j^\star A V_j - \frac{1}{2}[V_j^\star V_j, A]_+ \right), \tag{4.2.6}$$

where H is a self-adjoint element of $\mathcal{A}$ and V_j and $\sum_j V_j^\star V_j$ all belong to this algebra. In fact, this formula has been generalised to the generators of some quantum dynamical semi-groups satisfying weaker continuity conditions than the above one of norm continuity [Vh, Kho, Fa, CF, AlSe].

4.2.3 Conditional Expectations

Definition 4.2.3

(1) If $\mathcal{B}$ is a $C^\star$-subalgebra of a $C^\star$-algebra $\mathcal{A}$, then a *conditional expectation* (CE) of $\mathcal{A}$ with respect to $\mathcal{B}$ is defined to be a linear CP mapping $P : \mathcal{A} \to \mathcal{B}$, such that

    (i) $P(B) = B \ \forall B \in \mathcal{B}$, which signifies that P is both projective and identity preserving, that is, $P^2 = P$ and $P(I_{\mathcal{A}}) = P(I_{\mathcal{B}}) = I_{\mathcal{B}}$; and

    (ii) $P(B_1 A B_2) = B_1 P(A) B_2 \ \forall A \in \mathcal{A}$ and $B_1, B_2 \in \mathcal{B}$.

(2) The CE P is termed *compatible with a state* ρ on $\mathcal{A}$ if

$$\rho(A) = \rho(P(A)) \quad \forall A \in \mathcal{A}, \tag{4.2.7}$$

or equivalently, defining $\rho_{\mathcal{B}}$ to be the restriction of ρ to $\mathcal{B}$, if

$$\rho(A) = \rho_{\mathcal{B}}(P(A)) \ \forall A \in \mathcal{A}. \tag{4.2.7$'$}$$

(3) We define $S^{(P)}$ to be the set of states on $\mathcal{A}$ with which P is compatible, and $S^{(P)}_{\mathcal{B}}$ to be the restriction of $S^{(P)}$ to $\mathcal{B}$.

Examples of CEs

(1) The map f of Example (2) in Section 4.2.1 is a CE, and is compatible with any state of $\mathcal{A}$ that corresponds to a density matrix in $\mathcal{K}$.

(2) The map P of Example (3) in Section 4.2.1 is a CE and is compatible with any state of the form $\rho \otimes \rho_{\mathcal{R}}$.

Comments on CEs

(1) If P is compatible with ρ, then, by Definition 4.2.3 (2), $\rho(A)$ is obtained by first taking the conditional expectation of A with respect to the subalgebra, $\mathcal{B}$, of $\mathcal{A}$, and then taking the expectation value of $P(A)$ for the state $\rho_{\mathcal{B}}$. This precisely parallels the procedure for formulating classical expectation values of functions on a classical probability space in terms of conditional expectations.

(2) It follows easily from the above specifications that both $S^{(P)}$ and $S^{(P)}_{\mathcal{B}}$ are norm closed convex sets of states that are stable under the modifications $\rho \to \rho(B^\star(\cdot)B)/\rho(B^\star B)$ for $B \in \mathcal{B}$. Hence, $S^{(P)}_{\mathcal{B}}$ is a *folium*, as defined in Section 2.4.2.

4.3 INDUCED DYNAMICAL SUBSYSTEMS

We assume that Σ is a quantum dynamical system $(\mathcal{A}, S, \alpha)$, that $\mathcal{B}$ is a $C^\star$ subalgebra of $\mathcal{A}$, that is stable under the action of τ, and that P is a CE of $\mathcal{A}$ onto $\mathcal{B}$ that commutes with τ. We also assume that, in the case where Σ is a $W^\star$-dynamical system and $\mathcal{B}$ is a $W^\star$-algebra, P is normal, that is, ultra-weakly continuous.

Suppose now that Σ evolves from an initial state, ρ, with which P is compatible. Then it follows from Eq. $(4.2.7)'$ that the time-dependent expectation value, $\rho(\alpha(t)B)$, of an element B of $\mathcal{B}$ is $\rho_{\mathcal{B}}(\theta(t)B)$, where $\theta(t)$ is the CP_1-class transformation of $\mathcal{B}$ given by

$$\theta(t) = P\alpha(t)P \quad \forall t \in \mathbf{R}_+. \tag{4.3.1}$$

Hence, by the assumptions of the previous paragraph on E, $\theta(t)$ is norm or weakly continuous according to whether Σ is a $C^\star$ or $W^\star$ dynamical system.

These considerations on θ lead us to the following definition of an induced dynamical subsystem of Σ.

Definition 4.3.1   We define an *induced dynamical subsystem (IDS)*, Γ, of Σ to comprise a triple $(\mathcal{B}, \mathcal{F}, \theta)$, where

(i) $\mathcal{B}$ is a $C^\star$ or $W^\star$ subalgebra of $\mathcal{A}$, according to whether Σ is a $C^\star$ or $W^\star$ system;

(ii) $\{\theta(t) \mid t \in \mathbf{R}_+\}$ is a one-parameter family of identity preserving CP transformations of $\mathcal{B}$, induced by the automorphisms α according to Eq. (4.3.1) and possessing the continuity properties of $\alpha(\mathbf{R}_+)$; and

(iii) $\mathcal{F}$ $(\subset S_{\mathcal{B}}^{(P)})$ is a folium of states on $\mathcal{B}$ that is stable under $\tau^\star$ and separates the elements of $\mathcal{B}$; that is, if B $(\in \mathcal{B})$ is nonzero, then $\mathcal{F}$ contains a state ψ such that $\psi(B) \neq 0$.

We take $\mathcal{B}$ and $\mathcal{F}$ to be the algebra of observables and the state space, respectively, of Γ, and θ to represent its dynamics.

Example   The open system $(\mathcal{A}, \mathcal{N}(\mathcal{A}), \theta)$ of Section 2.7 is an IDS.

4.4 IRREVERSIBILITY

4.4.1 Irreversibility, Mixing and Markovian Dynamics

The evolution of the IDS Γ is governed by the dynamical transformations $\theta(t)$, and it can be seen readily from the presence of the factors P in the formula (4.3.1) that they do not neccessarily inherit either the invertibility or the MR property (4.1.7) of the automorphisms $\alpha(t)$. Thus, the possibility arises that the dynamics of Γ might, in some sense, be irreversible. To be precise, we define irreversibility as follows.

Definition 4.4.1   We term the dynamics of the IDS Γ *irreversible* if there is a functional, Λ, on the state space, $\mathcal{F}$, and a subset, $\mathcal{F}_0$, of $\mathcal{F}$, whose complement $\mathcal{F}_0^{(c)} := \mathcal{F} \backslash \mathcal{F}_0$ is $w^\star$-dense in $\mathcal{F}$, such that

   (i) Λ is τ-invariant;

   (ii) $\mathcal{F}_0$ is stable under the transformations $\theta^\star(t)$ and $\tau^\star$; and

   (iii) $\Lambda(\theta^\star(t)\psi)$ is a decreasing function of t for ψ in $\mathcal{F} \backslash \mathcal{F}_0$, and a constant one for ψ in $\mathcal{F}_0$.

Comments

   (1) Evidently, Λ is a Lyapunov functional, which serves as an "arrow of time".

   (2) The τ-invariance of Λ ensures that this functional is neutral with respect to past and future, and thus that the Lyapunov properties (iii) do represent a bona fide irreversibility.

   (3) The domain $\mathcal{F}_0$ is designed to comprise the "stable attractors" to which the states of Γ may evolve. Thus, as in the theory of classical dynamical systems [RT], they may consist of fixed points or periodic, or even "chaotic" orbits. We shall provide concrete examples of these various types of attractors in Chapter 11.

   (4) It follows from the above definition that the irreversibility of Γ implies that this subsystem cannot satisfy the MR condition of Section 4.1, with α replaced by θ. For that condition would imply that

$$\Lambda\left(\theta^\star(t)\tau^\star\theta^\star(t)\psi\right) = \Lambda(\psi) \quad \forall \psi \in \mathcal{F},$$

   and it is a simple consequence of Definition 4.4.1 that the left-hand side of this equation is strictly less than $\Lambda(\psi)$, for $\psi \in \mathcal{F} \backslash \mathcal{F}_0$, while the right-hand side is equal to this quantity.

   (5) The conditions for irreversibility given by Definition 4.4.1 could be weakened in various ways, and still preclude the possibility that Γ was microscopically reversible. For example, this could achieved by demanding only the monotonic decrease either of $\Lambda(\theta^\star(t)\psi)$, for $\psi \in \mathcal{F} \backslash \mathcal{F}_0$, as t increased over some infinite sequence of times $(t_1, t_2, ..., t_n, ...)$, or of its time averages over a sequence of intervals $\{[nb, (n+1)b] \mid n \in \mathbf{N}\}$, with $b > 0$.

   In Proposition 4.4.3, we shall relate irreversibility to the concepts of Markovian dynamics and mixing, which we define as follows.

Definition 4.4.2

   (1) We term the dynamics of Γ *Markovian* if the transformations θ form a

one-parameter semi-group, that is, if $\theta(t)\theta(s) = \theta(t+s)$ for all $t, s \geq 0$. In this case, $\theta(\mathbf{R}_+)$, which is evidently a quantum dynamical semi-group, is also termed Markovian.

(2) The dynamics of Γ is termed *mixing* if it is Markovian and if, further, all its states evolve to a (unique) element ϕ of $\mathcal{F}$, in that

$$w^\star : \lim_{t \to \infty} \theta^\star(t)\psi = \phi \quad \forall \psi \in \mathcal{F}. \tag{4.4.1}$$

Comments
(1) There is a wide class of open systems that enjoy Markov, mixing and irreversibility properties, either exactly [LTh] or in certain natural limits [Da]. Simple examples are provided in Appendices B and C.

(2) In the case where $\mathcal{B}$ is the algebra of bounded operators in a separable Hilbert space, $\mathcal{F}$ is the folium of normal states on $\mathcal{B}$, and $\theta(t)B$ is norm continuous in t, the infinitesimal generator, L, of θ is given by the Lindblad formula (4.2.6), that is,

$$LB = i[H, B] + \sum_j \left(V_j^\star B V_j - \frac{1}{2}[V_j^\star V_j, B]_+ \right) \quad \forall B \in \mathcal{B}.$$

Here the term $i[H, B]$ corresponds to a conservative evolution, and the remaining terms represent dissipative effects.

The following proposition establishes that the dynamics of a certain class of mixing IDS's is irreversible, and that their relative entropies serve as Lyapunov functionals.

Proposition 4.4.3 *If Γ is a mixing IDS $(\mathcal{B}, \mathcal{F}, \theta)$, with $\mathcal{B}$ the algebra of operators in a finite dimensional Hilbert space, $\mathcal{H}$, $\mathcal{F}$ the set of all states on $\mathcal{B}$, and ϕ the unique stationary state on $\mathcal{B}$, then the relative entropy $\hat{S}_{rel}(\phi \mid \theta^\star(t)\psi)$ decreases monotonically with t if $\psi \neq \phi$ and remains constant (and equal to zero) if $\psi = \phi$. Thus the dynamics of Γ is irreversible, and $\hat{S}_{rel}(\phi \mid \psi)$ serves as a Lyapunov functional.*

Proof. We use the properties of $\hat{S}_{rel}$ given by (S.3) and (S.5) in Section 3.2.2 and the CP property (CP.4), Section 4.2, of the conditional expectation P.

Since ϕ is invariant under $\theta^\star(t)$, it follows from (CP.4) and the semi-group property of θ, which ensues from the assumption of mixing, that

$$\hat{S}_{rel}(\phi \mid \theta^\star(t+s)\psi) \equiv \hat{S}_{rel}(\theta^\star(s)\phi \mid \theta^\star(s)\theta^\star(t)\psi) \leq \hat{S}_{rel}(\phi \mid \theta^\star(t)\psi)$$

$$\forall s, t \in \mathbf{R}_+.$$

$\hat{S}_{rel}(\phi \mid \theta^\star(t)\psi)$ is therefore a nonincreasing function of t. It is also a continuous function of this variable, by virtue of (S.5) and the finite dimensionality of $\mathcal{H}$. Hence the mixing property of Γ, represented by Eq. (4.4.1), implies that

$$\lim_{t\to\infty}\hat{S}_{\text{rel}}(\phi \mid \theta^\star(t)\psi) = \hat{S}_{\text{rel}}(\phi \mid \phi) = 0. \qquad (4.4.2)$$

Therefore, if $\psi \neq \phi$ and consequently, by (S.3), $\hat{S}_{\text{rel}}(\phi \mid \theta^\star(t)\psi)$ is strictly positive at $t = 0$, then this relative entropy is not merely a nonincreasing decreasing function of t, but a decreasing one; while, if $\psi = \phi$, it is zero, by virtue of (S.3) and the $\theta^\star(t)$-invariance of ϕ. $\square$

Comment The assumption of the finite dimensionality of $\mathcal{H}$ was used in this proof in order to establish the continuity of $\hat{S}_{\text{rel}}(\phi \mid \theta^\star(t)\psi)$ with respect to t. Otherwise, if $\mathcal{H}$ were infinite dimensional and $\mathcal{F}$ were the folium of normal states on the algebra $\mathcal{B}(\mathcal{H})$, the lower semicontinuity of $\hat{S}_{\text{rel}}$, specified by (S.3), would ensure that this function was continuous for finite nonzero values of t, but would not preclude the possibility that it might drop discontinuously in the limits $t \downarrow 0$ and $t \uparrow \infty$. However, since $\theta(0) = I$ and $\theta(t)$ satisfies (CP.4), the former discontinuity cannot occur. Hence, this relative entropy is continuous in t, except for a possible downward jump in the limit $t \uparrow \infty$. Therefore, in order to extend the proposition to the case where $\mathcal{H}$ is infinite dimensional, it suffices to add the condition that no such jump occurs. In fact, this condition is fulfilled by the Lyapunov functions of various models, including that of Appendix B.

4.5 NOTE ON CLASSICAL MACROSCOPIC CAUSALITY
From Quantum Stochastics to Classical Determinism

The central problem of nonequilibrium statistical mechanics is that of deriving the irreversible, deterministic macroscopic dynamical laws of many-particle systems, such as those of hydrodynamics or heat conduction, from their underlying quantum dynamics. We now sketch a rather general scheme for the connection between the macroscopic and the quantum pictures that lead to these laws. The scheme will be shown to be realised by a tractable model of an open system in Appendix C and will be further developed in Chapter 7, in the context of nonequilibrium thermodynamics, and in Chapters 10 and 11, in connection with the generations of ordered and chaotic evolutions far from equilibrium.

    We start by assuming, on standard empirical grounds, that the macroscopic picture is that of a *classical dynamical system* $\mathcal{M} = (\mathcal{Y}, T)$, where $\mathcal{Y}$ is a topological space and $\{T(t) \mid t \in \mathbf{R}_+\}$ is a one-parameter semi-group of transformations of $\mathcal{Y}$. Here, the points, Y, of $\mathcal{Y}$ represent the pure states of $\mathcal{M}$ and $T(t)Y$ represents the evolute of Y at time t. In the (usual) case that $\mathcal{M}$ is dissipative, we assume that its irreversibility is characterised by the existence of a closed subset, $\mathcal{Y}_0$, of $\mathcal{Y}$ and a function, $\Lambda_{\mathcal{M}}$, on $\mathcal{Y}$, such that (cf. [RT, Ru3])

($\mathcal{M}.1$) $\mathcal{Y}_0$ is stable under $T(\mathbf{R}_+)$,

($\mathcal{M}.2$) the complementary subset, $\mathcal{Y}_0^{(c)} = \mathcal{Y}\backslash\mathcal{Y}_0$, is dense in $\mathcal{Y}$, and

($\mathcal{M}.3$) for fixed Y, $\Lambda_{\mathcal{M}}(T(t)y)$ is a monotonically decreasing function of t if $Y \in \mathcal{Y}_0^{(c)}$ and is constant if $Y \in \mathcal{Y}_0$.

Under these assumptions, $\mathcal{Y}_0$ then comprises the attractors of $\mathcal{M}$, and the $\Lambda_{\mathcal{M}}$ is a Lyapounov function, that reveals the irreversibility of the dynamics of $\mathcal{M}$.

We now assume that the classical system $\mathcal{M}$ corresponds, in a sense that is specified below, to the dynamics of a particular set of observables, Y_Ω, of the quantum system $\Sigma = (\mathcal{A}, S, \alpha)$. Here, we assume, for simplicity, that Y_Ω consists of a finite set[4] of observables $(Y_\Omega^{(1)}, ..., Y_\Omega^{(k)})$ and that Ω is a "large", dimensionless, positive valued parameter, whose magnitude provides a measure of their macroscopicality. Thus, for example, in the case of a heavy projectile, $\mathcal{P}$, that moves through a material medium, a natural choice for Ω would be the mass of $\mathcal{P}$ in atomic units; while, in the context of hydrodynamics exhibited by Σ, Ω would be some mean number of atoms in a subvolume that is large from the microscopic standpoint but small from the macroscopic one.

We then assume that there is a set, Δ_Ω, of states of Σ, that satisfy the following conditions.

(a) For each state $\phi_\Omega \in \Delta_\Omega$, the means and dispersion of Y_Ω converge to limits, Y and 0, respectively, as $\Omega \to \infty$. We assume that, as ϕ_Ω runs through Δ_Ω, the resultant range of the values of Y is just the phase space, $\mathcal{Y}$, of $\mathcal{M}$. Thus, under the assumption that Y_Ω consists of k observables, $\mathcal{Y}$ is a susbset of $\mathbf{R}^k$.

(b) The dynamical semi-group, $T(\mathbf{R}_+)$, of $\mathcal{M}$ is induced by the evolution of Σ from Δ_Ω- class states on a "macroscopic" time scale Ω^γ, with $\gamma > 0$. Specifically, we assume that the mean and dispersion of $\alpha(\Omega^\gamma t)Y_\Omega$ for the state ϕ_Ω converge to $T(t)Y$ and 0, respectively, as $\Omega \to \infty$.

Thus, under these assumptions, the evolution of Σ from the Δ_Ω-class states induces a dynamics of the observables Y_Ω that reduces, on the macroscopic time scale Ω^γ, to that of the classical deterministic system $\mathcal{M}$ in the limit where Ω tends to infinity.

We remark here that characterisations of the macroscopicality parameter Ω, the observables Y_Ω and the initial states Δ_Ω seems to be limited to (a) particular models, such as those of Appendix C and Chapter 11, where the scheme presented here can be carried through, and (b) the nonequilbrium thermody-namical regime of Chapter 7, where the advent of general local conservation laws provides some guidance.

---

[4] A generalisation to the case where it consists of an infinite set of observables or a set of quantum fields (cf. Chapter 7) requires further topological considerations.

APPENDIX A: EXAMPLE OF A POSITIVE MAP THAT IS NOT COMPLETELY POSITIVE

Our example here is the spin reversal antiautomorphism, τ_{spin}, which was defined by Eq. (4.1.9). In fact, it is a simple case of a more general example provided by Stinespring [Sti].

In order to obtain the positivity properties of τ_{spin} within the terms of Section 4.2, we define $\mathcal{A}$ to be the algebra, M_2, of 2×2 matrices. Thus, $\mathcal{A}$ is the linear span of the identity and the Pauli matrices $(\sigma_x, \sigma_y, \sigma_z)$, defined by Eq. (2.2.8). The algebraic properties of these matrices are given by the well-known relations

$$\sigma_x \sigma_y = -\sigma_y \sigma_x = i\sigma_z, \text{ etc.,} \quad \sigma_x^2 = I, \text{ etc.} \tag{A.1}$$

The spin reversal antiautomorphism, which we now denote simply by τ, is just the linear, identity preserving transformation of $\mathcal{A}$ given by the formula

$$\tau \sigma_u = -\sigma_u \quad \text{for } u = x, y, z. \tag{A.2}$$

Our aim is to prove that τ is positive but not completely positive.

Proof of Positivity of τ. It follows from the definition of $\mathcal{A}$ that the self-adjoint elements of this algebra are of the form

$$A = a.\sigma + bI, \tag{A.3}$$

where σ is the spin vector $(\sigma_x, \sigma_y, \sigma_z)$, a and b are elements of $\mathbf{R}^3$ and $\mathbf{R}$, respectively, and the dot represents the scalar product in $\mathbf{R}^3$. Hence, by Eqs. (A.2) and (A.3),

$$\tau(A) = -a.\sigma + bI. \tag{A.4}$$

We now note that, by Eqs. (A.1) and (A.3), A is positive if and only if $b > 0$ and $b \geq | a |$; and by Eq. (A.4), these are also the neccessary and sufficient conditions for the positivity of $\tau(A)$. This signifies that τ preserves positivity and so is positive.

Proof that τ is not CP. We prove this by demonstrating that the linear transformation $\tau_2 := \tau \otimes I$ of $\mathcal{A}_2 := \mathcal{A} \otimes \mathcal{A} \equiv \mathcal{A} \otimes M_2$ is not positive. To this end, we recall that the single column vectors $\phi_{\pm 1}$, as defined by Eq. (2.2.9), form an orthonormal basis in the representation space of $\mathcal{A}$. Hence, the vectors $\{\phi_s \otimes \phi_{s'} \mid s, s' = \pm 1\}$ form such a basis for the representation space of $\mathcal{A}_2$, and, consequently, any vector in this latter space may be expressed in the form

$$\psi = \sum_{s=\pm 1}(a(s)\phi_s \otimes \phi_s + b(s)\phi_s \otimes \phi_{-s}), \tag{A.5}$$

where the coefficients $a(s)$ and $b(s)$ are complex numbers.

We define B to be the element of $\mathcal{A}_2$ given by the formula

$$B = \sigma_x \otimes \sigma_x - \sigma_y \otimes \sigma_y + \sigma_z \otimes \sigma_z + I \otimes I. \tag{A.6}$$

Hence, by Eq. (A.2) and the definition $\tau_2 := \tau \otimes I$,

$$\tau_2(B) = -\sigma_x \otimes \sigma_x + \sigma_y \otimes \sigma_y - \sigma_z \otimes \sigma_z + I \otimes I. \tag{A.7}$$

We now note that, by Eqs. (2.2.8)–(2.2.10), (A.1) and (A.5)–(A.7),

$$(\psi, B\psi) = 4 \mid a(1) + a(-1) \mid^2, \tag{A.8}$$

and

$$(\psi, \tau_2(B)\psi) = 2 \sum_{s=\pm 1} (\mid b(s) \mid^2 - \mid a(s) \mid^2). \tag{A.9}$$

From these last two equations, we see immediately that B is positive, but that, for certain choices of $a(s)$ and $b(s)$, $(\psi, \tau_2(B)\psi)$ is negative. Hence τ_2 is not positive, and therefore τ is not CP.

APPENDIX B: SIMPLE MODEL OF IRREVERSIBILITY AND MIXING

The model, Σ, which formed a part of the Hepp–Lieb laser [HL1], consists of a harmonic oscillator, Σ_B, coupled to a quantum field, Σ_C. Thus, Σ_B is an open system and Σ_C is its reservoir.

We formulate Σ_B in terms of creation and annihilation operators, $b^\star$ and b, acting in an irreducible representation space, $\mathcal{H}_B$, of the CCR

$$[b, b^\star] = I. \tag{B.1}$$

We assume that the algebra of observables of this system is the $W^\star$-algebra, $\mathcal{B}$, of bounded operators in $\mathcal{H}_B$. Thus, $\mathcal{B}$ is generated by the Weyl operators

$$W(z) = \exp\left(i(zb + \bar{z}b^\star)\right) \quad \forall z \in \mathbb{C}. \tag{B.2}$$

Expressed in terms of these operators, the CCR takes the form

$$W(z)W(z') = W(z + z')\exp(i\,\mathrm{Im}(\bar{z}z')). \tag{B.3}$$

We define ϕ to be the pure state on $\mathcal{B}$ corresponding to the eigenvector of the number operator $b^\star b$ with eigenvalue zero. Thus, by elementary quantum mechanics,[5]

$$\phi[W(z)] = \exp\left(-\frac{1}{2}\phi((zb + \text{h.c.})^2)\right) = \exp\left(-\frac{1}{2}\mid z \mid^2\right). \tag{B.4}$$

---

[5] This corresponds to the fact that the wave function for the ground state of a harmonic oscillator is a Gaussian function of its position, centered at the origin.

We take Σ_C to be a complex Bose field, $c(k)$, in one spatial dimension, where the variable k runs through $\mathbf{R}$ and indexes the normal modes. We assume that $c(k)$ and $c^\star(k)$ operate in a Fock space, $\mathcal{H}_C$, and are defined by the standard conditions that

(a) $\mathcal{H}_C$ contains a vacuum vector Φ_C, that is annihilated by the smeared fields $c(f) = \int_\mathbf{R} dk f(k)c(k)$, with $f \in L^2(\mathbf{R})$;

(b) $\mathcal{H}_C$ is generated by the application to Φ_C of the polynomials in the adjoints, $c(f)^\star$, of these smeared fields; and

(c) c and $c^\star$ satisfy the canonical commutation relations

$$[c(k), c^\star(k')] = \delta(k - k'), \quad [c(k), c(k')] = 0. \tag{B.5}$$

We assume that the algebra of observables for Σ_C is the $W^\star$-algebra, C, of bounded operators in $\mathcal{H}_C$. This is generated by the Weyl operators

$$W_C(f) = \exp\left(i(c(f) + c(f)^\star)\right) \quad \forall f \in L^2(\mathbf{R}). \tag{B.6}$$

We formulate the model Σ as a $W^\star$-dynamical system, whose algebra of observables, $\mathcal{A}$, is the $W^\star$ tensor product, $\mathcal{B} \otimes C$, and whose state space, S, consists of the normal states on $\mathcal{A}$. We generally identify $\mathcal{B}$ and C with the subalgebras $\mathcal{B} \otimes I$ and $I \otimes C$, respectively, of $\mathcal{A}$.

We assume that the formal Hamiltonian of Σ, in units where Planck's constant is unity, is

$$H = \omega b^\star b + \int_\mathbf{R} dk k c^\star(k)c(k) + g \int_\mathbf{R} dk(bc^\star(k) + b^\star c(k)), \tag{B.7}$$

where ω and g are positive constants. Thus, the coupling between $\Sigma_\mathcal{B}$ and Σ_C, as represented by the last integral in this formula, is *singular*, in that the coefficient of $(bc^\star(k) + \text{h.c.})$ is k-independent. For simplicity, we treat the dynamics of the model on a formal level. A mathematically rigorous treatment, leading to the same results, can be obtained *either* by introducing, and eventually removing, a cut-off at large $|k|$ in the last term in (B.7), *or* by formulating the model within the framework of the quantum stochastic calculus of Hudson and Parthasarathy [HP].

Thus, in our formal treatment, we express the dynamical automorphisms, α, of Σ by the formula

$$\alpha(t)A = \exp(iHt)A \exp(-iHt) \equiv A(t). \tag{B.8}$$

The triple $(\mathcal{A}, S, \alpha)$ constitutes the model of Σ.

We assume that the subsystems $\Sigma_\mathcal{B}$ and Σ_C are prepared independently of one another at time $t = 0$ in normal states ψ and ϕ_C, respectively, the latter being the Fock vacuum, that is, the pure state corresponding to the vector Φ_C. Thus, the initial state of Σ is

$$\rho = \psi \otimes \phi_C. \tag{B.9}$$

We note here that the action of ϕ_C on $W_C(f)$ has a Gaussian form analogous to that of ϕ on $W(z)$, as given by Eq. (B.4), that is, [BR]

$$\phi_C(W_C(f)) = \exp\left(-\frac{1}{2}\|f\|^2\right) \quad \forall f \in L^2(\mathbf{R}). \tag{B.10}$$

We define P, the conditional expectation of $\mathcal{A}$ with respect to $\mathcal{B}$, by the formula

$$P(B \otimes C) = \phi_C(C)B \quad \forall B \in \mathcal{B}, \; C \in \mathcal{C}. \tag{B.11}$$

Thus, P is compatible with all states ρ of the form given by Eq. (B.9).

It follows from the above construction that the *open* system $\Sigma_{\mathcal{B}}$ is given by $(\mathcal{B}, \mathcal{N}, \theta)$, where $\mathcal{N}$ is the folium of normal states on $\mathcal{B}$ and the transformations $\theta(t)$ of this algebra are given by

$$\theta(t) = P\alpha(t)P. \tag{B.12}$$

Our aim now is to obtain the specific form of the action of $\theta(t)$ on the Weyl operators $W(z)$. For this purpose, we note that, by Eqs. (B.1), (B.5), (B.7) and (B.8), the Heisenberg equations of motion for b and c are

$$\frac{db(t)}{dt} = -i\omega b(t) - ig \int_{\mathbf{R}} dk c(k, t) \tag{B.13}$$

and

$$\frac{dc(k, t)}{dt} = -ikc(k, t) - igb(t). \tag{B.14}$$

The solution of the latter equation, as expressed in terms of $b(t)$, is

$$c(k, t) = c(k)\exp(-ikt) - ig \int_0^t dsb(s)\exp(-ik(t-s)). \tag{B.15}$$

On inserting this formula into Eq. (B.13), we obtain the following equation for $b(t)$.

$$\frac{db(t)}{dt} = -\left(i\omega + \frac{\kappa}{2}\right)b(t) + \xi(t), \tag{B.16}$$

where

$$\kappa = 2\pi g^2 \tag{B.17}$$

and

$$\xi(t) = ig \int_{\mathbf{R}} dk c(k)\exp(-ikt). \tag{B.18}$$

Thus, Eq. (B.16) is a Langevin equation for a damped harmonic oscillator, with ξ a stochastic force, whose statistical properties are governed by the vacuum state, ϕ_C, of Σ_C. The solution of Eq. (B.16) is

$$b(t) = b \exp(-(i\nu + \kappa/2)t) + \int_0^t ds \xi(s) \exp(-(i\nu + \kappa/2)(t-s)). \quad \text{(B.19)}$$

To obtain the algebraic and statistical properties of ξ we note that, by Eqs. (B.5), (B.17) and (B.18), this force is ig times the Fourier transform of $c(k)$ and satisfies the CCR

$$[\xi(t), \xi^\star(t')] = \kappa\delta(t - t'), \quad [\xi(t), \xi(t')] = 0. \quad \text{(B.20)}$$

It also follows from those specifications and the definition of the vacuum state ϕ_C that

$$\phi_C(\xi(t)) = 0, \quad \text{(B.21)}$$

$$\phi_C\left(\xi(t)\xi^\star(t')\right) = \kappa\delta(t - t') \quad \text{and} \quad \phi_C\left(\xi(t)\xi^\star(t')\right) = 0. \quad \text{(B.22)}$$

Furthermore, the state ϕ_C has a Gaussian property analogous to that given by Eq. (B.4) for the ground state of the oscillator, that is, if $\xi(h)$ is the time-smoothed force

$$\xi(h) = \int_{\mathbf{R}} dt \xi(t) h(t) \quad \text{for } h \in L^2(\mathbf{R}), \quad \text{(B.23)}$$

and $Q = \xi(h) + \xi(h)^\star$, then

$$\phi_C(\exp(i(Q)) = \exp\left(-\frac{1}{2}\phi_C(Q^2)\right).$$

Thus, by Eqs. (B.22) and (B.23),

$$\left\langle \phi_C; \exp\left(i(\xi(h) + \xi(h)^\star)\right)\right\rangle = \exp\left(-\frac{1}{2}\|h\|^2\right). \quad \text{(B.24)}$$

The dynamics of Σ_B, as given by Eqs. (B.16) and (B.20–B.23), together with the Gaussian property of the fluctuating force $\xi(t)$, corresponds to a quantum Brownian motion. We remark here that the lack of "memory", both in the damping force and in the time-autocorrelation functions for the fluctuating force, stems from the singular property of the $\Sigma_B - \Sigma_C$ coupling, which we noted directly after Eq. (B.7).

We now infer the following Proposition from the above results.

Proposition B.1

(1) The dynamics of Σ_B is Markovian, the transformations $\theta(t)$ forming a one-parameter semi-group given by the formula

$$\theta(t)[W(z)] = W(z\exp(-(i\omega + \kappa/2)t))\exp\left(-\frac{1}{2}\mid z\mid^2 (1 - \exp(-\kappa t))\right).$$
$$(B.25)$$

This is, in fact, a particular example of a class of semi-groups obtained by Vanheuverszwijn [Vh].

(2) The system $\Sigma_{\mathcal{B}}$ is mixing, all its states evolving to the ground state ϕ.

(3) The dynamics of this system is irreversible, with the Lyapunov functional Λ given by

$$\Lambda(\psi) = \psi(b^\star b) \quad \forall \psi \in \mathcal{N}. \qquad (B.26)$$

Comments

(1) The functional Λ represents the energy of the oscillator, that is, its free energy at zero temperature, up to the factor ω. Hence, denoting the canonical equilbrium state and free energy at inverse temperature β by ϕ_β and $\Phi(\beta)$, respectively, it follows from Section 3.2.2, property (S.6), that Λ corresponds to the limiting form, as $\beta \to \infty$, of $(\beta\omega)^{-1}\left(\hat{S}_{\text{rel}}(\phi_\beta \mid \psi) + \Phi(\beta)\right)$.

(2) By Eq. (B.25), the infinitesimal generator, L, of the semi-group θ is given by the equation

$$LB = \omega[b^\star b, B]_- + \kappa b^\star Bb - \frac{\kappa}{2}[b^\star b, B]_+, \qquad (B.27)$$

and is thus *formally* of the Lindblad type (cf. Eq. (4.2.6)).

Proof of Proposition B.1

(1) Eq. (B.25) follows from Eqs. (B.2), (B.11), (B.12), (B.19) and (B.24). The semi-group property of θ then ensues from Eq. (B.25) and the Weyl form (B.3) of the CCR.

(2) By Eqs. (B.2) and (B.25),

$$w : \lim_{t\to\infty} \theta(t)W(z) = \exp\left(-\frac{1}{2}\mid z\mid^2\right)I,$$

and hence, by Eq. (B.4),

$$\lim_{t\to\infty}\langle\theta^\star(t)\psi; W(z)\rangle = \phi(W(z)) \quad \forall z \in C, \psi \in \mathcal{N}.$$

Since $\mathcal{B}$ is generated by the Weyl operators $W(z)$ and therefore, by Eq. (B.3), is the weakly closed linear span of these operators $W(z)$, this equation establishes the required result.

(3) By Eqs. (B.11), (B.12), (B.19), (B.21) and (B.22),

$$\theta(t)(b^{\star}b) = b^{\star}b\exp(-\kappa t).$$

Hence, by Eq. (B.26),

$$\Lambda(\theta^{\star}(t)\psi) = \psi(b^{\star}b)\exp(-\kappa t).$$

Since $\psi(b^{\star}b)$ vanishes if and only if $\psi = \phi$, it follows that the left-hand side of this equation remains constant in this case, and otherwise decreases monotonically with t. Therefore the dynamics of $\Sigma_{\mathcal{B}}$ is irreversible, and Λ serves as a Lyapunov function for it.

APPENDIX C: SIMPLE MODEL OF IRREVERSIBILITY AND MACROSCOPIC CAUSALITY

Here, we provide a treatment of a rather simple model that provides a concrete realisation of the scheme of Section 4.5 and also exhibits a macroscopic fluctuation process corresponding to a classical Brownian motion.

C.1 The Model

We take the model, Σ, to be a transversely vibrating, semi-infinite harmonic chain with nearest neighbor interactions, such that the first particle, B, has mass M and moves along a smooth straight wire, while the rest of the particles, which we sometimes call "atoms", are each of mass m and are confined to the plane of the wire and the axis of vibration[6] (cf. Figure 4.C.1). This is a type of model that has been studied extensively for various purposes (cf. [FKM, LTh, Se9, To]). Here, we are concerned with the dynamics induced by Σ on B in the situation where that particle is macroscopic, in the sense that $M \gg m$. Our main aim is to show that, under appropriate initial conditions and in a limit where M/m tends to infinity,

(a) the velocity of B evolves according to a classical, irreversible, deterministic law, and

(b) the fluctuations of that velocity, on an appropriate scale, execute a classical Brownian motion, that is related to the law (a) according to the fluctuation-dissipation relation, represented here by Eq. (C.1.14).

We formulate the model in units for which both m and Planck's constant are unity, and define ϵ to be $m/M = M^{-1}$: thus, ϵ^{-1} is a measure of the macroscopicality of B. We denote the displacement and momentum of B (respectively the nth atom) by Q and P (respectively q_n and p_n), and we assume that the only nonzero commutators between these observables are those given by

---

[6] Evidently, B is an open system, for which the chain of atoms serves as a reservoir.

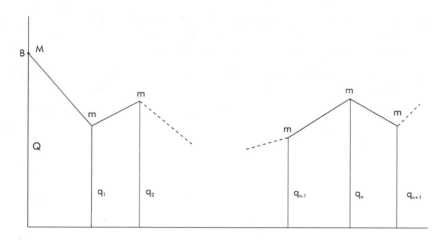

Figure 4.C.1   Particle B, of mass M, moving along smooth wire and coupled to an infinite chain of particles, each of mass m, that vibrates transversely in the direction of the wire.

the canonical commutation relations

$$[Q, P] = iI, \quad [q_m, p_n] = i\delta_{mn}I. \tag{C.1.1}$$

We assume that the formal Hamiltonian, governing the dynamics of the system, is

$$H = \frac{1}{2}\epsilon P^2 + \frac{1}{2}(Q - q_1)^2 + \frac{1}{2}\sum_{n=1}^{\infty}[p_n^2 + (q_{n+1} - q_n)^2]. \tag{C.1.2}$$

Since the mass of B is ϵ^{-1}, its velocity observable is

$$V = \epsilon P. \tag{C.1.3}$$

We denote by $\Sigma^{(0)}$ the *auxilliary* model of the atomic chain alone, in which its first particle is harmonically coupled to the origin, instead of B. Thus, the formal Hamiltonian for $\Sigma^{(0)}$ is

$$H^{(0)} = \frac{1}{2}\sum_{n=0}^{\infty}[p_n^2 + (q_n - q_{n+1})^2] \quad \text{with } q_0, p_0 \equiv 0. \tag{C.1.4}$$

Thus by Eqs. (C.1.1) and (C1.4), Σ comprises the harmonically coupled system formed by the particle B and the atomic chain $\Sigma^{(0)}$. For simplicity, we carry out the treatment of the model Σ on a formal basis[7].

In general, we denote the expectation values of observables R and S of Σ and $\Sigma^{(0)}$, respectively, with respect to the initial states of these systems, by $\langle R \rangle$

---

[7] The model could be formulated as a $W^{\star}$-dynamical system, expressed in terms of Weyl operators for both B and the chain Σ_0.

and $\langle S \rangle^{(0)}$; and we denote the evolutes of these observables at time t due to the conservative dynamics of these systems by $R(t)$ and $S^{(0)}(t)$.

The initial conditions we assume are the following.

(I.1) The observables of the particle B are uncorrelated to those of the atomic chain $\Sigma^{(0)}$ at time $t = 0$.

(I.2) The initial state of $\Sigma^{(0)}$ is one of thermal equilibrium, as characterised by the KMS condition,[8] which was specified in Section 2.5, and may be equivalently expressed in the following way [Ku]. Let

$$F_{\pm}(t) = \left\langle [S_1^{(0)}(t), S_2]_{\pm} \right\rangle^{(0)}, \tag{C.1.5}$$

where S_1 and S_2 are arbitrary observables of $\Sigma^{(0)}$, and let $F_{\pm}$ be the Fourier transforms of $F_{\pm}$, that is,

$$\hat{F}_{\pm}(\omega) = \int_{\mathbf{R}} dt F_{\pm}(t) \exp(-i\omega t).$$

Then the KMS condition is given by the formula

$$\hat{F}_{+}(\omega) = \hat{F}_{-}(\omega) \coth(\beta\omega). \tag{C.1.6}$$

(I.3) The initial state of the particle B yields Gaussian distribututions for all real linear combinations of Q and V, the means, dispersions and correlations of these observables being given by the formulae

$$\langle Q \rangle = Q_0, \quad \langle V \rangle = V_0, \tag{C.1.7}$$

$$\Delta Q \equiv \langle (Q - Q_0)^2 \rangle^{1/2} = \epsilon^{1/2} a, \quad \Delta V \equiv \langle (V - V_0)^2 \rangle^{1/2} = \epsilon^{1/2} b \tag{C.1.8}$$

and

$$\langle [Q - Q_0, V - V_0]_{+} \rangle = 0, \tag{C.1.9}$$

where a and b are ϵ-independent parameters that satisfy the Heisenberg inequality

$$ab \geq \frac{1}{2}.$$

We formulate the dynamics of B, induced by its interaction with the atomic chain, on a time scale for which the unit interval corresponds to that represented by ϵ^{-1} on the t-scale. We designate this the *macroscopic* time scale, and denote its time variable by τ. Thus, $\tau = \epsilon t$. We denote by $v_{\epsilon}(\tau)$ the evolute of V on the macroscopic scale, so that

---

[8] The physical significance of this condition is discussed in Chapter 5.

$$v_\epsilon(\tau) = V(\epsilon^{-1}\tau).$$ (C.1.10)

We also define $w_\epsilon(\tau)$ to be the fluctuation of $v_\epsilon(\tau)$ about its mean, as rescaled by the factor $\epsilon^{-1/2}$. Thus,

$$w_\epsilon(\tau) = \epsilon^{-1/2}(v_\epsilon(\tau) - \langle v_\epsilon(\tau)\rangle).$$ (C.1.11)

Our principal results, which are obtained in Sections C.4 and C.5, are that, under the conditions (I.1)–(I.3) and in the limit $\epsilon \to 0$, the velocity $v_\epsilon(\tau)$ and the fluctuation $w_\epsilon(\tau)$ reduce to classical variables $v(\tau)$ and $w(\tau)$, respectively, such that

(A) $v(\tau)$ is dispersion-free and evolves according to the deterministic, irreversible law

$$\frac{dv(\tau)}{d\tau} = -v(\tau),$$ (C.1.12)

and

(B) $w(\tau)$ executes a Brownian motion, conforming to the Langevin equation

$$\frac{dw(\tau)}{d\tau} = -w(\tau) + f(\tau),$$ (C.1.13)

where the force $f(\tau)$ is a Gaussian stochastic process, with zero mean and autocorrelation function

$$E(f(\tau)f(\tau')) = 2\beta^{-1}\delta(\tau - \tau'),$$ (C.1.14)

E being the expectation functional for the process.

Comment
(1) The macroscopicality of B is represented here in its sharpest form by the passage to the limit in which its mass becomes infinite, that is, where $\epsilon \to 0$. Evidently, ϵ and v_ϵ correspond to the Ω^{-1} and Y_Ω, respectively, of Section 4.5.

(2) Defining $\Lambda_{\mathcal{M}}(v) := v^2$, one sees immediately from Eq. (C.1.12) that $\Lambda_{\mathcal{M}}(v(\tau))$ decreases monotonically with τ for $v \neq 0$ and remains constant for $v = 0$. $\Lambda_{\mathcal{M}}$ is therefore a Lyapounov function for the deterministic evolution of $v(\tau)$.

(3) The damping of this velocity is due not to the contact between B and the wire, which is smooth, but to the loss of energy of the particle by "radiation" along the atomic chain.

(4) The damping force takes the same form (that is, minus the velocity) in the fluctuation dynamics, given by Eq. (C.1.13), as it does in the phenomenological law (C.1.12). This constitutes a realisation on Onsa-

ger's regression hypothesis [On], which we formulate in a more general context in Chapter 7.

C.2 EQUATIONS OF MOTION

By Eqs. (C.1.1)–(C.1.3), the Heisenberg equations of motion for the positions and momenta of the particles of Σ are

$$\frac{dQ(t)}{dt} = V(t), \quad \frac{dP(t)}{dt} = \epsilon(q_1(t) - Q(t)) \tag{C.2.1}$$

and

$$\frac{dq_n(t)}{dt} = p_n(t), \quad \frac{dp_n(t)}{dt} = q_{n+1}(t) - 2q_n(t) + q_{n-1}(t) + Q(t)\delta_{n,1}. \tag{C.2.2}$$

Correspondingly, by Eq. (C.1.4), the equations of motion for the positions and momenta of the particles of $\Sigma^{(0)}$ are

$$\frac{dq_n^{(0)}(t)}{dt} = p_n^{(0)}(t), \quad \frac{dp_n^{(0)}(t)}{dt} = q_{n+1}^{(0)}(t) - 2q_n^{(0)}(t) + q_{n+1}^{(0)}(t). \tag{C.2.3}$$

We now see from Eq. (C.2.1) that, in order to obtain the dynamics of B, we need to solve for $q_1(t)$ in terms of the evolutes of Q and V. For this purpose, we pass to the Fourier transforms of q_n and p_n given by

$$\hat{q}(k) = \sum_{n=0}^{\infty} q_n \sin kn, \quad \hat{p}(k) = \sum_{n=0}^{\infty} p_n \sin kn, \tag{C.2.4}$$

with k running from 0 to π. Conversely,

$$q_n = \frac{2}{\pi} \int_0^{\pi} dk \hat{q}(k) \sin kn, \quad p_n = \frac{2}{\pi} \int_0^{\pi} dk \hat{p}(k) \sin kn. \tag{C.2.5}$$

It follows now from Eq. (C.2.4) that the equations (C.2.2) transform to

$$\frac{d\hat{q}(k,t)}{dt} = \hat{p}(k,t), \quad \frac{d\hat{p}(k,t)}{dt} = -\omega(k)^2 \hat{q}(k,t) + Q(t) \sin k, \tag{C.2.6}$$

where

$$\omega(k) = 2 \sin(k/2). \tag{C.2.7}$$

Likewise, by Eqs. (C.2.3) and (C.2.4), the corresponding equations of motion for $\Sigma^{(0)}$ are

$$\frac{d\hat{q}^{(0)}(k,t)}{dt} = \hat{p}^{(0)}(k,t), \quad \frac{d\hat{p}^{(0)}(k,t)}{dt} = -\omega(k)^2 \hat{q}^{(0)}(k,t). \tag{C.2.8}$$

Hence, by Eqs. (C.2.6) – (C.2.8),

$$\hat{q}(k,t) = \hat{q}^{(0)}(k,t) + \cos(k/2) \int_0^t dsQ(t-s)\sin(\omega(k)s), \qquad \text{(C.2.9)}$$

where

$$\hat{q}^{(0)}(k,t) = \hat{q}(k)\cos(\omega(k)t) + \frac{\hat{p}(k)}{\omega(k)}\sin(\omega(k)t) \qquad \text{(C.2.10a)}$$

and

$$\hat{p}^{(0)}(k,t) = \hat{p}(k)\cos(\omega(k)t) - \omega(k)\hat{q}(k)\sin(\omega(k)t). \qquad \text{(C.2.10b)}$$

Therefore, by Eqs. (C.2.5), (C.2.9) and (C.2.10a),

$$q_1(t) = q_1^{(0)}(t) - \int_0^t ds\dot{\phi}(s)Q(t-s), \qquad \text{(C.2.11)}$$

where

$$\phi(s) = \frac{2}{\pi} \int_0^\pi dk \cos^2(k/2)\cos(\omega(k)s),$$

and $\dot{\phi}$ is the first derivative of ϕ. On putting $u = \sin(k/2)$ in the last equation, and using Eq. (C.2.7) and the evenness of cos, we obtain the following formula for ϕ.

$$\phi(s) = \frac{2}{\pi} \int_{-1}^1 du(1-u^2)^{1/2}\cos(2us). \qquad \text{(C.2.12)}$$

Further, since this implies that $\phi(0) = 1$, the integral in Eq. (C.2.11) is equal to

$$\phi(t)Q - Q(t) + \int_0^t ds\phi(s)\dot{Q}(t-s),$$

and consequently, by Eqs. (C.2.1) and (C.2.11), the equation of motion for $V(t)$ is

$$\frac{dV(t)}{dt} = -\epsilon \int_0^t ds\phi(s)V(t-s) + \epsilon q_1^{(0)}(t) - \epsilon\phi(t)Q. \qquad \text{(C.2.13)}$$

This is a generalised Langevin equation, the three terms on the right-hand side representing a retarded damping force, a stochastic force whose properties depend on the KMS state of $\Sigma^{(0)}$, and a "memory" of the initial position of B.

The stochastic properties of the second term are governed by the KMS condition for the state of $\Sigma^{(0)}$. Since this is a harmonic chain and thus corresponds to a *free* Bose field, the distribution of any real linear combination of the p_n and q_n, in an equilibrium state of this system is Gaussian (cf. [RST]). Furthermore, its mean is zero, for the following reasons. By Eqs. (C.2.10),

$$\langle \hat{q}^{(0)}(k,t)\rangle^{(0)} = \langle \hat{q}(k)\rangle^{(0)}\cos(\omega(k)t) + \frac{\langle \hat{p}(k)\rangle^{(0)}}{\omega(k)}\sin(\omega(k)t)$$

and

$$\langle \hat{p}^{(0)}(k, t)\rangle^{(0)} = \langle \hat{p}(k)\rangle^{(0)} \cos(\omega(k)t) - \omega(k)\langle \hat{q}(k)\rangle^{(0)} \sin(\omega(k)t),$$

and, in view of the stationarity of the equilibrium state, these equations imply that the mean values it yields for $\hat{q}(k)$ and $\hat{p}(k)$ are zero. Hence, so too are the means, for this state, of the observables given by linear combinations of the p_n and q_n. We conclude therefore that these observables have Gaussian distributions with zero means in the equilbrium states of $\Sigma^{(0)}$.

On applying this result to $q_1^{(0)}(t)$, which, by Eqs. (C.2.5) and (C.2.10), belongs to this category of observables, we infer that its stochastic properties are completely determined by its two-point autocorrelation functions with respect to the equilibrium state of $\Sigma^{(0)}$. To formulate these functions, we note that, by Eqs. (C.1.1), (C.2.4), (C.2.5), (C.2.7) and (C.2.10),

$$[q_1^{(0)}(t), q_1] = \frac{-4i}{\pi} \int_0^\pi \sin(k/2) \cos^2(k/2) \sin(2 \sin(k/2)t),$$

and therefore, putting $u = \sin(k/2)$ and noting Eq. (C.2.12) and the evenness of cos,

$$\langle [q_1^{(0)}(t), q_1]\rangle^{(0)} = \frac{-4i}{\pi} \int_{-1}^1 du\, u(1 - u^2)^{1/2} \sin(2ut) \equiv -i\dot{\phi}(t). \qquad (C.2.14)$$

Hence, by the KMS property (C.1.6),

$$\langle [q_1^{(0)}(t), q_1]_+\rangle^{(0)} \equiv \frac{4}{\pi} \int_{-1}^1 du\, u(1 - u^2)^{1/2} \coth(\beta u) \cos(2ut) := \psi(t).$$

$$(C.2.15)$$

Thus, Eqs. (C.2.14) and (C.2.15) represent the stochastic properties of $q_1^{(0)}(t)$, and consequently, it follows from these equations and Eq. (C.2.13) that the dynamics of B is governed by the forms of the functions ϕ and ψ. Most importantly, it follows from Eqs. (C.2.12) and (C.2.15) that, by the Riemann–Lebesgue lemma, $\phi(t)$ and $\psi(t)$ both tend to zero as $t \to \infty$. These decay properties are the key to the irreversibility of the motion of B.

C.3 Macroscopic Description of B

As remarked in Section C.1, we characterise the macroscopicality of B by the smallness of ϵ, and we idealise this by a passage to a limit where $\epsilon \to 0$ in a sense that we specify below.

Thus, we start by noting that, by Eq. (C.2.13), the rate of change of $V(t)$ is proportional to ϵ, which suggests that the time scale appropriate to the macroscopic description would be that whose unit corresponds to $t = \epsilon^{-1}$. Accordingly, we explore the dynamics of B on this scale. Thus, as stated in Section C.1, we define the time for the macroscopic description to be $\tau = \epsilon t$ and the velocity of B to be

$$v_\epsilon(\tau) = V(\epsilon^{-1}\tau). \tag{C.3.1}$$

We also define $\xi_\epsilon(\tau)$ to be the "stochastic force" $q_1^{(0)}(t)$, as represented on the macroscopic scale, that is,

$$\xi_\epsilon(\tau) = q_1^{(0)}(\epsilon^{-1}\tau). \tag{C.3.2}$$

It follows now from Eqs. (C.2.13), (C.3.1) and (C.3.2) that the generalised Langevin equation of motion for the velocity of B takes the following form in the macroscopic description.

$$\frac{dv_\epsilon(\tau)}{d\tau} = -\int_0^{\epsilon^{-1}\tau} ds\phi(s)v_\epsilon(\tau - \epsilon s) + \xi_\epsilon(t) - \phi(\epsilon^{-1}\tau)Q. \tag{C.3.3}$$

Further, by Eq. (C.3.2), $\xi_\epsilon(\tau)$, like $q_1^{(0)}(t)$, is a Gaussian process with zero mean. Its stochastic properties are therefore determined by its two point functions. In view of Eqs. (C.2.14) and (C.2.15), together with the stationarity of the equilibrium state of $\Sigma^{(0)}$, these are given by the formulae

$$\langle [\xi_\epsilon(\tau), \xi_\epsilon(\tau')] \rangle^{(0)} = i\dot\phi\left(\epsilon^{-1}(\tau - \tau')\right) \tag{C.3.4}$$

and

$$\langle [\xi_\epsilon(\tau), \xi_\epsilon(\tau')]_+\rangle^{(0)} = \psi\left(\epsilon^{-1}(\tau - \tau')\right). \tag{C.3.5}$$

Our aim now is to obtain the properties of $v_\epsilon(\tau)$ in the limit that ϵ tends to zero. For this purpose, we note the following key properties of the functions ϕ and ψ, which follow from Eqs. (C.2.12) and (C.2.15) by standard Fourier analysis and the Riemann–Lebesgue lemma.

(P.1) The functions ϕ, ψ and their derivatives are bounded, and tend to zero as their arguments tend to $\pm\infty$.

(P.2) $\int_0^T ds\phi(s)$ and $\int_0^T ds\psi(s)$ are bounded functions of T and tend to 1 and β, respectively, as T tends to infinity.

In order to control the evolution of v_ϵ in the limit where ϵ tends to zero, we integrate the Langevin equation (C.3.3) from 0 to τ, and observing that, by Eq. (C.3.1), $v_\epsilon(0) = V(0) \equiv V$. Thus we obtain the Volterra integral equation

$$v_\epsilon(\tau) + \int_0^\tau d\sigma\kappa_\epsilon(\tau - \sigma)v_\epsilon(\sigma) = V + \int_0^\tau d\sigma\left(\xi_\epsilon(\sigma) - \phi(\epsilon^{-1}\sigma)Q\right), \tag{C.3.6}$$

where

$$\kappa_\epsilon(\sigma) = \int_0^{\epsilon^{-1}\sigma} ds\phi(s). \tag{C.3.7}$$

It follows from this last equation and (P.2) that $\kappa_\epsilon(\sigma)$ is uniformly bounded with respect to both ϵ and σ, and that

$$\lim_{\epsilon \to 0} \kappa_\epsilon(\sigma) = 1 \quad \forall \sigma > 0. \tag{C.3.8}$$

By the standard theory of Volterra equations, the solution of Eq. (C.3.6) may be expressed in the form

$$v_\epsilon(\tau) = g_\epsilon(\tau, 0)V + \int_0^\tau d\sigma g_\epsilon(\tau, \sigma)\left(\xi_\epsilon(\sigma) - \phi(\epsilon^{-1}\sigma)Q\right), \tag{C.3.9}$$

where the Green function g_ϵ is given by

$$g_\epsilon(\tau, \sigma) + \int_\sigma^\tau d\sigma' \kappa_\epsilon(\tau - \sigma')g_\epsilon(\sigma', \sigma) = 1, \tag{C.3.10}$$

the uniqueness of the solution of this equation being guaranteed by the boundedness of κ_ϵ. Furthermore, as this boundedness is uniform, the solution for g_ϵ may be obtained by iteration as a uniformly convergent series, which is uniformly bounded with respect to ϵ and its arguments over the range $0 < \sigma \leq \tau \leq \tau_m$, for any finite τ_m. Moreover, by Eq. (C.3.8), the resultant series expansion for g_ϵ converges pointwise to the limit g_0, given by the equation

$$g_0(\tau, \sigma) + \int_\sigma^\tau d\sigma' g_0(\sigma', \sigma) = 1, \tag{C.3.11}$$

whose solution is simply

$$g_0(\tau, \sigma) = \exp(-(\tau - \sigma)). \tag{C.3.12}$$

Thus,

$$\lim_{\epsilon \to 0} g_\epsilon(\tau, \sigma) = \exp(-(\tau - \sigma)). \tag{C.3.13}$$

C.4 The Phenomenological Law

Since ξ_ϵ has zero expectation value, it follows from the initial condition (I.1) of Section C.1, together with Eqs. (C.1.7)–(C.1.9) and (C.3.9) that the mean and dispersion of $v_\epsilon(\tau)$ are given by the following formulae.

$$\langle v_\epsilon(\tau) \rangle = g_\epsilon(\tau, 0)V_0 - \int_0^\tau d\sigma g_\epsilon(\tau, \sigma)\phi(\epsilon^{-1}\sigma)Q_0 \tag{C.4.1}$$

and

$$(\Delta v_\epsilon(\tau))^2 \equiv \langle (v_\epsilon(\tau) - \langle v_\epsilon(\tau) \rangle)^2 \rangle = g_\epsilon(\tau, 0)\epsilon b^2 +$$

$$\int_0^\tau d\sigma \int_0^\tau d\sigma'' g_\epsilon(\tau, \sigma)g_\epsilon(\tau, \sigma')\left[\psi(\epsilon^{-1}(\sigma - \sigma')) + \phi(\epsilon^{-1}\sigma)\phi(\epsilon^{-1}\sigma')\epsilon a^2\right].$$

$$\tag{C.4.2}$$

Hence, by Eq. (C.3.13), together with the properties (P.1) and (P.2) of ϕ and ψ, specified in Section C.2,

$$\lim_{\epsilon \to 0} \langle v_\epsilon(\tau) \rangle = v(\tau) = V_0 \exp(-\tau)$$

and

$$\lim_{\epsilon \to 0} \Delta v_\epsilon(\tau) = 0.$$

These last two equations establish the following proposition.

Proposition C.4.1 *In the limit $\epsilon \to 0$, the velocity, $v(\tau)$, of B becomes dispersion-free and satisfies the classical phenomenological law*

$$\frac{dv(\tau)}{d\tau} = -v(\tau). \tag{C.4.3}$$

Thus the model exhibits irreversibility and macroscopic causality.

C.5 The Fluctuation Process

Since $\xi_\epsilon(\sigma)$ has zero mean, it follows from Eqs. (C.1.11) and (C.3.9) that the fluctuation observable w_ϵ takes the form

$$w_\epsilon(\tau) = \epsilon^{-1/2} g_\epsilon(\tau, 0)(V - \langle V \rangle) + \epsilon^{-1/2} \int_0^\tau d\sigma g_\epsilon(\tau, \sigma)$$

$$\times \left(\xi_\epsilon(\sigma) - \phi(\epsilon^{-1}\sigma)(Q - \langle Q \rangle) \right). \tag{C.5.1}$$

Hence, by the initial conditions (I.1)–(I.3) and the Gaussian property of ξ_ϵ, the process w_ϵ is also Gaussian, with zero mean, and therefore it is completely determined by its two-point correlation function. The following proposition establishes that the correlation functions defining the process w_ϵ converge to the corresponding ones for the classical stochastic process , w, introduced in Section C.1.

Proposition C.5.1 *Let $w(\tau)$ be the classical stochastic process defined by equations (C.1.13) and (C.1.14), subject to the conditions that*
 (i) the mean of the stochastic force $f(\tau)$ is zero,
 (ii) $w(0)$ is a Gaussian random variable with zero mean and dispersion b, and
 (iii) $w(0)$ and $f(\tau)$ are statistically independent of one another.
Then the quantum process w_ϵ converges to the classical one, w, as ϵ tends to

zero, in the sense that

$$\lim_{\epsilon\to 0}\langle w_\epsilon(\tau_1)\cdots w_\epsilon(\tau_n)\rangle = E(w(\tau_1)\cdots w(\tau_n)) \quad \forall \tau_1, ..., \tau_n \in \mathbf{R}, \; n \in \mathbf{N}.$$

$$(\text{C.5.2})$$

Note It follows easily from the above definition of w that this process, like w_ϵ, is Gaussian and that the mean of $w(\tau)$ is zero. Hence, in order to prove Proposition C.5.1, it suffices to establish Eq. (C.5.2) for $n = 2$.

Our proof of the proposition depends on the following lemma.

Lemma C.5.2 *Let F be a differentiable function on $[0, 1]$ that vanishes at the extremities of that interval and whose derivative, F', is absolutely integrable over it; and, for $\epsilon > 0$, let G_ϵ be the function on $\mathbf{R}$ defined by the formula*

$$G_\epsilon(x) = \epsilon^{-1} x \int_{-1}^{1} dy F(y) \exp(i\epsilon^{-1}xy). \qquad (\text{C.5.3})$$

Then the function G_ϵ is bounded, uniformly with respect to ϵ, and tends pointwise to zero as $\epsilon \to 0$.

The following corollary an immediate consequence of this lemma and the definitions of ϕ, $\dot{\phi}$ and ψ given by Eqs. (C.2.12), (C.2.14) and (C.2.15).

Corollary C.5.3 $\epsilon^{-1}\sigma\phi(\epsilon^{-1}\sigma)$, $\epsilon^{-1}\sigma\dot{\phi}(\epsilon^{-1}\sigma)$ *and* $\epsilon^{-1}\sigma\psi(\epsilon^{-1}\sigma)$ *are all uniformly bounded with respect to both σ and ϵ, and tend to zero as $\epsilon \to 0$.*

Proof of Lemma C.5.2. By Eq. (C.5.3),

$$G_\epsilon(x) = [-iF(y)\exp(i\epsilon^{-1}xy)]_{y=-1}^{1} + i\int_{-1}^{1} dy F'(y)\exp(i\epsilon^{-1}xy).$$

In view of our specifications of F and F', the required result follows immediately from this formula and the Riemann–Lebesgue lemma. $\square$

Proof of Proposition C.5.1. By the initial condition (I.1) and Eqs. (C.1.8), (C.1.9), (C.3.4), (C.3.5) and (C.5.1),

$$\langle w_\epsilon(\tau)w_\epsilon(\tau')\rangle = b^2 g_\epsilon(\tau, 0)g_\epsilon(\tau'', 0) \qquad (\text{C.5.4a})$$

$$+a^2 \int_0^\tau d\sigma \int_0^{\tau'} d\sigma' \phi(\epsilon^{-1}\sigma)\phi(\epsilon^{-1}\sigma') \qquad (\text{C.5.4b})$$

$$+\int_0^\tau d\sigma \int_0^{\tau'} d\sigma' (2\epsilon)^{-1} g_\epsilon(\tau, \sigma)g_\epsilon(\tau', \sigma')\left[\psi(\epsilon^{-1}(\sigma-\sigma')) + i\dot{\phi}(\epsilon^{-1}(\sigma-\sigma'))\right].$$

$$(\text{C.5.4c})$$

Now, by Eq. (C.3.13),

$$\text{Term } (C.5.4a) \rightarrow b^2 \exp(-\tau - \tau') \quad \text{as } \epsilon \rightarrow 0. \qquad (C.5.5a)$$

Also, by property (P.2),

$$\text{Term } (C.5.4b) \rightarrow 0 \quad \text{as } \epsilon \rightarrow 0. \qquad (C.5.5b)$$

Our treatment of the term (C.5.4c) is based on the following construction. For each $\eta > 0$, we denote by $I_{\epsilon,\eta}$ and $J_{\epsilon,\eta}$ the contributions to that integral from the inside and outside of the strip $| \sigma' - \sigma | \leq \eta$. It follows from this definition and Corollary C.5.3 that, in view of the uniform boundedness of g_ϵ, $J_{\epsilon,\eta}$ tends to zero as $\epsilon \rightarrow 0$.

Next, we observe that it follows from Eq. (C.3.10) and the uniform boundedness of κ_ϵ that $| g_\epsilon(\tau', \sigma') - g_\epsilon(\tau', \sigma) |$ is majorised by η times a constant that depends neither on ϵ nor on the arguments of g_ϵ. Hence, defining $I'_{\epsilon,\eta}$ to be the integral obtained by replacing $g_\epsilon(\tau', \sigma')$ by $g_\epsilon(\tau', \sigma)$ in the integrand for $I_{\epsilon,\eta}$, we see that $| I'_{\epsilon,\eta} - I_{\epsilon,\eta} |$ is majorised by a term of the form

$$c_1 \left| \epsilon^{-1}(\sigma - \sigma')\left(\psi(\epsilon^{-1}(\sigma - \sigma'))\right)\right| + c_2 \left| \epsilon^{-1}(\sigma - \sigma')\dot{\phi}\left(\epsilon^{-1}(\sigma - \sigma')\right)\right|$$

where c_1 and c_2 are finite ϵ-indpendent constants. Hence, by Corollary C.5.3, the difference between $I_{\epsilon,\eta}$ and $I'_{\epsilon,\eta}$ vanishes in the limit $\epsilon \rightarrow 0$.

In order to obtain the form of the latter integral in this limit, we note that, by Eq. (C.2.15), together with properties (P.1), (P.2) of Section C.2 and the evenness of ψ_ϵ, that

$$\lim_{\eta \rightarrow 0} \lim_{\epsilon \rightarrow 0} \epsilon^{-1} \int_{-\eta}^{\eta} d\sigma \psi(\epsilon^{-1}\sigma) = 4\beta^{-1}$$

and

$$\lim_{\eta \rightarrow 0} \lim_{\epsilon \rightarrow 0} \epsilon^{-1} \int_{-\eta}^{\eta} d\sigma \dot{\phi}(\epsilon^{-1}\sigma) = 0.$$

Hence, by the definition of $I'_{\epsilon,\eta}$,

$$\lim_{\eta \rightarrow 0} \lim_{\epsilon \rightarrow 0} I'_{\epsilon,\eta} = 2\beta^{-1} \int_{0}^{\min(\tau,\tau')} \exp(-\tau - \tau' + 2\sigma)$$

$$\equiv \beta^{-1}(\exp(- | \tau - \tau' |) - \exp(-\tau - \tau'))$$

the limit with respect to η serving to eliminate "end effects" of the strips $| \sigma' - \sigma |< \eta$. Therefore, since, as shown above, both $J_{\epsilon,\eta}$ and $I_{\epsilon,\eta} - I'_{\epsilon,\eta}$ vanish in the limit $\epsilon \rightarrow 0$, we conclude that

$$\text{Term } (C.5.4c) \rightarrow \beta^{-1}(\exp(- | \tau - \tau' |) - \exp(-\tau - \tau')) \quad \text{as } \epsilon \rightarrow 0.$$

$$(C.5.5c)$$

Eqs. (C.5.4) and (C.5.5) therefore yield the formula

$$\lim_{\epsilon \to 0} \langle w_\epsilon(\tau) w_\epsilon(\tau') \rangle = (b^2 - \beta^{-1}) \exp(-\tau - \tau') + \beta^{-1} \exp(-\mid \tau - \tau' \mid).$$

(C.5.6)

It now remains for us to prove that the right-hand side of this equation is equal to the two-point function of the process w, subject to the assumptions on the distribution of $w(0)$. To this end, we infer from Eq. (C.1.13) that

$$w(\tau) = w(0) \exp(-\tau) + \int_0^\tau d\sigma \exp(-\tau + \sigma) f(\sigma).$$

Hence, by Eq. (C.1.14),

$$E(w(\tau) w(\tau')) = E\left(w(0)^2\right) \exp(-\tau - \tau') + 2\beta^{-1} \int_0^{\min(\tau, \tau')} d\sigma \, \exp(-\tau - \tau' + 2\sigma).$$

(C.5.7)

Therefore, as the mean and dispersion of $w(0)$ are zero and b, respectively, we see that the right-hand side of this equation is identical to that of Eq. (C.5.6). In view of the note following the statement of Proposition C.5.1, this completes the proof of the proposition. □

Part II

From quantum statistics to equilibrium and nonequilibrium thermodynamics: prospectus

Part II is designed to provide a general quantum statistical basis for both classical thermodynamics and its extension to macroscopic irreversible processes of continuum mechanics.

Chapter 5 is devoted to a quantum statistical formulation of thermal equilibrium states, within the operator algebraic framework. There we review a variety of stability arguments that lead to the characterisation of equilibrium states and phases in terms of the KMS fluctuation-dissipation conditions. The resultant scheme, by contrast with its traditional statistical mechanical counterpart, can support different equilibrium states under the same thermodynamical conditions and can thus admit the phenomena of phase coexistence and spontaneous symmetry breakdown.

In Chapter 6, we present a quantum statistical formulation of classical thermodynamics in terms of the restriction of the equilibrium states to macroscopic observables, corresponding to densities of extensive conserved quantities, Q. This contains a precise characterisation of a *complete set of thermodynamical variables* and a justification of the standard assumption of classical thermodynamics that the singularities in the thermodynamic potentials arise at precisely the values of the control variables that admit phase coexistence.

In Chapter 7, we proceed to an approach to nonequilibrium thermodynamics, based on the evolution of the hydrodynamical observables given by local densities of the same extensive conserved quantities, Q, as those of the equilibrium theory. On this basis, we devise a scheme that contains a macrostatistical characterisation of local thermal equilibrium and a generalisation of the Onsager theory [On] to a nonlinear regime.

Chapter 5

Thermal equilibrium states and phases

5.1 INTRODUCTION

The statistical mechanical theory of equilibrium states serves to relate classical thermodynamics to the microscopic laws of motion governing many-particle systems. The traditional version of this theory pertains to large, but finite,
systems, and is based on the probabilistic assumptions underlying the standard
Gibbs ensembles [LL2], or, equivalently, on the still unproven ergodic
hypothesis[1] [Kh2, Far].

The recasting of statistical mechanics within the framework of algebraic
quantum mechanics offers radical advantages in that it yields both an enriched
picture of thermodynamic phase structure and some dynamical basis for the
Gibbsian assumptions, *even for finite systems*. In particular, an important work
of Kossakowski, Frigerio, Gorini and Verri [KFGV] has circumvented the old
problems of ergodic theory by establishing that the KMS conditions characterise the thermal states of both finite and infinite systems in the following
operational sense. Suppose that an infinite system, Σ, and a finite one, $\tilde{\Sigma}$, are
initially and independently prepared in states ρ and $\tilde{\rho}$, respectively, and then
weakly coupled together. Then, assuming that the time correlation functions
of Σ for the state ρ enjoy certain viable decay properties,

(a) $\tilde{\Sigma}$ will be driven to a terminal state, $\tilde{\rho}_0$, that is independent of both the
initial state of $\tilde{\Sigma}$ and the details of the $\Sigma - \tilde{\Sigma}$ coupling, if and only if ρ
satisfies the KMS condition for Σ at some inverse temperature β, and

(b) in this case, $\tilde{\rho}_0$ is the canonical Gibbs state for that inverse temperature
and so is also identified by the KMS condition.

Thus the KMS property characterises the states of both finite and infinite
sytems that are thermal in the sense of the the Zeroth Law of Thermodynamics.

---

[1] In fact, this latter hypothesis cannot be valid for finite conservative quantum systems,
since their Hamitonians have pure point spectra.

The object of this chapter is to review the theory of thermal states of conservative quantum systems within the terms of the algebraic framework of Chapter 2. Our account of this theory is essentially discursive: proofs of the relevant theorems can be found in the works cited here.

We start in Section 5.2 by noting that the canonical Gibbs states of finite systems are completely characterised by a thermodynamic variational principle or, equivalently, by the KMS conditions. In Section 5.3, we pass on to infinite systems and survey the operational grounds for characterising their equilibrium states by these conditions. Here, the essential argument consists of the following four points:

(I)   As discussed above, the KMS conditions characterise the thermal states in the sense of the Zeroth Law of Thermodynamics [KFGV].

(II) By contrast with the situation for finite systems, an infinite system may support different states satisfying the same KMS conditions. This set is convex, and its extremal elements may naturally be interpreted as pure phases [EKV, EK]. Thus, the generic model of infinite sytems provides the framework for a theory of coexisting phases. Moreover, some of the pure phases may possess ordering corresponding to symmetry breakdown, as in, say, ferromagnets or crystals.

(III) The KMS conditions are equivalent to a variational principle, representing thermodynamical stability against local changes of state [AS, Se7, Se8].

(IV) They are also closely related to a condition of global thermodynamical stability, corresponding to the absolute minimisation of the free energy density. To be precise, this latter condition implies that of KMS, and, for systems with short range forces, is equivalent to it [LaRo, Ar3, Se8]. On the other hand, there are models with long range interactions that support states with the KMS property that do not minimise the free energy density [Se2,8].

Our interpretation of these results in Section 5.4 is based on the view that true equilibrium states are those that meet the demands of both the Zeroth and the Second Laws of Thermodynamics. On this basis, we take these to be the states that satisfy the KMS conditions and minimise the global free energy density, this latter condition being redundant only for systems with short range interactions (cf. (IV)). For those with long range interactions, we interpret the KMS states that do not minimise the free energy density as metastable ones, since they have all the properties of equilibrium states except that of global thermodynamical stability [Se2, Se8]. We conclude the chapter in Section 5.5 with a brief discussion of the general picture of equilibrium and metastable states and phases, including the unsolved problem of the characterisation of the crystalline phase. In particular, we raise the question of whether crystals are, in general, fully thermodynamically stable or, as in the case of some allotropes, merely metastable.

5.2 FINITE SYSTEMS

Here we consider the standard model of a finite system, Σ, as formulated in Section 2.2.3. According to traditional statistical mechanics, the equilibrium state, ρ_β, of the model at inverse temperature β is given by the canonical density matrix

$$\hat\rho_\beta = \exp(-\beta H)/\mathrm{Tr}(idem). \tag{5.2.1}$$

Most importantly, this state may be equivalently characterised in dynamical and thermodynamical terms. Let us first look at the dynamical picture.

5.2.1 Equilibrium, Linear Response Theory and the KMS Conditions

We suppose that the system is prepared at time $t = 0$ in a stationary state ρ and then subjected to a perturbation corresponding to an interaction Hamiltonian $F(t)B$, where B is a bounded observable and F is a real-valued function of time. The evolute, $\rho(t)$, of the state, ρ, of the system at positive times t is then determined by the Von Neumann equation of motion for its density matrix, $\hat\rho(t)$, that is,

$$\frac{d}{dt}\hat\rho(t) = -\frac{i}{\hbar}[H + F(t)B, \hat\rho(t)]. \tag{5.2.2}$$

On developing $\hat\rho(t)$ as a perturbative series in F, subject to the initial condition that $\hat\rho(0) = \hat\rho$, one finds that the *linear* contribution to the time-dependent increment in the expectation value of an arbitrary observable A is (cf. [Ku])

$$\Delta\langle A\rangle_t = \int_0^t dt' L_{AB}(t - t')F(t'), \tag{5.2.3}$$

where

$$L_{AB}(t) = \frac{i}{\hbar}\rho([A(t), B]) \tag{5.2.4}$$

and

$$A(t) = \alpha(t)A = \exp(iHt/\hbar)A\exp(-iHt/\hbar), \tag{5.2.5}$$

the evolute of A for the unperturbed system Σ. Thus, the function L_{AB} governs the linear response of the system to perturbations corresponding to interaction Hamiltonians proportional to B.

On the other hand, the time-correlation function of A and B, governing spontaneous fluctuations of these observables, is

$$C_{AB}(t) = \rho([A(t), B]_+).$$ (5.2.6)

Remarkably, as first observed by Kubo [Ku], the functions L_{AB} and C_{AB} are simply related in the particular case where $\rho = \rho_\beta$. To be precise, it follows from Eqs. (5.2.1) and (5.2.4)–(5.2.6) that, in this case, the Fourier transforms[2] $\hat{L}_{AB}$ and $\hat{C}_{AB}$, of L_{AB} and C_{AB}, respectively, are related according to the formula

$$\hat{L}_{AB}(\omega) = \frac{i}{\hbar} \hat{C}_{AB}(\omega) \tanh\left(\frac{1}{2}\beta\hbar\omega\right).$$ (5.2.7)

This result is often termed the fluctuation-dissipation theorem, since it relates the dissipative properties of the system, as represented by the linear response function, L_{AB}, to the spontaneous equilibrium fluctuations, represented by C_{AB}. Moreover, Eq. (5.2.7) is just the Fourier transform of the following formula, due to Martin and Schwinger [MS], which was at the core of their Green function approach to statistical mechanics:

$$\rho(A(t)B) = \rho(BA(t + i\hbar\beta)).$$ (5.2.8)

Thus, the equilibrium state, ρ_β, satisfies the Kubo–Martin–Schwinger (KMS) relation (5.2.8), or equivalently (5.2.7). To be precise [HHW], it satisfies the KMS conditions (i)–(iii), specified at the end of Section 2.2.3, and moreover, it is the only state of the finite system that does so. In other words, these conditions are not only a property of the Gibbs state ρ_β, but also a *characterisation* of it.

5.2.2 Equilibrium and Thermodynamical Stability

To provide a thermodynamical characterisation of equilibrium, we invoke the result (S.6) of Section 3.2.2, which established that the free energy $\hat{F}_\beta(\rho)$ of the state ρ attains its minimum value at $\rho = \rho_\beta$ only. This signifies that the classical principle of thermodynamical stability, whereby equlilibrium corresponds to the absolute minimisation of free energy, extends right down to the microscopic level.

5.2.3 Résumé

We may summarise the contents of this section by saying that the canonical states of finite systems are completely characterised by the condition of thermodynamical stability, as represented by the minimisation of the free energy functional $\hat{F}_\beta$, or, equivalently, by the KMS fluctuation-dissipation condition.

---

[2] Here, we are employing the standard convention that the Fourier transform, $\hat{f}$, of a function, f, on $\mathbf{R}$ is defined by the formula $\hat{f}(\omega) = \int_{\mathbf{R}} dt f(t) \exp(-i\omega t)$.

However, despite the enormous *a posteriori* support for the standard assumption that these are the thermal equilibrium states, the traditional statistical mechanics of Gibbs and Boltzmann provides no *operational* basis for it.

5.3 INFINITE SYSTEMS

We now pass on to the operator algebraic model of an infinite system, Σ. Thus, Σ may be a $C^\star$ or $W^\star$ system $(\mathcal{A}, S, \alpha)$, with the quasi-local structure specified in Section 2.5. Our aim here is to provide an operational basis for characterising thermal equilibrium by KMS and thermodynamical stability conditions.

5.3.1 The KMS Conditions

As in the case of finite systems, the KMS conditions on a state ρ are given by specifications of Section 2.2.3, and we recall here that they take the following form.

(KMS) For each pair of elements A, B of $\mathcal{A}$, there is a function, F_{AB}, on the complex strip $S_\beta = \{z \in \mathbf{C} \mid \mathrm{Im}(z) \in [0, \hbar\beta]\}$, such that

  (i) F_{AB} is analytic in the interior of S_β and continuous on its boundaries;

 (ii) $F_{AB}(t) = \rho(B\alpha(t)A) \quad \forall t \in \mathbf{R}$, and

(iii) $F_{AB}(t + i\hbar\beta) = \rho(A\alpha(t)B) \quad \forall t \in \mathbf{R}$.

Moreover, as in the case of finite systems, the connection of these conditions with linear response theory is given by Eqs. (5.2.3)–(5.2.7) (cf. [PSSW]). However, the implications of these conditions differs from their counterparts for finite systems in the following two crucial respects.

(A) The infinite system may support more than one state that satisfies the same KMS conditions [EKV]. For example, a system in a ferromagnetic phase supports states with differently directed polarisations at the same temperature (cf. [EK; Se2, Chapter 3]).

(B) Under certain supplementary conditions on the decay of time-correlation functions, this KMS property characterises precisely those states for which Σ behaves as a thermal reservoir, in that it drives any finite system to which it is locally and weakly coupled into its canonical Gibbs state for the inverse temperature β [KFGV].

Let us now elaborate on these two features of the KMS states of infinite systems.

The Set of KMS States

We denote by S_β^{KMS} the set of states of Σ that satisfies the KMS condition (5.2.8), and we term these the KMS states of the system at inverse temperature β. Most importantly, this set is not empty, since it contains the infinite volume limits[3] of the Gibbs states $\rho_{\beta,\Lambda}$ of the finite systems Σ_Λ.

It follows immediately from the above definition that S_β^{KMS} is is a convex set. We denote the set of its extremal elements by $\mathcal{E}_\beta^{KMS}$.

The following general key properties of S_β^{KMS} and $\mathcal{E}_\beta^{KMS}$ have been established [EKV].

(KMS.1) The KMS states are stationary, that is, invariant under time translations.

(KMS.2) $\mathcal{E}_\beta^{KMS}$ consists precisely of the primary elements of S_β^{KMS}.

(KMS.3) Each element ρ of S_β^{KMS} has a *unique* decomposition into the extremals, that is, ρ induces a unique probability measure, μ, on $\mathcal{E}_\beta^{KMS}$, such that

$$\rho = \int_{\mathcal{E}_\beta^{KMS}} \sigma \, d\mu(\sigma). \qquad (5.3.9)$$

In fact, this corresponds to the central decomposition of ρ, discussed in Section 2.4.1, item (10).

(KMS.4) The set S_β^{KMS} is invariant under the action of any dynamical symmetry group, G. However, it may contain states that are not G-invariant. In this case, we have *symmetry breakdown*. The maximal subgroup, H, of G under which a symmetry breaking state ρ is invariant is then termed the residual symmetry of ρ. In particular, the phenomenon of crystallisation corresponds to the case where G is the Euclidean group and H is a crystallographic subgroup thereof.

The Thermal Reservoir Property

Suppose that Σ is prepared in a stationary state ρ, and then coupled, at time $t = 0$, to a finite system, $\tilde{\Sigma} = (\tilde{\mathcal{A}}, \tilde{S}, \tilde{\alpha})$. Then, at positive times, $\tilde{\Sigma}$ becomes an open system. We now sketch the argument that has established that the KMS condition serves to characterise those states, ρ, of Σ for which this system behaves as a *thermal* reservoir for $\tilde{\Sigma}$, in the sense described above in (B).

---

[3] To be precise, these limit states, ρ, are defined by the formula

$$\rho(A) = \lim_{\Lambda \uparrow X} \rho_\Lambda(A) \quad \forall A \in \mathcal{A}_L,$$

and it follows the compactness of the state space [Ru1] and the separability of the local Hilbert spaces $\mathcal{H}_\Lambda$ that at least one such limit exists.

We assume here that Σ and $\tilde{\Sigma}$ are both $C^{\star}$ or $W^{\star}$ dynamical systems and that the $\Sigma - \tilde{\Sigma}$ interaction energy is λV, where λ is a real-valued parameter and V is a self-adjoint element of $\mathcal{A} \otimes \tilde{\mathcal{A}}$ given by a finite sum of the form

$$V = \sum_{j=1}^{n} A_j \otimes \tilde{A}_j \quad \text{with } \rho(A_j) = 0. \qquad (5.3.10)$$

Thus, the composite, $\Sigma^{(c)}$, of Σ and $\tilde{\Sigma}$ is a $C^{\star}$ or $W^{\star}$ dynamical system $(\mathcal{A}^{(c)}, S^{(c)}, \alpha^{(c,\lambda)})$, where $\mathcal{A}^{(c)} = \mathcal{A} \otimes \tilde{\mathcal{A}}$, and the generator, $\delta^{(c,\lambda)}$, of the dynamical group, $\alpha^{(c,\lambda)}(\mathbf{R})$, is

$$\delta^{(c,\lambda)} = \delta \otimes \tilde{I} + I \otimes \tilde{\delta} + \frac{i}{\hbar}[\lambda V, \cdot], \qquad (5.3.11)$$

δ, and $\tilde{\delta}$ being the generators of α and $\tilde{\alpha}$, respectively. To formulate the evolution of $\tilde{\Sigma}$ induced by $\alpha^{(c,\lambda)}$, we make the canonical identifications of A $(\in \mathcal{A})$ and $\tilde{A}$ $(\in \tilde{\mathcal{A}})$ with $A \otimes \tilde{I}$ and $I \otimes \tilde{A}$, respectively, and then proceed as in Section 4.3 by defining the conditional expectation $\tilde{E}$ of $\mathcal{A}^{(c)}$ onto $\tilde{\mathcal{A}}$ and the one parameter family, $\{\tilde{\theta}^{(\lambda)}(t) \mid t \in \mathbf{R}_+\}$, of CP transformations of $\tilde{\mathcal{A}}$ by the formulae

$$\tilde{E}(A \otimes \tilde{A}) = \rho(A)\tilde{A} \quad \forall A \in \mathcal{A}, \ \tilde{A} \in \tilde{\mathcal{A}} \qquad (5.3.12)$$

and

$$\tilde{\theta}^{(\lambda)}(t) = \tilde{E}\tilde{\alpha}^{(c,\lambda)}(t)\tilde{E}. \qquad (5.3.13)$$

The dynamics of the open system $\tilde{\Sigma}$ is therefore given by the family $\{\tilde{\theta}^{(\lambda)}(t) \mid t \in \mathbf{R}_+\}$ of transformations of $\tilde{\mathcal{A}}$.

In order to extract the effect of the $\Sigma - \tilde{\Sigma}$ coupling on this dynamics, we employ an interaction representation corresponding to the time scale whose unit is λ^{-2}, which is the natural one for the Born approximation. Thus, we define

$$\tilde{\gamma}^{(\lambda)}(s) = \tilde{\alpha}(-\lambda^{-2}s)\tilde{\theta}^{(\lambda)}(\lambda^{-2}s) \quad \forall s \in \mathbf{R}_+. \qquad (5.3.14)$$

As proved by Davies [Da], the CP transformations $\{\tilde{\gamma}^{(\lambda)}(s) \mid s \in \mathbf{R}_+\}$ reduce to a Markovian semigroup, $\tilde{\gamma}(\mathbf{R}_+)$, in the Van Hove weak coupling limit $\lambda \to 0$, subject to suitable conditions on the decay properties of the time correlations of the observables $\{A_j\}$ in the state ρ. Specifically, in this case,

$$\text{norm: } \lim_{\lambda \to 0} \tilde{\gamma}^{(\lambda)}(s)\tilde{A} = \tilde{\gamma}(s)\tilde{A} \quad \forall \tilde{A} \in \tilde{\mathcal{A}}, \ s \in \mathbf{R}_+, \qquad (5.3.15)$$

and, moreover, the transformations $\tilde{\gamma}(s)$ commute with the time-translational automorphisms for isolated system $\tilde{\Sigma}$, that is,

$$[\tilde{\gamma}(s), \tilde{\alpha}(t)] = 0 \quad \forall s \in \mathbf{R}_+, \ t \in \mathbf{R}. \qquad (5.3.16)$$

The following proposition, which was proved by Kossakowski, Frigerio, Gorini and Verri [KFGV], establishes that, under the specified assumptions, the KMS relation is equivalent to the condition that ρ is a state of Σ for which this system behaves as a thermal reservoir, in the sense of the Zeroth Law of Thermodynamics. Earlier, and much more limited, versions of this result were obtained by Callen and Welton [CW], at a heuristic level, and by Sewell [Se9].

Proposition 5.3.1   *Assume that $\mathcal{A}$ has a dense[4] subalgebra $\mathcal{A}_0$, such that the above Markovian picture is realised, for any finite system $\tilde{\Sigma}$, if the operators $\{A_j\}$ of Eq. (5.3.10) belong to $\mathcal{A}_0$. Then the following conditions are equivalent.*

(1) ρ is a KMS state for some inverse temperature β.

(2)   There is precisely one state, $\tilde{\rho}_0$, of $\tilde{\Sigma}$ that is invariant under both $\alpha(\mathbf{R})$ and $\tilde{\gamma}(\mathbf{R}_+)$. In this case $\tilde{\rho}_0$ is the Gibbs canonical state of $\tilde{\Sigma}$ at the inverse temperature β. Moreover, there is a dense subset, $\tilde{\mathcal{A}}_0$, of $\tilde{\mathcal{A}}$, such that, if the operators $\tilde{A}_j$ of Eq. (5.3.10), lie in $\tilde{\mathcal{A}}_0$, then

$$\text{norm:} \lim_{s \to \infty} \gamma^\star(s)\tilde{\rho} = \tilde{\rho}_0 \quad \forall \tilde{\rho} \in \tilde{S}, \tag{5.3.17}$$

and hence, by Eqs. (5.3.14)–(5.3.16),

$$\lim_{s \to \infty} \lim_{\lambda \to 0} \tilde{\rho}\left(\tilde{\theta}^{(\lambda)}(\lambda^{-2}s)\tilde{A}\right) = \tilde{\rho}_0(\tilde{A}) \quad \forall \tilde{A} \in \tilde{\mathcal{A}}, \, \tilde{\rho} \in \tilde{S}. \tag{5.3.18}$$

Comments

(1)   The clustering conditions underlying the Markovian assumption of this Proposition imply that ρ is primary (cf. Section 2.6.3).

(2)   Hence, by (KMS.2), the states that satisfy conditions (1) and (2) of the proposition are extremal KMS states.

(3)   The proposition therefore carries the natural interpretation that the extremal KMS states are just those for which Σ behaves as a thermal reservoir at inverse temperature β; and that the equilibrium states of a finite system at the same temperature are just the canonical ones.

The next two points were first made by Emch and Knops [EK].

(4)   Given the above interpretation of the $\mathcal{E}_\beta^{\text{KMS}}$-class states as thermal ones, it follows from their primary character that they correspond to pure phases (cf Section 2.6.6).

(5)   Correspondingly, by (KMS.3), the nonextremal elements of S_β may be interpreted as mixtures of thermal states at the same inverse temperature β.

---

[4] The relevant topology here is the norm or weak one according to whether Σ is a $C^\star$ or $W^\star$ system.

To summarise, *the proposition provides a dynamical basis for the traditional assumption that the thermal states of a finite system are the canonical ones of Gibbs, as well as the more recent one [HHW] that the KMS conditions characterise the thermal states of infinite systems.* Here we remark that, in addition to the thermal equilibrium states of matter, there are metastable (e.g., superheated or supercooled) states that simulate those of true equilibrium except for the fact that their free energies are not minimal. Therefore, it is natural to ask whether the KMS conditions might characterise certain metastable states, as well as those of true equilibrium. In fact, we argue in Section 5.4 that they do so for certain classes of systems with long range interactions.

Note on the Chemical Potential

Although the KMS conditions are expressed in terms of only one thermodynamical parameter, namely β, a remarkable work by Araki, Haag, Kastler and Takesaki [AHKT] has established that they lead to the concept of a chemical potential in the following way. If ρ is a primary KMS state on the observable algebra, $\mathcal{A}$, of a system of indistinguishable particles, as formulated in Section 2.5.3, then its canonical extension[5] to the field algebra $\mathcal{A}^{(F)}$ satisfies the KMS conditions with respect to a dynamical automorphism group, $\alpha'(\mathbf{R})$, of the form

$$\alpha'(t) = \alpha(t)\gamma(-\mu t), \tag{5.3.19}$$

where γ is the gauge group given by Eq. (2.5.8) and μ is a real parameter that depends on ρ. Moreover, the dynamical group $\alpha'(\mathbf{R})$ corresponds to the family of local Hamiltonians

$$H_\Lambda^{\text{eff}} = H_\Lambda - \mu N_\Lambda, \tag{5.3.20}$$

where $N_\Lambda \ (= \int_\Lambda dx \psi^\star(x)\psi(x))$ is the local number operator. Evidently, Eq. (5.3.20) signifies that the parameter μ is a chemical potential. In other words, *the thermodynamical concept of the chemical potential emerges from the KMS conditions.* In view of this result, we may classify the primary, or extremal, KMS states in terms of β and μ. By Eq. (5.3.20), this entails the absorption of the term $-\mu N_\Lambda$ into the effective local Hamiltonian H_Λ^{eff}. The full set, $S_{\beta\mu}^{\text{KMS}}$ of KMS states at inverse temperature β and chemical potential μ is then given by the convex combinations of these extremals.

Note on Zero Temperature States

For zero temperature, that is, $\beta = \infty$, the KMS conditions reduce simply to the

---

[5] See the Note following Eq. (2.5.10) for a definition of this extension.

demands that ρ is a stationary and that $\rho(B\alpha(t)A)$ has an analytic continuation to the upper half of the complex t-plane, for any $A, B \in \mathcal{A}$. Moreover, these latter conditions are equivalent to the requirements that the GNS Hamiltonian, H_ρ, for the state ρ is a positive operator that vanishes on the cyclic vector, Φ_ρ[6] [BR]. This means simply that Φ_ρ is a ground state vector of the Hamiltonian H_ρ.

5.3.2 Thermodynamical Stability Conditions

For an infinite system, Σ, there are two distinct kinds of thermodynamical stability, corresponding to the nondecrease of free energy for localised and global changes of state. To formulate both of these kinds of stability, we need to invoke the quasi-local structure of the algebra of observables, $\mathcal{A}$, of Σ, the local normality of its physical states and the specification of the local Hamiltonians, or energy observables, H_Λ. In view of the above remarks on the chemical potential, μ, we assume that, in the case of systems of indistinguishable particles, formulated in Section 2.5.3, this parameter is absorbed into the effective local Hamiltonian, H_Λ^{eff}, according to the formula (5.3.20). On the other hand, for systems of distinguishable particles located on the sites of a lattice, the effective local Hamiltonian, H_Λ^{eff}, is simply H_Λ.

We assume that the interactions possess the properties[7] of stability and temperedness that ensure that the thermodynamic potentials, such as the free energy density $-\beta^{-1}\ln(\mathrm{Tr}\,\exp(-\beta H_\Lambda^{\mathrm{eff}})/|\Lambda|$, are well defined in the infinite volume limit.

For each state ρ of Σ, we define the local free energy of the region Λ at inverse temperature β to be that of the finite system Σ_Λ when its state is ρ_Λ. Thus,

$$\hat{F}_{\beta,\Lambda}(\rho) = \hat{E}_\Lambda(\rho) - \beta^{-1}\hat{S}(\rho_\Lambda), \tag{5.3.21}$$

where $\hat{E}_\Lambda$ and $\hat{S}_\Lambda$ are the local energy and entropy functionals defined by the formulae

$$\hat{E}_\Lambda(\rho) = \rho(H_\Lambda^{\mathrm{eff}}) \equiv \rho_\Lambda(H_\Lambda^{\mathrm{eff}}) \tag{5.3.22}$$

and, as in Eq. (3.2.22),

$$\hat{S}_\Lambda(\rho) = \hat{S}(\rho_\Lambda). \tag{5.3.23}$$

Here we remark that, in the case of continuous systems [Ro3], the effective

---

[6] This follows easily from the specification of H_ρ in Section 2.4.3, item (31).

[7] These are essentially the properties of short-range repulsion and long-range decay, which ensure stability against collapse and extensivity of energy, respectively (cf [Ru1]).

Hamiltonian H_Λ^{eff} must be subjected to Neumann, rather than Dirichlet, boundary conditions, in order that its kinetic energy component, T_Λ, satisfies the additivity condition

$$\rho(T_{\Lambda \cup \Lambda'}) = \rho(T_\Lambda) + \rho(T_{\Lambda'}) \quad \text{for } \Lambda \cap \Lambda' = \emptyset.$$

Local Thermodynamical Stability

In order to formulate stability against strictly localised modifications of state, we define an equivalence relation $\sim$ on the state space of Σ according to the specification that $\rho \sim \rho'$ signifies that the states ρ and ρ' coincide outside some bounded spatial region Λ_0, that is, that $\rho_{\Lambda_1} = \rho'_{\Lambda_1}$ if Λ_0 and Λ_1 are disjoint. We then define the increment in the free energy of Σ, corresponding to a transition from ρ to $\rho'(\sim \rho)$, to be

$$\Delta \hat{F}_\beta(\rho \mid \rho') = \lim_{\Lambda \uparrow X} \left(\hat{F}_{\beta,\Lambda}(\rho') - \hat{F}_{\beta,\Lambda}(\rho) \right) \quad \forall \rho' \sim \rho, \qquad (5.3.24)$$

assuming that this limit exists. In fact, this assumption is valid for lattice systems with short range forces [Se8] since, for such systems, the convergence of the entropic and energetic contributions to the right-hand side of Eq. (5.3.24) are guaranteed by the strong subadditivity of entropy and the short range of the interactions, respectively. It is also valid, at least for certain states ρ, in lattice systems with suitably tempered long range forces [Se8]. As for continuous systems, it is a simple matter to infer similar results, at least for those whose particles have hard cores, from Robinson's [Ro3] formulation of their thermodynamic functionals.

Assuming then that the right-hand side of Eq. (5.3.24) is well defined, we term the state ρ of Σ *locally thermodynamically stable* (LTS) if

$$\Delta \hat{F}_\beta(\rho \mid \rho') \geq 0 \quad \forall \rho' \sim \rho, \qquad (5.3.25)$$

and we denote the set of all LTS states at inverse temperature β by S_β^{LTS}.

The following proposition establishes that the equivalence between the KMS property and thermodynamical stability prevails not only for finite systems, but also, at the local level, for infinite lattice systems. This is perhaps not surprising, since the KMS conditions are essentially of a local nature, in that they pertains to time correlations of local, or quasi-local, observables.

Proposition 5.3.2 [AS, Se7, Se8] *For lattice systems, the locally thermodynamically stable states at inverse temperature β are precisely the KMS states at the same temperature, that is, $S_\beta^{\text{LTS}} = S_\beta^{\text{KMS}}$.*

No corresponding result has yet been established for continuous systems, though one imagines that it probably also prevails for these. The difficulties one faces in trying to prove it appear to be technical ones, arising from the unboundedness of the local Hamiltonians H_Λ^{eff} and the infinite dimensionality of the local Hilbert spaces $\mathcal{H}_\Lambda$. In the absence of a proof, we make the following conjecture.

Conjecture 5.3.3   *For continuous systems, too, the locally thermodynamically stable states at any given temperature are precisely the KMS states at the same temperature.*

Thus, according to Proposition 5.3.2 and Conjecture 5.3.3, the KMS and LTS conditions are equivalent.

Note

(1) The KMS and LTS conditions do not require space translational invariance of either the state or the interactions.

(2) Furthermore, these conditions do not require the global density of the system to be sharply defined, that is, dispersion-free. In fact, they may satisfied by mixtures of KMS states of different densities, and such mixtures evidently carry global density fluctuations.

Global Thermodynamical Stability

Assuming now that the interactions of Σ are invariant with respect to the space translation group X, we define the energy and free energy density functionals, $\hat{e}$ and $\hat{f}_\beta$, on the set, S_X, of translationally invariant states by the following formulae.

$$\hat{e}(\rho) = \lim_{\Lambda \uparrow X} \hat{E}_\Lambda(\rho) \tag{5.3.26}$$

and

$$\hat{f}_\beta(\rho) = \lim_{\Lambda \uparrow X} |\Lambda|^{-1} \hat{F}_{\beta,\Lambda}(\rho). \tag{5.3.27}$$

Here the limits are well defined both for lattice systems [Ru1] and for continuous ones with stable and tempered interactions[8] [Ro3]. It follows immediately from Eqs. (3.2.25), (5.3.21), (5.3.26) and (5.3.27) that

$$\hat{f}_\beta = \hat{e} - \beta^{-1}\hat{s}, \tag{5.3.28}$$

where $\hat{s}(\rho)$ is the entropy density of ρ. The following properties have been established for $\hat{f}_\beta$ [Ru1, Ro3].

---

[8] In fact, the treatment of [Ro3], which was restricted to systems of particles with hard cores, was generalized by A. Pflug (Thesis, University of Vienna, 1979) to ones with strong short range repulsive forces, which included those of the Lenard–Jones type.

(f.1) $\hat{f}_\beta$ is affine, that is,

$$f(\lambda\rho_1 + (1 - \lambda)\rho_2) = \lambda f(\rho_1) + (1 - \lambda)f(\rho_2) \quad \forall \rho_1, \rho_2 \in S_X, \ 0 < \lambda < 1.$$

(f.2) $\hat{f}_\beta$ is lower semi-continuous with respect to the $w^\star$ topology and hence attains its lower bound.

(f.3) The minimum value of $\hat{f}_\beta$ is the Gibbs potential, $\phi(\beta)$, of Σ at inverse temperature β, that is,

$$\min_{\rho \in S_X} \hat{f}_\beta(\rho) = \phi(\beta), \tag{5.3.29}$$

where

$$\phi(\beta) = \lim_{\Lambda \uparrow X} \ln\left(\operatorname{Tr} \exp(-\beta H_\Lambda^{\mathrm{eff}})\right)/|\Lambda|, \tag{5.3.30}$$

with Dirichlet boundary conditions on the right-hand side in the case of continuous systems.[9]

We term the set of states that minimise $\hat{f}_\beta$ *globally thermodynamically stable* (GTS). We denote this set by $S_{X,\beta}^{\mathrm{GTS}}$ and note that, by Eq. (5.3.29), it consists of the elements, ρ, of S_X that satisfy the equation

$$\hat{f}_\beta(\rho) = \phi(\beta). \tag{5.3.31}$$

It follows immediately from condition and (f.1) that the set $S_{X,\beta}^{\mathrm{GTS}}$ is convex. We denote the set of its extremals by $\mathcal{E}_{X,\beta}^{\mathrm{GTS}}$. The following key properties of the GTS states have been established.

Proposition 5.3.4 [Ru1, Ro3] *There is a unique decomposition of the GTS states into extremals, that is, for $\rho \in S_{X,\beta}^{\mathrm{GTS}}$, there is a unique probability measure μ_ρ on $\mathcal{E}_{X,\beta}^{\mathrm{GTS}}$, such that*

$$\rho = \int_{\mathcal{E}_{X,\beta}^{\mathrm{GTS}}} \omega \, d\mu_\rho(\omega). \tag{5.3.32}$$

Moreover, this is just the spatial ergodic decomposition, as restricted to $S_{X,\beta}^{\mathrm{GTS}}$.

Comment It follows immediately from this proposition that the extremal GTS states are those for which the space-averages of the local observables are sharply defined (cf. Section 2.6.2).

Proposition 5.3.5 *The following relationships between the KMS and GTS properties hold for quantum lattice systems:*

---

[9] Cf. Robinson [Ro3] and A. Pflug (Thesis, University of Vienna, 1979) for discussions of these boundary conditions.

(1) [LaRo, Se8] The GTS states possess the KMS property.

*(2) [Ar3] Conversely, for systems with short range interactions, the trans-
     lationally invariant KMS states are GTS.*

*(3) [Se8] However, certain systems[10] with long range interactions support
     translationally invariant KMS states that are not GTS.*

Discussion

There is, as yet, no general proof that these results hold for continuous
systems. However, it is not difficult to prove that the method of [LaRo] to
establish Proposition 5.3.5 (1) can be extended to continuous systems of
particles with hard cores and suitably regular interactions, as formulated by
Miracle-Sole and Robinson [MR]. The same thing can also be proved for a
significant class of continuous systems by a method devised by Narnhofer and
Sewell [NS, Appendix 2] for the ideal Fermi gas. The problems confronting
generalisations of these results to the full counterpart of Proposition 5.3.5 for
continuous systems appear to be technical ones, stemming from the unboundedness of the local Hamiltonians, H_Λ, and the infinite dimensionality of the
local Hilbert spaces, $\mathcal{H}_\Lambda$. In the absence of general proven results, we make
the following conjecture.

Conjecture 5.3.6   *The results of Proposition 5.2.5 are valid also for continuous systems.*

Relationship between the GTS and KMS decompositions

By the property (KMS.3), specified in Section 5.3.1, the KMS decomposition
is just the central one; and further, Ruelle [Ru2] has proved that the latter is
finer than that of GTS. Hence, by Proposition 5.3.5 and Conjecture 5.3.6, it
serves to decompose any element of $S_{X,\beta}^{\mathrm{GTS}}$ into extremal (primary) KMS states,
that is, for $\rho \in S_{X,\beta}^{\mathrm{GTS}}$, there is a unique probability measure ν_ρ on $\mathcal{E}_\beta^{\mathrm{KMS}}$ such
that

$$\rho = \int_{\mathcal{E}_\beta^{\mathrm{KMS}}} d\nu_\rho(\sigma)\sigma \qquad\qquad (5.3.33)$$

and, moreover, the extremals occurring here are not neccessarily translationally invariant. However, as emphasised by Emch, Knops and Verboven
[EKV], they may be invariant under some crystallographic group, and
hence the KMS decomposition may provide the seeds of a theory of crystallisation.

---

[10] Examples are the Fisher droplet model [Fi1] and the Heisenberg-Weiss ferromagnet
     (cf. [Se8]).

Extension of GTS Condition to Crystalline States

In view of the last remark, we now seek to extend the above definition of global stability to states that are invariant under crystallographic subgroups, Y, of X, rather than the full translation group X. For this purpose, we define S_Y to be the (convex) set of states of Σ that are invariant under such a group, Y.

It follows easily from these specifications that the above definition of $\hat{f}_\beta$ extends to S_Y. Further, the proofs of (f.1)–(f.3) [Ru1, Ro3] remain valid when X is replaced by Y. Hence, by the modification of (f.3) thereby obtained,

$$\min_{\rho \in S_Y} \hat{f}_\beta(\rho) = \phi(\beta),$$

and therefore the elements of S_Y for which $\hat{f}_\beta$ attains its minimum are those that satisfy Eq. (5.3.31). We term these states the GTS elements of S_Y. Since $\hat{f}_\beta$ is affine, these form a convex set, which we denote by $S_{Y,\beta}^{GTS}$. We define $\mathcal{E}_{Y,\beta}^{GTS}$ to be the set of its extremals.

We now remark that the proofs of Propositions 5.3.4 and 5.3.5 remain valid when X is replaced by Y; and we assume that Conjecture 5.3.6 is also valid then. Thus, we conclude that the above propositions and conjecture extend from the translationally invariant states to those with crystalline symmetry. Hence, by the resulting extended version of Proposition 5.3.5 and Conjecture 5.3.6, we arrive at the following proposition.

Proposition 5.3.7 *Under the above assumptions,*

(1) the crystalline GTS states possess the KMS property; and

(2) the crystalline KMS states are GTS in the case of short range interactions, but are not necessarily so in the case of long range ones.

5.4 EQUILIBRIUM AND METASTABLE STATES

5.4.1 Equilibrium States

We assume that theses are the states that meet the demands of the Zeroth and Second Laws of Thermodynamics. Now, by Proposition 5.3.1 and the ensuing discussion, the KMS conditions characterise the states that satisfy the requirements of the Zeroth Law, while the GTS condition of minimum free energy is precisely that for the fufillment of the Second Law, at least if either full translational or crystallographic symmetry prevails. In this case, it follows from Proposition 5.5, Conjecture 5.6 and Proposition 5.7 that the GTS states satisfy the KMS condition, and consequently we take these to be the thermal equilibrium states. Moreover, as noted in Comment (4) following Proposition 5.3.1, the extremal KMS states are the primary ones, which in turn correspond to the pure phases. Accordingly, we assume that the pure equilibrium phases comprise the globally stable extremal invariant KMS states: in the case of short range interactions, the global stability requirement is redundant, by Proposition 5.3.7. Most importantly (cf. (KMS.4) of Section 5.3.1), some of the pure equilibrium phases may be symmetry breaking states, as exemplified by those of ferromagnets, antiferromagnets and crystals.

We remark here that the characterisation of thermal equilibrium by the GTS condition can be extended to certain states with neither full translational nor crystallographic symmetry. For example, globally stable states formed from spatially separated phases of Ising spin systems have been constructed by Dobrushin [Dob] and Van Beijeren [VB] as infinite volume limits of Gibbs states with boundary conditions that lead to equal and opposite polarisations on the two sides of a certain surface. These states evidently satisfy the KMS condition, since they are infinite volume limits of canonical ones.

5.4.2 Metastable States

These are the states that simulate those of thermal equilibrium, except for the fact that they do not minimise the free energy of the system. Such states are quite ubiquitous, classic examples being supercooled or superheated phases of matter, supercurrent carrying states of superconductors, and even certain crystalline allotropes of materials with higher free energy than others of the same material: a dramatic example of this is diamond!

Now a class of states that do have the hallmark of metastability, namely the simulation of thermal equilibrium except for the minimisation of free energy, are those that satisfy the KMS, but not the GTS condition. For, notwithstanding their lack of global stability, these states possess the reservoir property demanded by the Zeroth Law (by Proposition 5.3.1), are thermodynamically stable against local modifications (by Proposition 5.3.2 and Conjecture 5.3.3) and, as a consequece, have infinite lifetimes when locally coupled to thermal reservoirs of the same temperature [Se8]. For these reasons, we take the KMS states that are not GTS to be metastable. Examples of states that have been proved to satisfy these metastability conditions are the following:

(a) A quantum lattice system [Se2,8] with many-body interactions of the type favoring the formation of "clusters", analogous to those of the Fisher model [Fi1]. The metastable states of the model correspond to a superheated liquid phase.

(b) Various mean field theoretic models. In particular, the Ising–Weiss ferromagnet supports a superheated ferromagnetic phase [Se2,8], while a certain mean field theoretic bipolaronic system, designed to represent a ceramic superconductor, supports certain metastable phases [GMR].

Of course the metastable KMS states are rather special, in that they require suitably long range forces and have infinite lifetimes. Presumably, the more usual metastable states have very long, but finite, lifetimes [PL, Se2, Se8] and can occur even in systems with short range forces. A proven prototype example of such metastability is the Ising ferromagnet with polarisation opposed to an external magnetic field [SS].

5.5 FURTHER DISCUSSION

The above formulation of thermal states of infinite systems provides a framework for the theory of the phase structure of matter. Moreover, this framework is not an empty shell, since various models have been shown to exhibit the phenomena of phase transitions, phase coexistence and metastability.[11]

There are of course many outstanding problems in the area, and we now specify just two of them. The first is that of proving Conjectures 5.3.3 and 5.3.6, and thereby raising the theory of continuous systems to the same level as that of lattice systems. This is certainly a tough mathematical problem, and it remains to be seen whether it hides some deep physics.

A second big problem is the phenomenon of crystallisation, and that is certainly a deep one! In particular, as various crystalline allotropes, including diamond, are metastable, it is natural to ask which crystals, if any, are truly GTS. The states of those that are correspond to absolute minima of the free energy density and thus, by Proposition 5.3.7 satisfy the KMS conditions. They may be described within the framework of Emch, Knops and Verboven [EKV], wherein crystals correspond to extremal KMS states of systems that break the Euclidean symmetry of their interactions in favour of the residual symmetry of a crystallographic group Y. As regards metastable crystals, it is natural to conjecture that these correspond to states that minimise the free energy density, subject to the constraint of invariance under the relevant crystallographic groups, and that they satisfy KMS conditions relative to a reduced dynamics under which their normal folia are stable.[12] However, we still lack models that could throw light on these conjectures, although there are some interesting results concerning the breakdown of space translational symmetry in favor of a reduced lattice symmetry in (a) classical low-dimensional systems of interacting particles at zero temperature [Ra2,3], and (b) quantum lattice systems, such as the Heisenberg antiferromagnet [DLS] and a version of the Hubbard model [KL], in their condensed phases.

Finally we emphasise that the idealisation of macroscopic systems as infi-

---

[11] Cf. [Se2] and works cited there, as well as the treatment of metastability by Schonmann and Shlosman [SS].

[12] This dynamics should presumably be an approximant to the full dynamics of the systems, corresponding to a limit in which the lifetimes of the metastable states are infinite.

nite ones is really essential to any mathematically sharp theory of the phase structure of matter, since a finite system has but one Gibbs state at any fixed temperature and therefore cannot support the phenomenon of phase coexistence. Nor can it support states that are metastable in the sense of satisfying the KMS condition without minimising the free energy, since, as we remarked in Section 5.3.1, the KMS and (global) thermodynamical stability properties are equivalent for finite systems. This accords with Fisher's observation [Fi2] that there is no natural statistical thermodynamical characterisation of metastable states of finite systems.

Chapter 6

Equilibrium thermodynamics and phase structure

6.1 INTRODUCTION

Classical thermodynamics presents a purely macroscopic picture of the equilibrium states and phase structures of large systems. We now provide a brief exposé of this picture and show how it emerges from the quantum statistical theory of equilibrium states and phases, as described in the previous chapter. Evidently, our treatment is radically different from that of traditional statistical thermodynamics, since that is based on the theory of finite systems, which cannot support different equilibrium states under the same external macroscopic constraints.

We start by recalling that the structure of classical equilibrium thermodynamics may be encapsulated by the following scheme (cf. [Ca], [Bu]). The independent thermodynamic variables of a system are its internal energy, E, a finite set of other extensive conserved quantities $(Q_1, ..., Q_n)$, and the volume V. The work done by the system against external constraints in order to effect infinitesimal changes $(dQ_1, ..., dQ_n, dV)$ in $(Q_1, ..., Q_n, V)$ is then a differential form

$$\mathcal{W} = pdV + \sum_{k=1}^{n} y_k dQ_k,$$

where p, the pressure, and the coefficients $(y_1, ..., y_n)$ are all intensive variables. Hence, by energy conservation, the heat (that is, the energy) required to effect these changes, together with one of dE in E, is

$$\mathcal{H} = dE + pdV + \sum_{k=1}^{n} y_k dQ_k,$$

which is the First Law of Thermodynamics. The Second Law[1] then asserts that, in an adiabatic change of state, $\mathcal{H}$ is of the form TdS, where $T = \beta^{-1}$ is the temperature[2] and S is an extensive variable, termed the thermodynamic

---

[1] This stems from Caratheodory's principle [Bu], which represents the irreversibilty in nature by the assertion that, for each macroscopic state $(E, Q_1, ..., Q_n, V)$, there are adiabatically inaccessible neighbouring states $(E', Q_1', ..., Q_n', V')$.

[2] Recall that, as noted in the remark following Eq. (3.2.10)', we are employing units in which Boltzmann's constant is unity.

entropy, that is a function of E, the Q_k's and V. Thus, T arises here as an integrating factor for the form $\mathcal{H}$, and S satisfies the fundamental equation

$$TdS = dE + pdV + \sum_{k=1}^{n} y_k dQ_k. \qquad (6.1.1)$$

Equivalently, putting $E = Q_0$, $T^{-1} = \theta_0$, $T^{-1}y_k = \theta_k$, for $k = 1, ..., n$, and $T^{-1}p = \pi$,

$$dS = \pi dV + \sum_{k=0}^{n} \theta_k dQ_k. \qquad (6.1.2)$$

We term θ_k the *thermodynamical conjugate* of Q_k and π the *reduced pressure*. We denote the sets of variables $(Q_0, Q_1, ..., Q_n)$ and $(\theta_0, ..., \theta_n)$ by Q and θ, respectively.

In view of the extensivity of Q and S, we may express $S(Q, V)$ in the form $Vs(q)$, where $q = (q_0, ..., q_n) = V^{-1}Q$. Thus, q and the entropy density, $s(q)$, are intensive variables, and Eq. (6.1.2) takes the form

$$V(ds - \theta.dq) + (s - \theta.q - \pi)dV = 0$$

where the dot denotes the $\mathbf{R}^{n+1}$ scalar product. It follows immediately from this last equation that

$$ds = \theta.dq, \qquad (6.1.3)$$

that is,

$$\theta_k = \frac{\partial s}{\partial q_k} \quad \text{for } k = 0, 1, ..., n, \qquad (6.1.4)$$

and

$$\pi = s - \theta.q. \qquad (6.1.5)$$

The equilibrium thermodynamics of the system is governed by the form of s. A key property of this function, which ensues from the demand of thermo-dynamical stability, is that it is concave.[3] Hence, by Eqs. (6.1.3) and (6.1.5), the reduced pressure, π, is the Legendre transform of s [Ca], that is,

$$\pi(\theta) = \max_q (s(q) - \theta.q). \qquad (6.1.6)$$

The explicit form of this function constitutes the *equation of state* of the system. We note that it follows from Eq. (6.1.6) that π is a convex function, since it implies that, for any values $\theta^{(1)}$ and $\theta^{(2)}$ of θ, any value of q and $0 < \lambda < 1$,

$$\lambda \pi(\theta^{(1)}) + (1 - \lambda)\pi(\theta^{(2)}) \geq \lambda \left(s(q) - \theta^{(1)}.q \right) + (1 - \lambda)\left(s(q) - \theta^{(2)}.q \right)$$

$$\equiv s(q) - \left(\lambda \theta^{(1)} + (1 - \lambda)\theta^{(2)} \right).q$$

---

[3] A rudimentary sketch of convexity and concavity is provided in Section 6.2.

and hence the left-hand side of this inequality is greater than or equal to the maximum value, with respect to q, of its right-hand side, that is, by Eq. (6.1.6), π satisfies the convexity condition

$$\lambda\pi(\theta^{(1)}) + (1-\lambda)\pi(\theta^{(2)}) \geq \pi\left(\lambda\theta^{(1)} + (1-\lambda)\theta^{(2)}\right).$$

The phase structure of the system is determined by the singularities of π. The empirically based picture of it is that the singularities form smooth surfaces in θ-space that are the boundaries of open regions $R_1, R_2, R_3, \ldots$, where π is analytic (cf. Figure 6.1). This is taken to signify that the R_j's correspond to the different phases of the system and that their boundaries represent the phase transitions, or the regions of phase coexistence.

Thus, the equilibrium thermodynamics and phase structure of a system stem from the specification of the variables q and the form of the entropy function $s(q)$ or, equivalently, the reduced pressure $\pi(\theta)$. At the level of classical thermodynamics, this imput has to be empirically determined on the basis of the measurability of the variables q and the experimentally obtained equation of state. Here, it is important to appreciate that the choice of q and even the number, $(n+1)$, of its components depend on the system in question. For example, in the case of a simple gas, q consists of the densities of energy and particle number; whereas, in the case of a compressible magnetic material, it consists of these variables together with the polarisation.

It follows from these considerations that the first task of statistical thermodynamics is to provide a quantum characterisation of the variables Q. The fact

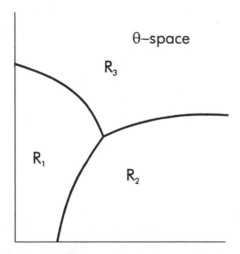

Figure 6.1 Example of a phase diagram. R_1, R_2 and R_3 are single phase regions, and the boundaries separating them are two- or multi-phase regions.

that they are extensive conserved quantities is completely in line with ergodic theory, which tells us that the ergodic equilibrium states are labeled by such variables [Far]. However, there remains the problem of specifying them in terms of the quantum structure of the system. Here, although obvious candidates are the extensive variables that satisfy universal conservation laws, such as those for energy, charge and linear momentum, it is not *a priori* clear which of these are the natural thermodynamical variables of a particular system, or even whether there might be others that are peculiar to that system.[4]

In this chapter we present an approach to quantum statistical thermodynamics that provides a characterisation of the variables q and a formulation of the entropy function s, along lines that may be summarised as follows. We assume that q corresponds to macroscopic quantum observables $\hat{q} = (\hat{q}_0, ..., \hat{q}_n)$, representing the global densities of extensive conserved quantities; in particular, $\hat{q}_0$ is the energy density, $\hat{e}$. Furthermore, we assume that q is a *complete* set of thermodynamical variables, in that $\hat{q}$ constitutes a minimal set of functionals that separate the pure equilibrium phases of the system.[5] As for the thermodynamical potentials, we define $s(q)$ to be the supremum of the quantum statistical entropy density $\hat{s}(\rho)$, taken over the set of translationally invariant states, ρ, for which the expectation value of $\hat{q}$ is q. On this basis, we derive the classical thermodynamical laws (6.1.3)–(6.1.5) from quantum statistics; and, furthermore, we substantiate the classical thermodynamical hypothesis that the values of the control variable, θ, at which the system supports different equilibrium phases are just those where the reduced pressure, π, has a discontinuous derivative (cf. Proposition 6.4.2).

We present our scheme as follows. In Section 6.2, we note some rudimentary properties of convex functions, which we employ in Section 6.3 to show that the classical thermodynamical states, as represented by q, correspond to the tangents to the reduced pressure function π.

In Section 6.4, we pass to the quantum statistical description of the system. There we formulate our basic assumptions concerning the macroscopic observables $\hat{q}$, and express the thermodynamic potentials s and π in terms of them. We then show that our quantum statistical formalism yields the above-described classical thermodynamical picture. In particular, we show that the classical thermodynamical characterisation of phase transitions by singularities in π coincides with the quantum statistical one given by the existence of different equilibrium states with the same values of θ. To be precise, we prove that the discontinuities in the derivatives of π occur at precisely those values

---

[4] Certainly some models, such as that of Ising, have numerous extensive conserved quantities that do not fall into the category covered by universal conservation laws.

[5] Thus, for example, in the case of the two-dimensional Ising ferromagnet, it follows from the solution of the model [MM] that, according to this scheme, the energy density and polarization constitute a complete set of thermodynamic variables.

of θ that support different pure phases, in the sense defined in Chapter 5. Thus, these discontinuities correspond to phase transitions of the first kind in Landau's classification [LL2]. By contrast, phase transitions of the second kind correspond to values of θ where π and its first derivatives are continuous but singular.

In Section 6.5, we extend the scheme of Section 6.4 to systems in ordered phases, characterised by order parameters, η, that are densities of extensive, *nonconserved* quantities. The thermodynamical variables then comprise both q and η.

We conclude in Section 6.6 with a brief discussion of the factors that lead macroscopic systems, with enormous (ideally infinite) numbers of degrees of freedom, to conform to self-contained thermodynamical laws, involving just a few variables.

Appendix A is devoted to the proof of two propositions of Section 6.4, while Appendix B prepares the ground for the following chapter by refining our formulation of the observables $\hat{q}$ for the important case where these are space averages of locally conserved quantum fields.

6.2 PRELIMINARIES ON CONVEXITY[6]

Convex Sets

A convex set, K, is defined to be a region of a vector space V, such that, if $v_1, v_2 \in K$ and $\lambda \in (0, 1)$, then $\lambda v_1 + (1 - \lambda)v_2 \in K$. In other words, straight lines joining pairs of points in K lie wholly in K (cf. Figure 6.2a).

It follows by induction from this definition that, if $v_1, ..., v_n \in K$ and $c_1, ..., c_n$ are positive numbers whose sum is unity, then $\sum_{j=1}^{n} c_j v_j \in K$. We term this such weighted sums *convex combinations* of the v_j's.

Extremal Elements of K

An extremal element of a convex set, K, is one that cannot be expressed as a convex combination of different elements of K. In other words, v is extremal if and only if there is no pair v_1, v_2 of different points in K such that, for some $\lambda \in (0, 1)$, $v = \lambda v_1 + (1 - \lambda)v_2$. We denote the set of extremal points of K by $\mathcal{E}(K)$. Thus, assuming that K is a closed domain of a finite-dimensional vector space V, $\mathcal{E}(K)$ is its boundary, as illustrated by Figure 6.2a.

---

[6] For a general treatment of convexity, see, for example, the book by Roberts and Varberg [RV].

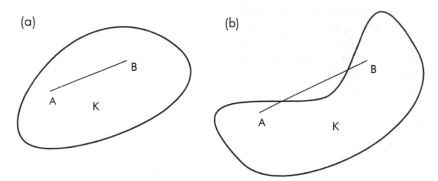

Figure 6.2 (a) An oval-shaped region K is convex, since the chord joining two of its points A and B lies wholly in K. (b) The region K here is not convex, since the chord joining its points A and B lies partially outside K.

The Krein–Millman Theorem

This is a classic theorem, which asserts that any element, v, of a convex, compact set is a convex combination of its extremal points, that is, that there is a probability measure, μ, on $\mathcal{E}(K)$, such that

$$v = \int_{\mathcal{E}(K)} w\,d\mu(w).$$

Here, μ is not necessarily unique. In those cases where it is, K is termed a Choquet simplex.

Convex, Concave and Affine Functions

A real-valued function, g, on a convex domain, K, of V is termed *convex* if

$$g(\lambda v_1 + (1-\lambda)v_2) \geq \lambda g(v_1) + (1-\lambda)g(v_2) \quad \forall v_1, v_2 \in K \text{ and } \lambda \in (0,1).$$

$$(6.2.1)$$

In other words, the chord joining any two points on the graph of g lies above, or else coincides with, the arc of the graph joining those points (cf. Figure 6.3). The function g is termed *strictly convex* if the strict inequality $>$ prevails in Eq. (6.2.1).

On the other hand, a real-valued function, g, on a convex set, K, is termed *concave* if $-g$ is convex, that is, if the inequality sign is reversed in Eq. (6.2.1). Further, it is termed *affine* if it is both convex and concave, that is, if the inequality sign in Eq. (6.2.1) is replaced by that of equality.

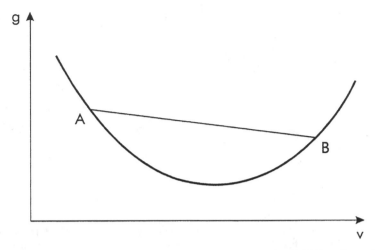

Figure 6.3   Graph of a convex function g. Note that the chord AB lies above the arc joining A to B.

Continuity and Differentiability Properties

Assume now that the convex set, K, is finite-dimensional. Then a convex function, g, on K is continuous in the interior of its domain of definition, K, that is, it cannot support discontinuities there (cf. Figure 6.4a). Its derivative, however, might have discontinuities in the interior of K (cf. Figure 6.4b).

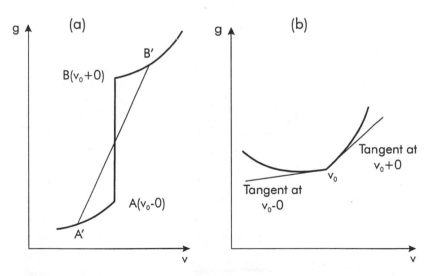

Figure 6.4   (a) Function g with discontinuity at v_0 is not convex since A lies above the chord $A'B'$. (b) Convex function with discontinuous derivative at v_0.

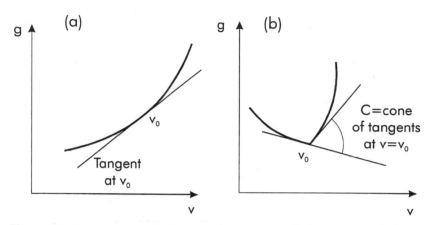

Figure 6.5 Tangents to graph of a convex function, $g(v)$, lie below the graph. In (a), where g is differentiable at v_0, it has a unique tangent there. In (b), where it is not differentiable at v_0, the tangents at that point form a cone $\mathbf{C}$.

Tangents to Convex Functions

Assume still that g is a convex function on a finite-dimensional convex set K. Then, if g is differentiable at the point v_0, say, it follows easily from Eq. (6.2.1) that

$$g(v) - g(v_0) \geq \phi.(v - v_0) \quad \forall v \in K, \tag{6.2.2}$$

where

$$\phi = g'(v_0), \tag{6.2.3}$$

the derivative, that is, the gradient, of g at v_0. Thus, the tangent to the graph of g at v_0 lies below, or possibly level with, that graph (cf. Figure 6.5a).

On the other hand, if g is not differentiable at v_0, then there is still a convex set, $\mathcal{T}_{v_0}$, of vectors, ϕ, that satisfy Eq. (6.2.2); these form a cone, C, with vertex at v_0 (cf. Figure 6.5b). We term $\mathcal{T}_{v_0}$ the set of *tangents* to g at v_0. The extremals of this set correspond to the generators of C.

In fact, we may relate these extremals to certain limits of derivatives of g at sequences of points where this function is differentiable. Specifically, each extremal tangent, ϕ_α, at v_0 is the limit of the derivative of g at a sequence of points $\{v_\alpha^{(n)}\}$ that approaches v_0 along a curve to which ϕ_α is tangent at v_0, that is,

$$\phi_\alpha = \lim_{n \to \infty} g'(v_\alpha^{(n)}). \tag{6.2.4}$$

6.3 THERMODYNAMIC STATES AS TANGENTS TO THE REDUCED PRESSURE FUNCTION

In order to express the classical thermodynamic picture of Section 6.1 in a precise form, we now stipulate that the variables q and θ range over convex subsets of $\mathbf{R}^{n+1}$, which we denote by Q and Θ, respectively. We term Q the *thermodynamic state space* and Θ the *control space*. Correspondingly, we term the elements of Q and Θ the thermodynamic states and the control variables, respectively.

We assume that the entropy function s is both differentiable and concave.[7] Here, the differentiability is a prerequisite for the fundamental formula (6.1.3), and the concavity is required for thermodynamical stability. On the other hand, as discussed in Section 6.1, there may be discontinuities, corresponding to phase transitions, in the derivative of the reduced pressure, π. We recall that the convexity of this function was established directly after Eq. (6.1.6), and we shall denote by $\mathcal{T}_\theta$ the set of all tangents to π at the point θ.

We assume that the thermodynamical equilibrium states, q, corresponding to the value, θ, of the control variable are those for which the maximum of Eq. (6.1.6) is attained. Thus, we take the equilibrium condition for q to be that

$$\pi(\theta) = s(q) - \theta.q. \tag{6.3.1}$$

We denote by Q_θ the set of these states.

Now suppose that $q \in Q_\theta$. Then Eq. (6.3.1) holds good and also, by Eq. (6.1.6),

$$\pi(\theta') \geq s(q) - \theta'.q \quad \forall \theta' \in \Theta.$$

It follows from the last two equations that

$$\pi(\theta') - \pi(\theta) \geq (\theta - \theta').q \quad \forall \theta' \in \Theta, \tag{6.3.2}$$

which signifies that $-q$ is tangent to π at θ.

We now prove the converse, subject to the assumption that the tangents to π all lie in the domain $-Q := \{-q \mid q \in Q\}$. In this case, if ϕ is an extremal tangent to π at θ, we can approach this value of the control variable by a sequence $\{\theta^{(n)}\}(\subset \Theta)$ at which π is differentiable and $\theta^{(n)}$ and $\pi'(\theta^{(n)}) = \phi^{(n)}$ converge to θ and ϕ, respectively, as n tends to infinity. Further, since $\phi^{(n)}$ is the unique tangent to π at $\theta^{(n)}$, it follows from the above that $q^{(n)} = -\pi'(\theta^{(n)})$ must belong to $Q_{\theta^{(n)}}$, that is, that

$$\pi(\theta^{(n)}) = s(q^{(n)}) - \theta^{(n)}.q^{(n)}.$$

On passing to the limit $n \to \infty$, it follows from the above specifications that q satisfies Eq. (6.3.1). This signifies that, if ϕ is an extremal tangent to π at θ,

---

[7] These assumptions are substantiated, on a quantum statistical basis, in Section 6.4.

then $q = -\phi$ lies in $\mathcal{Q}_\theta$. Consequently, by the Krein–Millman theorem, the same is true for all tangents to π at that point.

Thus, we have established the following proposition.

Proposition 6.3.1   *Assuming that all tangents to π lie in the convex set $-\mathcal{Q}$, there is a one-to-one correspondence, $q = -\phi$, between the thermodynamical equilibrium states, q, and the tangents, ϕ, to π, at each value of the control variable, θ.*

6.4 QUANTUM STATISTICAL BASIS OF THERMODYNAMICS

Our scheme for the passage from quantum statistics to classical thermodynamics is formulated within the algebraic framework of Chapter 5, and we employ the same notation here as in that chapter. Here again we assume that the interactions of the infinitely extended quantum system, Σ, are translationally invariant.

We base our treatment of the relationship between the quantum statistics and classical thermodynamics of Σ on a set, $\hat{Q}$, of extensive conserved observables that are designed to be the counterparts of the thermodynamical variables Q of Section 6.1. Here our main objectives are (a) to provide a characterisation of these observables by a condition of *thermodynamical completeness*, whereby they serve to label the pure equilibrium phases and (b) to derive the classical picture of Sections 6.1 and 6.3 from the resultant quantum statistical description of these phases.

We formulate the observables $\hat{Q}$, or more precisely the global densities thereof, in terms of their finite volume versions, $\hat{Q}_\Lambda$, for the bounded open regions Λ of X. Specifically, we assume that $\hat{Q}_\Lambda$ comprises $(n+1)$ linearly independent observables, $(\hat{Q}_{0,\Lambda}, ..., \hat{Q}_{n,\Lambda})$ of the finite system Σ_Λ and possesses the following properties:

($\hat{Q}$.1) $\hat{Q}_\Lambda$ transforms covariantly with respect to space translations, that is, $\xi(x)\hat{Q}_\Lambda = \hat{Q}_{\Lambda+x}$ for all $x \in X$ and $\Lambda \in L$, the set of bounded open subsets of X.

($\hat{Q}$.2) $\hat{Q}_{0,\Lambda}$ is the local Hamiltonian, H_Λ, which we assume to be lower bounded. We denote by $S^{(0)}$ (respectively $S_X^{(0)}$) the set of all (respectively translationally invariant) states, ρ, of Σ for which

$$\rho(H_\Lambda) < \infty \quad \forall \Lambda \in L$$

and we assume that $\rho(\hat{Q}_\Lambda)$ is well defined[8] on these states.

---

[8] This can be taken to mean that the expectation values of the positive operators $(\hat{Q}_{k,\Lambda}^2)^{1/2}$ for the state ρ are all finite.

($\hat{Q}$.3) The observables $\hat{Q}_{1,\Lambda}, ..., \hat{Q}_{n,\Lambda}$ are extensive conserved quantities, in that[9]

(a) they commute with H_Λ, and

(b) if $\rho \in S_X^{(0)}$ and Λ, Λ' are disjoint, then

$$\rho(\hat{Q}_{k,\Lambda \cup \Lambda'}) = \rho(\hat{Q}_{k,\Lambda}) + \rho(\hat{Q}_{k,\Lambda'}) \quad \text{for } r = 1, ..., n. \quad (6.4.1)$$

It follows from this equation that, for $r = 1, ..., n$, we may define the functional $\hat{q}_k$ on $S_X^{(0)}$ by the formula

$$\hat{q}_k(\rho) = \lim_{\Lambda \uparrow} \frac{\rho(\hat{Q}_{k,\Lambda})}{|\Lambda|} \quad \text{for } r = 1, ..., n. \quad (6.4.2)$$

Further, in accordance with our definition $\hat{Q}_{0,\Lambda} := H_\Lambda$, we define $\hat{q}_0$ to be the global energy density functional[10]

$$\hat{q}_0(\rho) = \lim_{\Lambda \uparrow} \frac{\rho(H_\Lambda)}{|\Lambda|}. \quad (6.4.3)$$

Thus, the functionals $\hat{q}_k$ are affine. We assume that they are linearly independent[11].

We denote $(\hat{q}_0, ..., \hat{q}_n)$ by $\hat{q}$ and define the *thermodynamic state space*, Q, to be the range of $\hat{q}$ as ρ runs through the domain $S_X^{(0)}$. This space is convex, since $S_X^{(0)}$ is convex and $\hat{q}$ is affine.

We define the thermodynamical entropy function, s, in terms of its quantum statistical counterpart, $\hat{s}$, by the formula

$$s(q) := \sup\{\hat{s}(\rho) \mid \hat{q}(\rho) = q\} \quad \forall q \in Q. \quad (6.4.4)$$

To show that his function is concave, we note that it follows from the above definitions that, if $q^{(1)}, q^{(2)} \in Q$, then, for any $\epsilon > 0$, there are states $\rho^{(1)}, \rho^{(2)}$ in $S_X^{(0)}$ such that

$$q^{(j)} = \hat{q}(\rho^{(j)}) \quad \text{and} \quad \hat{s}(\rho^{(j)}) \geq s(q^{(j)}) - \epsilon \quad \text{for } j = 1, 2.$$

Hence, by Eq. (6.4.4) and the affine property of $\hat{s}$,

---

[9] In fact, the conditions ($\hat{Q}$.3a, b) could be weakened by the incorporation of "surface effects", without changing any of their consequences. Thus, for example, one could replace the equation in ($\hat{Q}$.3b) by the assertion that the difference in its two sides was majorized by a term proportional to the total surface area of Λ and Λ'.

[10] This is equivalent to the energy density functional, $\hat{e}$, defined by Eqs. (5.3.20), (5.3.22) and (5.3.26) for the case where $\mu = 0$.

[11] This is a slightly stronger assumption than that of the linear independence of the observables $(\hat{Q}_{0,\Lambda}, ..., \hat{Q}_{n,\Lambda})$, since it involves a limit where Λ becomes infinite.

$$s(\lambda q^{(1)} + (1-\lambda)q^{(2)}) \geq \hat{s}(\lambda \rho^{(1)} + (1-\lambda)\rho^{(2)})$$

$$= \lambda \hat{s}(\rho^{(1)}) + (1-\lambda)\hat{s}(\rho^{(2)})$$

$$\geq \lambda s(q^{(1)}) + (1-\lambda)s(q^{(2)}) - \epsilon \quad \forall \lambda \in (0,1),$$

and therefore, since ϵ is an arbitrary positive number,

$$s(\lambda q^{(1)} + (1-\lambda)q^{(2)}) \geq \lambda s(q^{(1)}) + (1-\lambda)s(q^{(2)}) \quad \forall \lambda \in (0,1),$$

that is, s is concave.

The Control Space Θ

In the classical picture of Section 6.1, Q_0 is the internal energy and $Q_0 + \sum_{k=1}^{n} y_k Q_k$ is the effective energy when $y_1, ..., y_n$ are fixed. Correspondingly, we define the effective local Hamiltonian of Σ to be

$$H_\Lambda^{\text{eff}}(y) := H_\Lambda + \sum_{k=1}^{n} y_k \hat{Q}_{k,\Lambda}, \tag{6.4.5}$$

where $y_1, ..., y_n$ are real-valued variables. Further, as in Section 6.1, we define the thermodynamical variables $\theta = (\theta_0, ..., \theta_n)$ by the prescription that θ_0 is the inverse temperature, β, and $\theta_k = \beta y_k$ for $k = 1, ..., n$. Thus, by Eq. (6.4.5), $\theta.Q_\Lambda \equiv \beta H_\Lambda^{\text{eff}}(y)$. We define the control space, Θ, to be the subset of $\mathbf{R}^{n+1}$ whose points θ satisfy the condition that $\theta.Q_\Lambda$ is lower bounded for all $\Lambda \in L$. Evidently, Θ is convex. We remark that, in the case of the lattice model of Section 2.5.2, the finite dimensionality of the local Hilbert spaces, $\mathcal{H}_\Lambda$, ensures that $\Theta = \mathbf{R}^{n+1}$; whereas in the case of the generic continuum model, where the local Hamiltonians are bounded from below, but not from above, the inverse temperature θ_0 is restricted to positive values, at least when the other θ_j's vanish.[12]

$(\hat{Q}.4)$ We assume that the conditions (f.2) and (f.3) of Section 5.3.2 still prevail when H_Λ^{eff} is replaced by the effective local Hamiltonian $H_\Lambda^{\text{eff}}(y)$. Thus, the functional $(\hat{s} - \theta.\hat{q})$ attains its supremum, namely

$$\pi(\theta) = \max_{\rho}(\hat{s}(\rho) - \theta.\hat{q}(\rho)) \quad \forall \theta \in \mathbf{R}_+ \times \mathbf{R}^n, \tag{6.4.6}$$

and, furthermore,

$$\pi(\theta) = \lim_{\Lambda \uparrow} |\Lambda|^{-1} \ln\left(\text{Tr}_{\mathcal{H}_\Lambda} \exp(-\theta.\hat{Q}_\Lambda)\right). \tag{6.4.7}$$

It follows now from Eqs. (6.4.4) and (6.4.6) that π is the Legendre transform of s (cf. Section 6.1), that is,

[12] In fact, under further, realistic conditions on $\hat{Q}$, $\Theta \subset \mathbf{R}_+ \times \mathbf{R}^n$.

$$\pi(\theta) = \sup_{q \in Q} (s(q) - \theta.q). \tag{6.4.8}$$

We term π the *reduced pressure*, as in Section 6.1, and note that, as shown there, it is convex.

Generalised Version of GTS

We now generalise our earlier definition of global thermodynamical stability, given in Section 5.3.2, by defining the GTS states corresponding to θ to be the translationally invariant ones that maximise $\hat{s} - \theta.\hat{q}$, or, equivalently, that minimise the Gibbs free energy. Thus, by Eq. (6.4.6), they are the translationally invariant states ρ that satisfy the condition

$$\pi(\theta) = \hat{s}(\rho) - \theta.\hat{q}(\rho). \tag{6.4.9}$$

Hence, as both $\hat{s}$ and $\hat{q}$ are affine, $S_{X,\theta}^{\mathrm{GTS}}$ is a convex set. We denote the subset of its extremals by $\mathcal{E}_{X,\theta}^{\mathrm{GTS}}$. It follows from these specifications that $S_{X,\theta}^{\mathrm{GTS}}$ possesses the spatial ergodic decomposition property formulated for $S_{X,\beta}^{\mathrm{GTS}}$ in Chapter 5, that is, that Proposition 5.3.4 remains valid when the β there is replaced by θ.

Note Since, by Eqs. (6.4.6) and (6.4.9),

$$\pi(\theta') - \pi(\theta) \geq -(\theta' - \theta).\hat{q}(\rho) \quad \forall \theta, \theta' \in \Theta, \ \rho \in S_{X,\theta}^{\mathrm{GTS}},$$

it follows from the convexity of π that, for each $\theta \in \Theta$, the set of equilibrium expectation values of $\hat{q}$, namely $\{\hat{q}(\rho) \mid \rho \in S_{X,\theta}^{\mathrm{GTS}}\}$, are tangents to $-\pi$ at the point θ.

($\hat{Q}$.5) We supplement ($\hat{Q}$.4) with the assumption that, if $\theta \in \mathbf{R}^{n+1}\backslash\Theta$, then the functional $\hat{s} - \theta.\hat{q}$ does not attain its supremum. Hence, in general, continuous systems cannot support equilibrium states at negative or infinite temperatures.

($\hat{Q}$.6) We assume that the formulation of the dynamics of Σ, as presented in Section 2.5, prevails when H_Λ is replaced by the effective local Hamiltonian $H_\Lambda^{\mathrm{eff}}(y)$. Thus, on a time scale for which $\beta(\equiv \theta_0)$ is the unit, this dynamics is given by the one-parameter group $\alpha_\theta(\mathbf{R})$ of automorphisms of $\mathcal{A}$, corresponding to the effective local Hamiltonians $\theta.\hat{Q}_\Lambda = \sum_{j=0}^n \theta_j \hat{Q}_{j,\Lambda}$. Further, we assume that $\alpha_\theta \neq \alpha_{\theta'}$ if $\theta \neq \theta'$. This assumption is closely related to that of the linear independence of the $\hat{Q}_{k,\Lambda}$, since the generator of α_θ is of the form $\sum_{k=0}^n \theta_k \delta_k$, where δ_k is the contribution due to $\hat{Q}_k$.

Since θ_0 is absorbed into the time variable here, the KMS condition corresponding to the value θ of the control variable takes the following form, in units where $\hbar=1$:

$$\langle \rho; [\alpha_\theta(t)A]B \rangle = \langle \rho; B[\alpha_\theta(t+i)A] \rangle \quad \forall A, B \in \mathcal{A}, \ t \in \mathbf{R}. \qquad (6.4.10)$$

We denote by S_θ^{KMS} the (convex) set of states that satisfy this condition, and we denote the subset of its extremals by $\mathcal{E}_\theta^{\text{KMS}}$.

(Q̂.7) We assume that the relationships between the GTS and KMS states formulated in Chapter 5 survive the replacement of the effective Hamiltonian H_Λ^{eff} by $H_\Lambda^{\text{eff}}(y)$. Thus, we assume that Proposition 5.3.5 and Conjecture 5.3.6 prevail when the thermodynamical parameter β is replaced by θ and that the KMS decomposition for the dynamical group α_θ is finer than the ergodic one of $S_{X,\theta}^{\text{GTS}}$. This latter assumption is just the canonical analogue of Eq. (5.3.33), namely that, for each element ρ of $S_{X,\theta}^{\text{GTS}}$, there is a unique probability measure, ν_ρ, on $\mathcal{E}_\theta^{\text{KMS}}$ such that

$$\rho = \int_{\mathcal{E}_\theta^{\text{KMS}}} d\nu_\rho(\sigma)\sigma. \qquad (6.4.11)$$

The extremal KMS states in the support of ν_ρ are then the pure equilibrium phases. We further assume that, in the case where the KMS decomposition of ρ is nontrivial, that is, that ν_ρ is not the Dirac measure δ_ρ, its pure phase components have the symmetry of some crystallographic subgroup, Y, of X. Thus, in this case, Eq. (6.4.11) reduces to the form

$$\rho = | X/Y |^{-1} \int_{X/Y} dx \xi^\star(x)\rho_0, \qquad (6.4.12)$$

where ρ_0 is Y-invariant and the coset X/Y is a crystallographic cell.

(Q̂.8) We assume that $\hat{q}$ satisfies the following conditions of *thermodynamical completeness*. The second of these condition serves to extend the relationship between the tangents to $-\pi$ and the equilibrium expectation values of $\hat{q}$ specified in the Note at the end of (Q̂.4).

(a) $\hat{q}$ is a minimal set of functionals that separates the extremal GTS states. Thus, if ρ_1, ρ_2 are elements of $\bigcup_{\theta \in \Theta} \mathcal{E}_{X,\theta}^{\text{GTS}}$ and $\hat{q}(\rho_1) = \hat{q}(\rho_2)$, then $\rho_1 = \rho_2$; and, further, the same is not true for the proper subsets of $\hat{q} = (\hat{q}_0, ..., \hat{q}_n)$.

(b) The tangents to $-\pi$ belong to the set of equilibrium expectation values of $\hat{q}$, namely $\mathcal{Q}_{\text{equ}} := \bigcup_{\theta \in \Theta} \{\hat{q}(\rho) \mid \rho \in S_{X,\theta}^{\text{GTS}}\}$.

Note   These completeness conditions do *not* discriminate between different pure crystalline phases that are merely space translates of one another since it follows from the translational invariance of $\hat{q}$ that if σ is any crystalline state, then the canonical extension of $\hat{q}$ to the crystalline states satisfies the equation

$$\hat{q}\!\left(\xi^\star(x)\sigma \right) \equiv \hat{q}(\sigma) \ \forall x \in X.$$

Note The restriction to *extremal* GTS states here is essential,[13] as one can see from the following example. Suppose that the system Σ is an isotropic ferromagnet, and that $\hat{q}$ consists of the energy density $\hat{e}$ and the Cartesian components $\hat{m}_1, \hat{m}_2, \hat{m}_3$ of the magnetic polarisation vector $\hat{m}$. Suppose further that the pure phases at a certain fixed temperature below the critical point are $\{\rho_u\}$, with the index u running through the unit vectors in $\mathbf{R}^3$; and that the expectation value of the energy density $\hat{e} = \hat{q}_0$ and $\hat{m}$ for the state ρ_u are e and mu respectively, where e and m are u-independent scalars. Then it follows immediately from the affine property of $\hat{s}$, $\hat{e}$ and $\hat{m}$ that the action of $\hat{q} = (\hat{e}, \hat{m})$ on the GTS mixtures $1/2(\rho_u + \rho_{-u})$ all yield the same result. Hence $\hat{q}$ does not separate these mixtures.

This completes our basic definitions and assumptions pertaining to the passage from the quantum statistical to the thermodynamical picture of Σ.

Thermodynamical Consequences of $(\hat{Q}.1)$–$(\hat{Q}.8)$

As we have already seen in the treatments of $(\hat{Q}.3)$ and $(\hat{Q}.4)$, s and π have the concavity and convexity properties, respectively, required by classical thermodynamics. The following proposition, which is proved in Appendix A, provides a quantum statistical substantiation of the classical formulae (6.1.3) and (6.1.5), which were shown in Section 6.1 to be equivalent to the fundamental equation (6.1.1), subject to the assumption of the extensivity of $S, E, Q_1, ..., Q_n$.

Proposition 6.4.1 *The following properties of π and s ensue from the assumptions $(\hat{Q}.3)$–$(\hat{Q}.6)$.*

  (a) *The supremum of Eq. (6.4.8) is attained, that is,*

$$\pi(\theta) = \max_{q \in Q}(s(q) - \theta.q) \quad \forall \theta \in \Theta. \qquad (6.4.13)$$

  (b) *The function s is differentiable in Q_{equ}, and, in this domain,*

$$ds = \theta.dq. \qquad (6.4.14)$$

Macroscopic Degeneracy and Thermodynamic Discontinuity

Since different equilibrium states can prevail at the same value of θ, it follows from $(\hat{Q}.8a)$ that each such value of this control variable supports more than one equilibrium value of the thermodynamical variable q. We refer to this

---

[13] This point was overlooked in our earlier formulation of the minimal completeness condition in [Se2, Chapter 4].

latter as *macroscopic degeneracy*. A basic assumption of classical thermodynamics is that this occurs at precisely those values of θ at which the derivative
of π is discontinuous. The next proposition, which we prove in Appendix A,
provides a quantum statistical substantiation of this assumption.

Proposition 6.4.2 *Under the assumption of* $(\hat{Q}.1)–(\hat{Q}.8)$, *the condition for
macroscopic degeneracy at the value,* θ, *of the control variable is simply that
the thermodynamical potential* π *has a discontinuous derivative at that point.*

We supplement this proposition with the following assumption of classical
thermodynamics, which is satisfied by solvable models (cf. [Sta, Dor]).

(A) The domain of the (n+1)-dimensional space Θ where π is not differentiable consists of a finite number of surfaces of dimensionality n. Thus, the
interiors of the regions bounded by these surfaces correspond to the pure
phases of the system.

Single Phase Region

By Proposition 6.4.2, this is the domain of Θ where π is differentiable. Thus
since, by Eq. (6.4.13),

$$\pi(\theta) = s(q) - \theta.q,$$

where the value of q is the one that maximises $s - \theta.q$, it follows from Eq.
(6.3.2) that $-q$ is the *unique* tangent to π at the point θ, that is, that

$$d\pi = -q.d\theta. \tag{6.4.15}$$

Résumé   We have established here that the scheme $(\hat{Q}.1)–(\hat{Q}.8)$ leads from
quantum statistics to classical thermodynamics, as represented by the concavity and differentiability of the entropy function, s, the convexity of the reduced
pressure, π, the key relationships (6.4.14) and (6.4.15), and the identification,
by Proposition 6.4.2, of macroscopic degeneracy with discontinuities in the
derivative of π. The supplementary assumption (A), stated after Proposition
6.4.2, is needed for the picture represented by the standard phase diagrams of
classical thermodynamics. In Appendix B, we extend this scheme to the
situation where the functionals $\hat{q}$ are global space averages of locally
conserved quantum fields $\hat{q}(x) = (\hat{q}_0(x), ..., \hat{q}_n(x))$.

6.5 AN EXTENDED THERMODYNAMICS WITH ORDER
PARAMETERS

Landau [LL2, Section 134] observed that the order structures of certain states
can be naturally represented by the values of certain classical variables,

usually termed order parameters. Typical examples of these are the polarisation of a ferromagnet and the staggered polarisation of an antiferromagnet. We remark that these two examples are radically different, in that the former is the global density of an extensive conserved quantity, whereas, as we argue below, the latter is not. Thus, the general statistical thermodynamical scheme of Section 6.4 presents a picture of the ordering in ferromagnets, but not antiferromagnets.

We now extend the scheme of the previous section by the incorporation of additional variables that are required for the description of the order structures of certain systems. We start by considering the prototype case of an antiferromagnet.

The Heisenberg Antiferromagnet

This is the system consisting of spin vectors, $\sigma(x)$, situated at the sites, x, of the three-dimensional simple cubic lattice, $X = \mathbf{Z}^3$, with local Hamiltonian

$$H_\Lambda = J {\sum_{x,y\in\Lambda}}' \sigma(x).\sigma(y), \qquad (6.5.1)$$

where J is positive and the prime indicates that summation is confined to nearest neighboring sites. Here, each spin, $\sigma(x) = (\sigma^{(1)}(x), \sigma^{(2)}(x), \sigma^{(3)}(x))$ is a three dimensional vector, which satisfies the standard angular momentum commutation relations, that is,

$$[\sigma^{(1)}(x), \sigma^{(2)}(x)] = i\sigma^{(3)}(x), \text{ etc.} \qquad (6.5.2)$$

It follows from these specifications that the model is isotropic, since H_Λ is manifestly invariant under rotations. Furthermore, in the case where the spin value is unity, that is, where $\sigma(x)^2 = 3$, it undergoes a phase transition at a certain positive temperature, T_c, below which it is antiferromagnetically ordered (cf. [DLS]). This ordering may be naturally described in terms of the spin configurations on the two interlocking sublattices, $X_1 = (2\mathbf{Z})^3$ and $X_2 = (2\mathbf{Z}+1)^3$, whose union is X. To be precise, there is a breakdown of rotational symmetry, in that the expectation value of $\sigma(x)$ for any pure phase, ρ, at a temperature, T, below T_c takes values $\pm m$, according to whether x is in X_1 to X_2, where m is vector whose magnitude and direction depend on T and ρ, respectively. Thus,

$$\langle\rho; \sigma(x)\rangle = (-1)^{j+1}m \quad \text{for } x \in X_j, \ j = 1, 2. \qquad (6.5.3)$$

Hence, reverting to the terminology of Section 3.3, the rotational symmetry of the model is broken and σ is a G-field, with G the three-dimensional rotation group.

We formulate the order parameter of the system in terms of the staggered magnetic moment, namely

$$M_{\Lambda,s} = \sum_{x \in \Lambda \cap X_1} \sigma(x) - \sum_{x \in \Lambda \cap X_2} \sigma(x). \tag{6.5.4}$$

This is evidently an extensive quantity, and its density, in the infinite volume limit, is the staggered polarisation, $\hat{m}_s$. Thus,

$$\hat{m}_s(\rho) = \lim_{\Lambda \uparrow} \frac{\rho(M_{\Lambda,s})}{|\Lambda|}. \tag{6.5.5}$$

It follows from this formula, together with Eqs. (6.5.3) and (6.5.4), that

$$\hat{m}_s(\rho) = m, \tag{6.5.6}$$

which signifies that $\hat{m}_s$ provides a measure of the antiferromagnetic order of the state ρ.

However, the staggered magnetic moment is *not* a conserved quantity, since, by Eqs. (6.5.1), (6.5.2) and (6.5.4), $M_{\Lambda,s}$ does not commute with H_Λ. Correspondingly, the staggered polarisation observable, $\hat{m}_s$, does not belong to the category of thermodynamical observables formulated in Section 6.4.

Nevertheless, we can extend the scheme of that section by supplementing the set of macroscopic variables, $\hat{q}$, by $\hat{m}_s$, thereby incorporating the ordering of the state into the thermodynamical picture. It is a straightforward matter to confirm that all the results of Section 6.4, as summed up in the Résumé there, carry through to this extended description. For this, the thermodynamical conjugate of $\hat{m}_s$ is simply T^{-1} times a staggered magnetic field that takes equal and opposite constant values on the sublattices X_1 and X_2.

The General Extended Thermodynamic Picture

The above considerations pertaining to the antiferromagnetic model apply also to other systems. This, for example, the characterisation of a superfluid phase requires not only the standard thermodynamical variables but also an order field, Φ, discussed in Section 3.3.4, whose integral is not a conserved quantity. A similar situation arises also in Landau's picture of the ordering in certain crystalline phases [LL2, Section 134].

Thus, in general, the thermodynamical picture can be extended by supplementing the macroscopic observables, $\hat{q}$, by certain global densities of nonconserved quantities, that represent the order structure of the system under consideration.

6.6 CONCLUDING REMARKS ON THE PAUCITY OF THERMODYNAMICAL VARIABLES

The miracle of thermodynamics is that the equilibrium states of systems of enormously many particles can be classified in terms of just a few macroscopic variables. According to the scheme of the previous two sections, these latter variables are determined by the global conservation laws and the broken dynamical symmetries of the system, together with the restrictions imposed by the thermodynamical completeness conditions ($\hat{Q}$.8). Now for real physical systems, the known *universal* conservation laws and dynamical symmetries are rather few in number. This would therefore account for the paucity of the thermodynamical variables in real physical systems, provided that there were no hidden global conservation laws or symmetries, which might not be universal: at present, the problem of their existence is unresolved. Apart from these considerations, the conditions ($\hat{Q}$.8) can cut down the number of thermodynamical variables to just a few, and this is important even for the success of certain models which are but caricatures of real systems. For example, in the case of the two-dimensional Ising model, which supports an infinity of global conserved quantities, the energy density and polarisation consitute a complete set of thermodynamic variables (cf. [MM]).

Thus, one sees how the self-contained macroscopic description of complex systems, represented by classical thermodynamics, can arise as a result of (a) the paucity of universal global conservation laws and dynamical symmetry principles and (b) the restrictive character of the conditions ($\hat{Q}$.8).

APPENDIX A: PROOFS OF PROPOSITIONS 6.4.1 AND 6.4.2

Proof of Proposition 6.4.1. (a) Let $\rho \in S_{X,\theta}^{\text{GTS}}$. Then, by Eq. (6.4.9),

$$\pi(\theta) = \hat{s}(\rho) - \theta.\hat{q}(\rho)$$

$$\leq s(\hat{q}(\rho)) - \theta.\hat{q}(\rho) \quad \text{(by Eq. (6.4.4))}$$

$$\leq \pi(\theta) \quad \text{(by Eq. (6.4.8))}.$$

It follows immediately from these inequalities that

$$s(\hat{q}(\rho)) = \hat{s}(\rho) \tag{A.1}$$

and

$$\pi(\theta) = s(\hat{q}(\rho)) - \theta.\hat{q}(\rho). \tag{A.2}$$

Thus, by Eqs. (6.4.8) and (A.2), $s(q) - \theta.q$ attains its supremum at $q = \hat{q}(\rho)$.

(b) Let $q \in \mathcal{Q}_{\text{equ}}$, which signifies, by the definition of this set, that $q = \hat{q}(\rho)$ for some $\rho \in S_{X,\theta}^{\text{GTS}}$, where $\theta \in \Theta$. Hence the argument leading to Eq. (A.1) is again applicable, and so

$$s(q) = \hat{s}(\rho) \quad \text{and} \quad q = \hat{q}(\rho). \tag{A.3}$$

Suppose now that s is not differentiable at q. Then it follows from the concavity of s that it has two different extremal tangents, $\theta^{(1)}$ and $\theta^{(2)}$, at this point, that is, that

$$s(q) - \theta^{(j)}.q \geq s(q') - \theta^{(j)}.q' \quad \forall q' \in \mathcal{Q}, j = 1, 2.$$

Hence, putting $q' = \hat{q}(\rho')$, where ρ' is an arbitrary element of $S_X^{(0)}$, and using Eq. (A.3),

$$\hat{s}(\rho) - \theta^{(j)}.\hat{q}(\rho) \geq \hat{s}(\rho') - \theta^{(j)}.\hat{q}(\rho') \quad \forall \rho' \in S_X^{(0)}, j = 1, 2.$$

Thus, ρ maximises the functionals $\hat{s} - \theta^{(j)}.\hat{q}$ for $j = 1, 2$ and consequently, by (Q̂.5)–(Q̂7), $\theta^{(1)}$ and $\theta^{(2)}$ belong to Θ and ρ satsfies the KMS condition (6.4.10) both for $\theta = \theta^{(1)}$ and for $\theta = \theta^{(2)}$. However, it follows from (Q̂.7) that the latter condition serves to define α_θ uniquely in terms of ρ [Ta]; and therefore since, by (Q̂.6), $\alpha_\theta \neq \alpha_{\theta'}$ if $\theta \neq \theta'$, it also defines θ uniquely in terms of this state. Hence, since the assumption of the nondifferentiability of s at q implied that there were at least two different tangents there, this assumption is untenable. We conclude, therefore, that s is differentiable at all points of $\mathcal{Q}_{\text{equ}}$. $\square$

Proof of Proposition 6.4.2. Since $\hat{q}$ is affine, it follows from Proposition 6.3.1 and assumptions (Q̂8a, b) that there is a one-to-one correspondence between $\mathcal{E}_{X,\theta}^{\text{GTS}}$ and the extremal tangents to π at the point θ. Consequently there is just one tangent to π at that point, that is, π is differentiable there, if and only if $\mathcal{E}_{X,\theta}^{\text{GTS}}$, and hence also $S_{X,\theta}^{\text{GTS}}$, consists of just one state. Therefore the values of θ at which π is not differentiable are precisely those where macroscopic degeneracy prevails. $\square$

APPENDIX B: FUNCTIONALS $\hat{q}$ AS SPACE AVERAGES OF LOCALLY CONSERVED QUANTUM FIELDS

We now refine the scheme of Section 6.4 in order to express the thermodynamical functionals $\hat{q}$ as space averages of quantum fields, $\hat{q}(x)$, that satisfy local conservation laws. We concentrate our argument on the case where Σ is a system of identical particles of mass m that interact via two-body forces, for which the potential energy of interaction between particles at the points x and y is $V(x - y)$. The generalisation of the argument to systems with many-body interactions, whether on a lattice or in a continuum, is straightforward. We present our treatment here on a rather formal level, leaving the discussion of its mathematical structure to Section 7.2.

Thus, we assume the system Σ is represented by the generic model of Section 2.5.3, with formal Hamiltonian

$$H = \frac{\hbar^2}{2m} \int_X dx \nabla \psi^\star(x).\nabla \psi(x) + \int_X dx \int_X dy \psi^\star(x)\psi^\star(y)V(x - y)\psi(y)\psi(x),$$

$$(\text{B.1})$$

which is then the spatial integral of the energy density, $\hat{q}_0(x)$, defined by the formula

$$\hat{q}_0(x) = \frac{\hbar^2}{2m} \nabla \psi^*(x) \cdot \nabla \psi(x) + \frac{1}{2} \int_X dy \psi^*(x) \psi^*(y) V(x-y) \psi(y) \psi(x). \quad (B.2)$$

On the other hand, the local Hamiltonian H_Λ is the quantity obtained by replacing the domain of integration, X, by Λ in Eq. (B.1) and imposing Neumann boundary conditions (cf [Ro3]). Thus, assuming that the interactions are suitably tempered, the difference between H_Λ and $\int_\Lambda dx \hat{q}_0(x)$ reduces to a surface term, and so the replacement of H_Λ by the latter integral on the right-hand side of Eq. (6.4.3) leads to no change in its value. Hence [Ro3], that formula is equivalent to

$$\hat{q}_0(\rho) = \lim_{\Lambda \uparrow} |\Lambda|^{-1} \int_\Lambda dx \rho(\hat{q}_0(x)). \quad (B.3)$$

Furthermore, it follows from Eqs. (B.1) and (B.2) that the commutator of H with $\hat{q}_0(x)$ is i times the divergence of a certain, rather complicated, current density $\hat{j}_0(x)$ (cf. [WH]), and hence, assuming that the generator of the dynamical automorphism group of the system is $i/\hbar[H, \cdot]$, the evolutes, $\hat{q}_{0,t}$ and $\hat{j}_{0,t}$, of $\hat{q}_0(x)$ and $\hat{j}_0$, respectively, satisfy the local conservation law

$$\frac{\partial}{\partial t} \hat{q}_{0,t}(x) + \nabla \hat{j}_{0,t}(x) = 0. \quad (B.4)$$

Turning now to the functionals $\hat{q}_1, \ldots, \hat{q}_n$, we strengthen the conditions $(\hat{Q}.3)$, concerning the extensivity and conservation of $\hat{Q}_{1,\Lambda}, \ldots, \hat{Q}_{n,\Lambda}$, by the assumption that they are the space integrals over Λ of locally conserved quantum fields, $\hat{q}_1(x), \ldots, \hat{q}_n(x)$, respectively. In fact, it is easily checked that this assumption is fulfilled in standard cases such as those where $\hat{Q}$ represents particle number or magnetic moment, the $\hat{q}(x)$ fields then being $\psi^*(x)\psi(x)$ or $\psi^*(x)\sigma\psi(x)$.

Thus, our strengthened form of assumption $(\hat{Q}.3)$ is that

$$\hat{Q}_{k,\Lambda} = \int_\Lambda dx \hat{q}_k(x) \quad \text{for } k = 0, \ldots, n, \quad (B.5)$$

and that the evolute, $\hat{q}_{k,t}(x)$, of $\hat{q}_k(x)$ satisfies a local conservation law

$$\frac{\partial}{\partial t} \hat{q}_{k,t}(x) + \nabla \hat{j}_{k,t}(x) = 0 \quad \text{for } k = 1, \ldots, n, \quad (B.6)$$

where $\hat{j}_k(x)$ is the corresponding current. It follows then from Eqs. (6.4.1) and (B.5) that, formally,

$$\hat{q}_k(\rho) = \lim_{\Lambda \uparrow} |\Lambda|^{-1} \int_\Lambda dx \rho(\hat{q}_k(x)) \quad \text{for } k = 1, \ldots, n. \quad (B.7)$$

We conclude from Eqs. (B.3), (B.4), (B.6) and (B.7) that the $(n + 1)$-component functional $\hat{q} = (\hat{q}_0, ..., \hat{q}_n)$ is the global space average of the $(n + 1)$-component quantum field $\hat{q}(x) = (\hat{q}_0(x), ..., \hat{q}_n(x))$. As we remarked at the beginning of this appendix, this result may readily be generalised to systems with many-body interactions, whether on a lattice or in a continuum.

Chapter 7

Macrostatistics and nonequilibrium thermodynamics

7.1 INTRODUCTION

The extension of classical thermodynamics to nonequilibrium phenomena is concerned with the dynamics of systems composed of spatially disjoint macroscopic parts that are not in equilibrium with one another (cf. [deGM]). Here, the remarkable, empirically established fact is that such systems generally evolve according to classical, deterministic, irreversible, macroscopic laws, given by autonomous equations of motion for the set of time-dependent thermodynamical variables, q_t, of its component parts. Thus, they are of the form

$$\frac{dq_t}{dt} = \mathcal{F}(q_t). \tag{7.1.1}$$

In general, these laws stem from conservation principles, together with constitutive equations that render the dynamics of the variables q_t autonomous. Their irreversibility is dictated by the Second Law of Thermodynamics and, in Boltzmann's picture [Bol], it originates at the microscopic level from the stochasticity (molecular chaos) generated by collisions of its constituent particles. In order to relate this irreversibility to that discussed in Chapter 4, we note that the macroscopic variables here correspond to observables of subsystems of a conservative many-particle system.

An important class of nonequilibrium thermodynamical laws are those of macroscopic continuum mechanics, as exemplified by hydrodynamics and heat conduction. There, q_t is a set of classical fields, representing the position-dependent densities of the same extensive conserved quantities that constitute the thermodynamical variables of the system in its equilibrium states. Most importantly, this macroscopic description is a large scale, or hydrodynamical, one in which the field variables at a point represent, in essence, extensive conserved observables of a region large enough to contain a thermodynamically viable system comprising an enormous number of particles.

An essential general feature of the macroscopic laws of nonequilibrium thermodynamics is that their time scales are enormously long by comparison with those of the underlying microscopic processes (cf. [On], [VK2],

[Bog]).[1] This is absolutely crucial for the autonomy of those laws, which requires the almost instantaneous adjustment of the values of certain microscopic variables to those of the thermodynamical ones. Thus, to express this observation more sharply, the strict autonomy of nonequilibrium thermodynamical evolution arises only in a limit in which the ratio of the macroscopic to the microscopic time scales becomes infinite.[2] Furthermore, since the macroscopic observables undergo both thermal and quantum fluctuations, it is clear that some macroscopic limit is also required to reduce their dynamics to the classical deterministic form (7.1.1).

It is evident from these considerations that nonequilibrium thermodynamics emanates from a subtle interplay of physical processes at different levels of macroscopicality. Its central problem is that of determining the restrictions imposed by microphysics on the form of the phenomenological laws (7.1.1). A seminal treatment of this problem was provided by Onsager [On], who employed a statistical thermodynamical argument to obtain certain general relationships between the transport coefficients appearing in those laws. The essential input into Onsager's argument consisted of

(a) the assumption of linear macroscopic equations of motion for the regime close to thermal equilibrium;

(b) the assumption that the time scale for these equations is enormously long by microscopic standards;

(c) the so-called "regression hypothesis",[3] which signifies that these equations have precisely the same form as those governing the regression of the spontaneous fluctuations of the relevant macroscopic variables about their mean equilibrium values;

---

[1] For example, suppose that Σ consists of two bodies, Σ_1 and Σ_2, each of volume V, that are placed in contact with one another at parts of their surfaces. Suppose further that their energies are their only extensive conserved observables. Then the rate of energy transfer between them is proportional to their surface areas, that is, to $V^{2/3}$. Hence, as their energies are proportional to V, the time scale for their macroscopic evolution is proportional to $V^{1/3}$, that is, to the cube root of the number, N, of their constituent particles. Thus, the ratio of the macroscopic scale to a characteristic microscopic one is of the order of $N^{1/3}$, which is generally enormous.

[2] This parallels the situation in classical thermodynamics in which all changes of state are quasi-static.

[3] This hypothesis may be verified in some simple examples such as the model of Chapter 4, Appendix C (cf. Comment (4) in Section C.1). In general, however, as Onsager himself pointed out, the hypothesis is far from innocuous, since it implies a strict parallel between processes on two quite different scales. For example, the deviation from its equilibrium value of an extensive variable of a subsystem of N particles is of the order of N in the deterministic process, whereas it is only of the order of $N^{1/2}$ in a fluctuation.

(d) the reversibility of the underlying microscopic dynamics; and

(e) the Einstein formula,

$$P = \text{const.}\exp(S), \tag{7.1.2}$$

which relates the equilibrium probability distribution, P, for the macroscopic variables to the thermodynamical entropy S.

Thus, the Onsager theory is essentially of *macrostatistical* character, since its only microscopic imput is the principle of microscopic reversibility.

We now remark that, although that theory has vast ramifications in both physics and chemistry (cf. [deGM]), it has the following deficiencies. Firstly, although it is essential that both its thermodynamical variables and its time scale are macroscopic, the theory carries no explicit mathematical characterisation of their macroscopicality and consequently does not admit the limiting procedures required for the emergence of the phenomenological law (7.1.1). Secondly, it is restricted to a regime close to thermal equilibrium, where the macroscopic dynamics is linear. Thirdly, it lacks a statistical mechanical formulation of the key concept of local equilibrium, which is central to the nonequilibrium thermodynamics of continuous media (cf. [deGM]). In our view, these limitations were virtually inevitable, since Onsager's theory was devised prior to the advent of algebraic statistical thermodynamics, when there was no natural framework within which the concepts of macroscopic variables and local equilibrium could be formulated.

The object of this chapter is to formulate a general macrostatistical mechanical approach to nonequilibrium thermodynamics that is cast within the operator algebraic framework and incorporates Onsager's kcy assumptions of microscopic reversibility and the regression hypothesis. Here we confine ourselves to the case where the macroscopic law (7.1.1) is one of continuum mechanics, with q_t an $(n + 1)$-component field whose value at a spatial point x is a local version, $q_t(x) = (q_{t,0}(x), ..., q_{t,n}(x))$, of the thermodynamical variables q of the previous chapter.

Our strategy may be summarised as follows. We first note that, by Eq. (7.1.1), the linearised deterministic evolution of a small perturbation, z_t, of q_t is given by the equation

$$\frac{\partial z_t(x)}{\partial t} = \frac{\delta \mathcal{F}(q_t(x))}{\delta(q_t(x))} z_t(x). \tag{7.1.3}$$

In view of the t-dependence of $\delta \mathcal{F}(q_t(x))/\delta q_t(x)$, the solution of this equation may be expressed in the form

$$z_t = T(t, t_0) z_{t_0} \quad \text{for } t \geq t_0, \tag{7.1.4}$$

where $\{T(t, t') \mid t \geq t'\}$ is a two-parameter family of transformations satisfying the condition

$$T(t, t')T(t', t'') = T(t, t''), \quad T(t, t) = I. \tag{7.1.5}$$

In order to relate the phenomenological equations (7.1.1) and (7.1.3)–(7.1.5) to the underlying quantum dynamics of the system, we assume that the classical field $q_t(x)$ is the expectation value, in a certain large scale limit, of the quantum field, $\hat{q}_t(x) = (\hat{q}_{t,0}(x), ..., \hat{q}_{t,n}(x))$, introduced in Chapter 6, Appendix B as the position dependent density of the extensive conserved observables $\hat{Q}$. Similarly, we formulate the dynamics of the fluctuations, $\tilde{q}_t(x)$, of $\hat{q}_t(x)$ about its mean, in the same large scale limit, as a classical Gaussian Markov process, $\mathcal{M}_{\text{fluct}}$. We then assume that the regression of the fluctuations, $\tilde{q}_t(x)$, follow the same dynamical law as the small perturbations, $z_t(x)$, of $q_t(x)$; this is our generalisation of the original Onsager regression hypothesis, which pertained to perturbations and fluctuations of equilibrium states. In the present theory, we express this hypothesis in terms of the conditional expectation value, $E(\tilde{q}_t \mid \tilde{q}_{t_0})$, of $\tilde{q}$ at time t, given this field at an earlier time t_0. Specifically, we take the regression hypothesis to signify that the evolution of $\mathcal{M}_{\text{fluct}}$ is governed by the stochastic counterpart of Eq. (7.1.4), namely

$$E(\tilde{q}_t \mid \tilde{q}_{t_0}) = T(t, t_0)\tilde{q}_{t_0}. \tag{7.1.6}$$

By employing this hypothesis, together with the principle of microscopic reversibility and a macrostatistical characterisation of local equilibrium, we are able to extend the Onsager theory to a nonlinear regime.

Our specific treatment is centred on the case where the phenomenological law (7.1.1) is a nonlinear diffusion of the form

$$\frac{\partial q_{t,k}(x)}{\partial t} = \sum_{l=0}^{n} \nabla \cdot (K_{kl}(\theta_t(x))\nabla \theta_{t,l}(x)) \quad \forall k, l \in \{0, 1, ..., n\}, \tag{7.1.7}$$

where the q, or more precisely their quantum counterparts, $\hat{q}$, are even functions of the particle velocities, where the K_{kl} are scalars and where $\theta_t(x) = (\theta_{t,0}(x), ..., \theta_{t,n}(x))$ is the set of canonical space-time dependent versions of the control variables, θ, of the previous chapter. On the basis of the strategy described above, we obtain the following Onsager relations for this nonlinear dynamics.

$$K_{kl}(\theta_t(x)) = K_{lk}(\theta_t(x)). \tag{7.1.8}$$

Our treatment of the relationship between the microscopic and macroscopic pictures of Σ is organised as follows. In Section 7.2, we formulate our basic assumptions for the quantum field $\hat{q}$, and in Section 7.3, we specify those for the autonomous classical dynamical system, $\mathcal{M}$, constituted by the field q. In Section 7.4, we interrelate $\hat{q}$ and q by expressing the latter field as a certain large scale limit of the expectation value of the former. Correspondingly, in Section 7.5, we provide a quantum statistical formulation of $\mathcal{M}_{\text{fluct}}$ as a classical Gaussian Markov process, that satisfies the regression hypothesis.

We then employ the constructions of Sections 7.2–7.5, as reinforced by the principle of microscopic reversibility, to obtain the strictures imposed by microphysics on the phenomenological dynamics of the system. We start, in Section 7.6, with a treatment of the linear regime, close to thermal equilibrium; and for that we derive both the standard Onsager relations and a macrostatistical characterisation of equilibrium from the properties of $\mathcal{M}$ and $\mathcal{M}_{\text{fluct}}$. To generalise these results to the nonlinear regime, we proceed, in Section 7.7, to extend that characterisation of thermal equilibrium from the global setting to one that is local, in the hydrodynamical sense. In this way, we arrive at a definition of local thermodynamical equilibrium (LTE), and employ an LTE assumption to extend the relations (7.1.8) to the nonlinear regime.

We conclude this chapter in Section 7.8 with considerations regarding the generalisation of the theory to the Galilean covariant[4] nonequilibrium thermodynamics of continuous media, including fluid mechanics.

Appendices A and B are devoted to simple exposés of the rudiments of the theories of tempered distributions and classical stochastic processes, which we require for our treatments of the fields $\hat{q}$, q and $\tilde{q}$: Appendix B also contains the proofs of two key propositions, concerning the construction and Markov property of the process $\mathcal{M}_{\text{fluct}}$. Appendix C provides a derivation of a counterpart, for infinite systems, of the Einstein fluctuation formula (7.1.2).

Throughout this chapter, we employ a notation appropriate to quantum systems in a Euclidean continuum, with the understanding that it can also be employed for lattice systems via the simple expedient of interpreting spatial integrals as sums, etc.

7.2 THE QUANTUM FIELD $\hat{q}(x)$

We adopt the assumptions of Chapter 6, Appendix B, which led to the picture of the thermodynamical functional $\hat{q} = (\hat{q}_0, ..., \hat{q}_n)$ as the global space average of a locally conserved quantum field $\hat{q}(x) = (\hat{q}_0(x), ..., \hat{q}_n(x))$. We now formulate the mathematical structure of this field, based on the assumption that it is an operator-valued tempered distribution, in accordance with the standard formulation of quantum field theory (cf. [StWi]). Here, as in Appendix A, where we provide a rudimentary account of these distributions, we denote the Schwartz space of smooth fast decreasing functions[5] on X by $\mathbf{S}(X)$, its mth topological power by $\mathbf{S}^{(m)}(X)$, and the duals of $\mathbf{S}(X)$ and $\mathbf{S}^{(m)}(X)$ by $\mathbf{S}'(X)$ and $\mathbf{S}^{(m)\prime}(X)$, respectively. Thus, the latter two spaces consist of tempered distributions. We indicate the canonical vector field counterparts of the various $\mathbf{S}$

---

[4] The equations (7.1.7) are, of course, not Galilean covariant.

[5] In the case of a lattice system, this is just the space of functions on X that decrease faster than any inverse power of $|x|$.

and $\mathbf{S}'$ spaces by the subscript V, and the real subspaces of any of these spaces by the subscript R.

We formulate the field $\hat{q}$ in the GNS representation space of a primary state ρ, whose normal folium is stable under time translations. Thus, denoting the GNS triple of this state by $(\mathcal{H}, \pi, \Phi)$, we assume that each component $\hat{q}_k$ of $\hat{q}$ is a mapping $f_k \rightarrow \hat{q}_k(f_k)$ of $\mathbf{S}(X)$ into the unbounded affiliates of $\pi(\mathcal{A})''$, $\hat{q}_k(f_k)$ thus being a smeared field. Correspondingly, $\hat{q}$ is a mapping of $\mathbf{S}^{(n+1)}(X)$ into the affiliates of $\pi(\mathcal{A})''$, with the rule that, for $f = (f_0, ..., f_n) \in \mathbf{S}^{(n+1)}(X)$,

$$\hat{q}(f) = \sum_{k=0}^{n} \hat{q}_k(f_k). \tag{7.2.1}$$

We denote the time translate of $\hat{q}(f)$ for the conservative system Σ by $\hat{q}_t(f)$. We define $\hat{Q}$ to be the algebra of polynomials in $\{\hat{q}_t(f) \mid f \in \mathbf{S}^{(n+1)}(X), t \in \mathbf{R}\}$, we assume that Φ lies in the domain of these polynomials, and we denote by $\mathbf{D}_{\hat{Q}}$ the linear manifold $\hat{Q}\Phi$.

We assume that the field $\hat{q}$ possesses the following properties (q̂.1)–(q̂.6).

(q̂.1) The field $\hat{q}(x)$ is localised within a sphere whose center is x and whose radius, R, is fixed and finite.[6] This may be taken to be a consequence of Eqs. (B.2) and (B.5) of Appendix B in Chapter 6, together with the further assumption that the interactions are of finite range, R. The algebraic expression of the assumption is that if f is a test function of bounded support, then $\hat{q}(f)$ is affiliated to the local algebras $\pi(\mathcal{A}_\Lambda)''$ for which Λ contains that support and the distance between the boundaries of Λ and $\mathrm{supp}(f)$ exceeds R everywhere.

(q̂.2) $\hat{q}$ transforms covariantly with respect to space translations, that is, if ξ is their representation in $\mathrm{Aut}(\pi(\mathcal{A})'')$, then, formally,

$$\xi(y)\hat{q}(x) = \hat{q}(x + y), \tag{7.2.2}$$

that is,

$$\xi(y)\hat{q}(f) = \hat{q}(f_y) \quad \text{where } f_y(x) = f(x - y). \tag{7.2.2}'$$

(q̂.3) Both the field $\hat{q}$ and the state ρ are invariant under time reversals.

Note This assumption implies that $\hat{q}$ contains neither the local momentum density nor the local polarisation, since both of these change sign under the action of time reversals. Consequently, the present treatment is not applicable to either fluid mechanics or magnetic relaxation. We discuss the problems involved in generalising the theory to such areas in Section 7.8.

---

[6] This assumption could certainly be weakened to a less stringent localisation condition.

(q̂.4) The components of $\hat{q}$ satisfy the following commutation relations, which signify that their space integrals over finite volumes intercommute, up to "surface effects":

$$[\hat{q}_k(x), q_l(x')] = -i\hbar \nabla.\hat{p}_{k,l}(x)\delta(x - x'), \qquad (7.2.3)$$

that is,

$$[\hat{q}_k(g), q_l(h)] = i\hbar p_{k,l}(\nabla(gh)) \quad \forall g, h \in S(X), \qquad (7.2.3)'$$

where $p_{k,l}$ is an operator-valued tempered distribution and $(gh)(x) \equiv g(x)h(x)$.

(q̂.5) On the domain $\mathbf{D}_{\hat{Q}}$ defined above, the operator $\hat{q}_t(f)$ is strongly continuous in f and differentiable in t. Thus, denoting by ρ the canonical extension of that state to unbounded affiliates of $\pi(\mathcal{A})''$, the correlation function $\langle \rho; \hat{q}_{t_1}(f^{(1)}), ..., \hat{q}_{t_m}(f^{(m)}) \rangle$ is continuous in the $f^{(k)}$'s and differentiable with respect to the t.

(q̂.6) $\hat{q}_t$ satisfies a local conservation law of the form

$$\frac{\partial}{\partial t}\hat{q}_t(x) + \nabla.\hat{j}_t(x) = 0, \qquad (7.2.4)$$

that is,

$$\frac{\partial}{\partial t}\hat{q}_t(f) = \hat{j}_t(\nabla f), \qquad (7.2.4)'$$

where the current vector field $\hat{j}_t = (\hat{j}_{t,0}, ..., \hat{j}_{t,n})$ is an operator-valued tempered distribution, such that the action of $\hat{j}_t(v)$ on $\mathbf{D}_{\hat{Q}}$ is continuous in both v ($\in \mathbf{S}_V(X)$) and $t(\in \mathbf{R})$.

7.3 THE MACROSCOPIC MODEL, $\mathcal{M}$

We take $\mathcal{M}$ to be a continuum mechanical model consisting of a classical $(n + 1)$-component field, corresponding to a space-time dependent version of the thermodynamical variables q of the previous chapter.

We assume that $\mathcal{M}$ occupies a Euclidean space, X, of the same dimensionality as that occupied by Σ; in the case where that is a lattice, X would be its large scale continuum limit. Thus, here the field q is a function on $X \times \mathbf{R}_+$. We denote its value at position x and time t by $q_t(x) = (q_{t,0}(x), ..., q_{t,n}(x))$, and assume that each of its components is real-valued and satisfies a local conservation law of the standard form, that is,

$$\frac{\partial q_{t,k}}{\partial t} + \nabla.j_{t,k} = 0 \quad \text{for } k = 0, ..., n. \qquad (7.3.1)$$

We note here that, in view of this equation, the currents j_k are odd with respect to time-reversals and therefore do not belong to the set comprising q.

We define the time dependent entropy density at the point x to be

$$s_t(x) = s(q_t(x)), \tag{7.3.2}$$

where s is the equilibrium entropy function of the previous chapter; and correspondingly, we define the $(n+1)$-component field $\theta = (\theta_0, ..., \theta_n)$, conjugate to q, by the natural analogue of Eq. (6.1.4), namely

$$\theta_{t,k}(x) = \frac{\partial s(q_t(x))}{\partial q_{t,k}(x)} \quad \text{for } k = 0, ..., n. \tag{7.3.3}$$

We assume that the system is confined to a single phase, that is, that $q_t(x)$ and $\theta_t(x)$ lie in domains where the entropy density, s, and the reduced pressure, π, are infinitely differentiable.

Since the definitions (7.3.2) and (7.3.3) imply that the increment in $s_t(x)$ due to an infinitesimal change $\delta q_t(x)$ in $q_t(x)$ is $\sum_{k=0}^{n} \theta_{t,k}(x) \delta q_{t,k}(x)$, we define the entropy current density to be

$$c_t(x) = \sum_{k=0}^{n} \theta_{t,k}(x) j_{t,k}(x). \tag{7.3.4}$$

It follows now from Eqs. (7.3.1)–(7.3.4) that

$$\frac{\partial s_t(x)}{\partial t} + \nabla . c_t(x) = \sigma_t(x), \tag{7.3.5}$$

where

$$\sigma_t(x) = \sum_{k=1}^{n} j_{t,k}(x) . \nabla \theta_{t,k}(x). \tag{7.3.6}$$

Thus, by Eq. (7.3.5), $\sigma_t(x)$ is the local entropy source, that is, the entropy production density of at the position x and time t. The Second Law of Thermodynamics requires that it be nonnegative [dGM], since its essential demand is that the sum of the entropies of the macroscopic constituents of a system, rather than the entropy of quantum state of the composite, increases with time.[7]

We now assume that the currents j_k are related to the field θ, and thus to q, by the constitutive equation

$$j_{t,k}(x) = \sum_{l=0}^{n} K_{kl}(\theta_t(x)) \nabla \theta_{t,l}(x), \tag{7.3.7}$$

where K_{kl} is scalar and the matrix $[K_{kl}(\theta_t(x)]$ is negative definite. Thus, by Eqs.

---

[7] For example, in the case of a system, Σ, composed of two coupled bodies Σ_1 and Σ_2 at different temperatures, the Second Law asserts that the sum of the entropies of the latter systems increases with time. On the other hand, the entropy of the state, ρ, of Σ is just $-\text{Tr}(\rho \ln \rho)$, which is a conserved quantity.

(7.3.1) and (7.3.7), the dynamics of the model is given by the nonlinear diffusion equation (7.1.7), which we now write in the following more compact form.

$$\frac{\partial q_t(x)}{\partial t} = \nabla.(K(\theta_t)\nabla\theta_t), \qquad (7.3.8)$$

where $K(\theta_t) = [K_{kl}(\theta_t)]$ and, by Eq. (7.3.3), θ_t is the single column matrix valued function of position given by the derivative of $s(q_t)$ with respect to q_t, that is,

$$\theta_t(x) = s'(q_t(x)). \qquad (7.3.9)$$

The assumption of the constitutive equation (7.3.7) for the currents j, with K negative definite, serves to ensure that the dynamics of $\mathcal{M}$ possesses the following properties.[8]

(a) The state represented by the field q is one of equilibrium, that is, time-independent, if and only if it is spatially uniform.

(b) The entropy production σ is zero at equilibrium and otherwise positive.

(c) The resultant equations of motion for q are covariant with respect to space-translations and rotations.

(d) They are also covariant with respect to scale transformations $x \to Lx$, $t \to L^2 t$.

(e) $j_{t,k}(x)$ depends only on the local value of θ_t and its gradient at the point x and time t. In fact, the argument, $\theta_t(x)$, of K in Eq. (7.3.8) and its gradient, $\nabla\theta_t(x)$, may naturally be interpreted as the local macroscopic state and driving force, respectively.

The Perturbed Process $\mathcal{M}_{\text{pert}}$

Suppose now that δq_t is a "small" perturbation of q_t. Then it follows easily from Eqs. (7.3.8) and (7.3.9) that the linearised equation of motion for δq_t is

$$\frac{\partial}{\partial t}\delta q_t = L_t \delta q_t, \qquad (7.3.10)$$

where

$$L_t \delta q_t = \nabla.(K(\theta_t)s''(q_t)\nabla\delta q_t + K'(\theta_t)s''(q_t)\delta q_t\nabla\theta_t) \qquad (7.3.11)$$

and the primes denote derivatives: thus $s''(q_t)$ is the Hessian $[\partial^2 s(q_t)/\partial q_{t,k}\partial q_{t,l}]$. We assume that δq_t is a tempered distribution, and thus, in view of the t-dependence of L_t, that the solution of Eq. (7.3.10) is of the form

---

[8] In fact, under further mild assumptions, these even imply a constitutive equation of the form (7.3.7), with K negative definite.

$$\delta q_t = T_{t,t_0} \delta q_{t_0} \quad \text{for } t \ge t_0, \tag{7.3.12}$$

where $\{T_{t,t'} \mid t \ge t'\}$ is a two-parameter family of transformations of $\mathbf{S}^{(n+1)\prime}(X)$ that satisfies the deterministic condition

$$T_{t,t'} T_{t',t''} = T_{t,t''}, \quad T_{t,t} = I. \tag{7.3.13}$$

Evidently, as q, and hence δq, is a real field, the transformations $T_{t,t'}$ must leave $\mathbf{S}_R^{(n+1)\prime}$, the real subspace of $\mathbf{S}^{(n+1)\prime}(X)$, invariant. We denote by $\mathcal{M}_{\text{pert}}$ the deterministic process governed by the transformations $\{T_{t,t'}\}$ of $\mathbf{S}^{(n+1)\prime}(X)$.

We define $T_{t,t'}^{\star}$ to be the dual of $T_{t,t'}$. Thus,

$$[T_{t,t'} \delta q_{t'}](f) = \delta q_{t'}(T_{t,t'}^{\star} f) \tag{7.3.14}$$

and the conditions dual to those of Eq. (7.3.13) are

$$T_{t,t''}^{\star} = T_{t',t''}^{\star} T_{t,t'}^{\star}, \quad T_{t,t}^{\star} = I. \tag{7.3.15}$$

We note here that the stability of $\mathbf{S}_R^{(n+1)\prime}(X)$ under $T_{t,t'}$ implies that of $\mathbf{S}_R^{(n+1)}(X)$, the real subspace of $\mathbf{S}^{(n+1)}(X)$, under $T_{t,t'}^{\star}$; or, equivalently, it implies that

$$\overline{T_{t,t'}^{\star} f} = T_{t,t'}^{\star} \bar{f} \quad \forall f \in S^{(n+1)}(X), \tag{7.3.16}$$

where the bar denotes complex conjugation. Further, the dual, $\mathcal{L}_t^{\star}$, of $\mathcal{L}_t$, is defined by the formula

$$[\mathcal{L}_t \delta q_t](f) \equiv \delta q_t(\mathcal{L}_t^{\star} f), \tag{7.3.17}$$

and by Eqs. (7.3.10), (7.3.12) and (7.3.14), it is the infinitesimal generator of the two-parameter family of transformations $\{T_{t,t'}^{\star}\}$, that is,

$$\mathcal{L}_t^{\star} f = \lim_{h \to +0} h^{-1}[T_{t+h,t}^{\star} f - f]. \tag{7.3.18}$$

By Eqs. (7.3.11) and (7.3.17), the explicit form of $\mathcal{L}_t^{\star}$ is given by the formula

$$\mathcal{L}_t^{\star} f = s''(q_t)\Big(\nabla.(K^{\star}(\theta_t)\nabla f) - K'^{\star}(\theta_t)\nabla f.\nabla \theta_t\Big), \tag{7.3.19}$$

where $K^{\star}$ and $K'^{\star}$ are the duals of K and K', respectively.

7.4 RELATIONSHIP BETWEEN THE CLASSICAL FIELD q AND THE QUANTUM FIELD $\hat{q}$

We now aim to relate the macroscopic model $\mathcal{M}$ to its underlying quantum mechanics by representing the classical field q as the expectation value of the quantum field $\hat{q}$ in a suitable large scale limit. To this end, we first note that the phenomenological equation (7.3.8) is invariant under scale transformations $x \to \lambda x,\ t \to \lambda^2 t$, and interpret this invariance as signifying that a length scale λ corresponds to a time scale λ^2. Accordingly, we introduce a length parameter, L, and define the quantum field $\hat{q}^{(L)}$ to be the rescaled version of $\hat{q}$ for which the units of length and time are L and L^2, respectively, that is,

$$\hat{q}_t^{(L)}(x) = \hat{q}_{L^2t}(Lx), \tag{7.4.1}$$

or, in distribution theoretic terms,

$$\hat{q}_t^{(L)}(f) = \hat{q}_{L^2t}(f^{(L)}), \tag{7.4.1}'$$

where

$$f^{(L)}(x) = L^{-d}f(x/L). \tag{7.4.2}$$

It follows from these definitions that the local conservation law (7.2.4)$'$ rescales to the form

$$\frac{\partial}{\partial t}\hat{q}_t^{(L)}(f) = \hat{j}_t^{(L)}(\nabla f), \tag{7.4.3}$$

where, formally,

$$\hat{j}_t^{(L)}(x) = L\hat{j}_{L^2t}(Lx). \tag{7.4.4}$$

To relate the quantum picture of $\hat{q}$ to the macroscopic model, $\mathcal{M}$, of the classical field q, we make the following assumptions.

(I) The initial state, $\rho^{(L)}$, of Σ carries an L-dependence, representing the length scale of its spatial variations, in that the functions $\langle\rho^{(L)}; \hat{q}_{t_1}^{(L)}(f^{(1)})\cdots\hat{q}_{t_k}^{(L)}(f^{(k)})\rangle$ tend to definite limits as L tends to infinity. Thus, we may define the classical time-dependent field q_t by the equation

$$q_t(f) = \lim_{L\to\infty} \rho^{(L)}\big(\hat{q}_t^{(L)}(f)\big), \tag{7.4.5}$$

or, formally, using Eq. (7.4.1),

$$q_t(x) = \lim_{L\to\infty} \rho^{(L)}(\hat{q}_{L^2t}(Lx)). \tag{7.4.5}'$$

Hence, by condition (q̂.5) and the completeness of S' spaces, q_t is a tempered distribution.

(II) We assume that, for any $\mathbf{S}_V(X)$-class vector field, v, $\rho^{(L)}(\hat{j}_t^{(L)}(v))$ converges to a limit $j_t(v)$, as L tends to infinity. We also assume that this is continuous in t; its continuity in f follows from assumption (q̂.6) and the completeness of $\mathbf{S}'(X)$.

Note Since, by Eq. (7.4.3) and assumptions (q̂.5, q̂.6),

$$\langle\rho^{(L)}; \hat{q}_t^{(L)}(f)\rangle - \langle\rho^{(L)}; \hat{q}_{t_0}^{(L)}(f)\rangle = \int_{t_0}^{t} ds\, \langle\rho^{(L)}; \hat{j}_s^{(L)}(\nabla f)\rangle, \tag{7.4.6}$$

it follows from (I) and (II) that

$$q_t(f) - q_{t_0}(f) = \int_{t_0}^{t} ds\, j_s(\nabla f).$$

and hence, by the assumed continuity of $j_t(v)$ in t, $q_t(f)$ is differentiable with respect to this variable and

$$\frac{\partial}{\partial t} q_t(f) = j_t(\nabla f). \qquad (7.4.7)$$

(III)   We assume that the distributions q_t and j_t are continuously differentiable functions on X. This implies that the distribution theoretic version, (7.4.7), of the continuity equation reduces to the standard classical form, that is,

$$\frac{\partial q_t}{\partial t} + \nabla.j_t = 0. \qquad (7.4.8)$$

(IV)   We assume that j_t satisfies the constitutive equation (7.3.7) for $t \geq 0$. In view of Eq. (7.4.8), this implies that q evolves according to the phenomenological law (7.3.8) for positive times, *subject to the initial condition given in (I)*.

(V)   We further assume that the state, $\rho^{(L)}$, of the quantum field $\hat{q}^{(L)}$ becomes coherent[9] in the limit $L \to \infty$, that is, that

$$\lim_{L\to\infty} \langle \rho^{(L)} ; \hat{q}_{t_1}^{(L)}(f^{(1)})\cdots\hat{q}_{t_k}(f^{(k)}) \rangle = \prod_{r=1}^{m} q_{t_r}(f^{(r)}). \qquad (7.4.9)$$

This implies that the field $\hat{q}^{(L)}$ becomes dispersion-free in this limit, and thus that the dynamical law (7.3.8) is deterministic.

Comments

(1)   In view of assumption ($\hat{q}$.1), of Section 7.2, the scaling limit of Eq. (7.4.5)$'$ ensures that the macroscopic field $q_t(x)$ is strictly localised at the point x.

(2)   The constitutive equation (7.3.7), and hence also the quantum model Σ, is *not* Galilean invariant. The breakdown of Galilean situation prevails if, for example, the system is permeated by random scattering centres that are fixed relative to a particular inertial frame of reference. In this case, the state of Σ must incorporate the probability measure governing the statistical properties of the scatterers.

(3)   The assumptions (I) and (II) would be quite compatible with a consti-

---

[9] Cf. Section 3.3.2 for a definition of coherence.

tutive equation of the more general form $j_t(x) = J(q_t, \nabla q_t; x)$ and correspondingly, with the phenomenological equation

$$\frac{\partial}{\partial t} q_t(x) + \nabla \cdot J(q_t, \nabla q_t; x) = 0.$$

(4) As noted in (IV), the assumed initial condition renders the phenomenological equation of motion (7.3.8) valid for positive times only. For negative times, the principle of microscopic reversibility ensures that, for $t < 0$, the time-reversed version of that equation, obtained by replacing t by $-t$, would prevail. *Hence, the macroscopic model $\mathcal{M}$ of Section 7.3 is viable for positive times only.*

7.5 THE MODEL $\mathcal{M}_{\text{fluct}}$

We define the quantum fluctuation field $\tilde{q}^{(L)}$ by the formula

$$\tilde{q}_t^{(L)}(f) = L^{d/2}\left(\hat{q}_t^{(L)}(f) - \langle \rho^{(L)}; \hat{q}_t^{(L)}(f) \rangle \right). \tag{7.5.1}$$

Our aim now is to reduce the dynamics of $\tilde{q}^{(L)}$, in the limit $L \to \infty$, to a classical stochastic process, $\mathcal{M}_{\text{fluct}}$: a rudimentary account of such processes, which suffices for the present purposes, is presented in Appendix B.

We express the dynamics of the fluctuation field $\tilde{q}^{(L)}$ that stems from the above-prescribed initial conditions, in terms of the correlation functions

$$W_m^{(L)}(f^{(1)}, ..., f^{(m)}; t_1, ..., t_m) = \left\langle \rho^{(L)}; \tilde{q}_{t_1}^{(L)}(f^{(1)}) \cdots \tilde{q}_{t_m}^{(L)}(f^{(m)}) \right\rangle$$

$$\forall f^{(1)}, ..., f^{(m)} \in \mathbf{S}^{(n+1)}(X); \ t_1, ..., t_m \in \mathbf{R}_+. \tag{7.5.2}$$

Since $\tilde{q}$ is a Hermitian field, it follows from this equation that

$$\overline{W_m^{(L)}(f^{(1)}, ..., f^{(m)}; t_1, ..., t_m)} = W_m^{(L)}(\bar{f}^{(m)}, ..., \bar{f}^{(1)}; t_m, ..., t_1)$$

$$\forall f^{(1)}, ..., f^{(m)} \in \mathbf{S}^{(n+1)}(X); \ t_1, ..., t_m \in \mathbf{R}_+. \tag{7.5.3}$$

Our treatment of the functions $W_m^{(L)}$ is based on the following assumptions:

(W.1) Each $W_m^{(L)}$ converges pointwise to a limit, W_m, as $L \to \infty$, that is,

$$W_m(f^{(1)}, ..., f^{(m)}; t_1, ..., t_m) = \lim_{L \to \infty} W_m^{(L)}(f^{(1)}, ..., f^{(m)}; t_1, ..., t_m)$$

$$\forall f^{(1)}, ..., f^{(m)} \in \mathbf{S}^{(n+1)}(X); \ t_1, ..., t_m \in \mathbf{R}_+. \tag{7.5.4}$$

Furthermore, we assume that the W_m are continuous in all their arguments.

Comments

(1) The pointwise convergence of $W_m^{(L)}$ to a limit, W_m, as $L \to \infty$ is a standard property to be expected of normal fluctuations that is well established for the static case [GVV].

(2) The continuity of W_m in f follows directly from Eqs. (7.5.2) and (7.5.4), condition (q̂.5) and the completeness of $\mathbf{S}'$ spaces. On the other hand, the continuity of $W_m^{(L)}$ in t implies only the measurability of W_m in these variables, and therefore the assumption that W_m is continuous in them is an additional imput into the theory.

(3) It follows immediately from Eqs. (7.5.3) and (7.5.4) that

$$\overline{W_m(f^{(1)}, ..., f^{(m)}; t_1, ..., t_m)} = W_m(\bar{f}^{(m)}, ..., \bar{f}^{(1)}; t_m, ..., t_1)$$

$$\forall f^{(1)}, ..., f^{(m)} \in \mathbf{S}^{(n+1)}(X); \ t_1, ..., t_m \in \mathbf{R}_+. \tag{7.5.5}$$

(W.2) The field $\tilde{q}$ is Gaussian, that is, since $W_1 = 0$, by Eqs. (7.5.1) and (7.5.2), the functions W_m are all determined by W_2 according to the following formulae.

$$W_m = 0 \quad \text{if } m \text{ is odd,} \tag{7.5.6a}$$

and

$$W_m(f^{(1)}, ..., f^{(m)}; t_1, ..., t_m) = \sum_P \Pi_{(k,l) \in P} W_2(f^{(k)}, f^{(l)}; t_k, t_l), \tag{7.5.6b}$$

if m is even,

where $\sum_P$ denotes the sum over all partitions, P, of $(1, 2, ..., m)$ into pairs (k, l), with $k < l$.

Comment This Gaussian property is also expected to prevail for normal fluctuations, and its counterpart for static equilibrium correlation functions has been established by [GVV].

(W.3) $W_2(f^{(1)}, f^{(2)}; t_1, t_2)$ is invariant under the permutations $(f^{(1)}, t_1) \rightleftharpoons (f^{(2)}, t_2)$; and consequently, by (W.2), $W_m(f^{(1)}, ..., f^{(m)}; t_1, ..., t_m)$ is invariant under the permutations $(f^{(k)}, t_k) \rightleftharpoons (f^{(l)}, t_l)$.

Comments

(1) In view of (W.1), this is essentially an assumption of space-time asymptotically abelian properties of Σ in the normal folium of the primary state $\rho^{(L)}$.

(2) In the particular case where $\rho^{(L)}$ is an L- independent equilibrium state, the validity of (W.3) follows from (W.1) and the KMS conditions (cf. [Se10]). The essential reason for this is that the inverse temperature β is rescaled to $L^{-2}\beta$ in the macroscopic description and therefore vanishes

in the limit $L \to \infty$. Consequently, by Eqs. (7.5.2) and (7.5.4), the vanishing of the effective inverse temperature in the KMS condition (5.2.8) in this limit leads to the commutativity required for (W.3).

The following proposition is now a simple consequence of assumptions (W.1)–(W.3) (cf. Section B.3 of Appendix B).

Proposition 7.5.1   *Under the above assumptions, the functions $\{W_m \mid m \in \mathbf{N}\}$ define a Gaussian stochastic process, $\mathcal{M}_{\mathrm{fluct}}$, executed by a real classical field $\tilde{q}$, indexed by $\mathbf{S}^{(n+1)}(X) \times \mathbf{R}_+$, according to the formula*

$$E(\tilde{q}_{t_1}(f^{(1)}) \cdots \tilde{q}_{t_m}(f^{(m)})) = W_m(f^{(1)}, \ldots, f^{(m)}; t_1, \ldots, t_m)$$

$$\forall f^{(1)}, \ldots, f^{(m)} \in \mathbf{S}^{(n+1)}(X); \; t_1, \ldots, t_m \in \mathbf{R}_+, \tag{7.5.7}$$

where E is the expectation functional for the process.

Comment   Evidently, the positivity of E ensues from (W.1) and Eq. (7.5.2), while the classical and Gaussian properties of $\mathcal{M}_{\mathrm{fluct}}$ stem from (W.3) and (W.2), respectively.

Since, by Proposition 7.5.1 and assumption (W.2), $E(\tilde{q}_t(f)) \equiv 0$, the Gaussian process $\mathcal{M}_{\mathrm{fluct}}$ is completely specified by its two-point function $E(\tilde{q}_t(f)\tilde{q}_{t_0}(g))$. Further, by assumption (W.1), this function is continuous in f, g, t and t_0.

Our next assumption is the following generalisation of Onsager's regression hypothesis [On].

(W.4) The two-point function $E(\tilde{q}_t(f)\tilde{q}_{t_0}(g))$ is governed by the same dynamical law as that given by Eq. (7.3.13)–(7.3.15) for $\mathcal{M}_{\mathrm{pert}}$, in that

$$E\left(\tilde{q}_t(f)\tilde{q}_{t_0}(g)\right) = E\left(\tilde{q}_{t_0}(T^{\star}_{t,t_0}f)\tilde{q}_{t_0}(g)\right) \quad \forall f, g \in \mathbf{S}^{(n+1)}(X), \; t \geq t_0 \geq 0.$$

$$\tag{7.5.8}$$

As we shall prove in Appendix B (Section B.3), (W.4) reinforces the content of Proposition 7.5.1 by rendering $\mathcal{M}_{\mathrm{fluct}}$ Markovian. Thus, we have the following result.

Proposition 7.5.2   *Under the assumptions (W.1)–(W.4), $\mathcal{M}_{\mathrm{fluct}}$ is a Markov process with respect to time.*

We now define $E\left(\tilde{q}_t \mid \tilde{q}_{t_0}\right)$ to be the conditional expectation of $\tilde{q}_t$, given $\tilde{q}_{t_0}$, and $D\tilde{q}_t$ to be Nelson's [Ne] forward stochastic derivative of $\tilde{q}_t$), given by the formula

$$D\tilde{q}_t = \lim_{h \to +0} h^{-1} E(\tilde{q}_{t+h} - \tilde{q}_t \mid \tilde{q}_t). \tag{7.5.9}$$

The following corollary is an immediate consequence of Proposition 7.5.2 and Eqs. (7.3.18) and (7.5.9).

Corollary 7.5.3 *Under the above assumptions,*

$$E\left(\tilde{q}_t(f) \mid \tilde{q}_{t_0}\right) = \tilde{q}_{t_0}(T^{\star}_{t,t_0}f) \quad \forall f \in \mathbf{S}^{(n+1)}(X); \ t \geq t_0 \geq 0 \qquad (7.5.10)$$

and

$$D\tilde{q}_t(f) = \tilde{q}_t(\mathcal{L}^{\star}_t f) \quad \forall f \in \mathbf{S}^{(n+1)}(X), \ t \in \mathbf{R}_+. \qquad (7.5.11)$$

7.6 THE LINEAR REGIME: MACROSCOPIC EQUILIBRIUM CONDITIONS AND THE ONSAGER RELATIONS

Macroscopic Equilibrium Conditions

Let us now examine the consequences of the above scheme in the case where $\rho^{(L)}$ is an L-independent equilibrium state, ρ_θ, corresponding to a pure phase for the value θ of the control variable. We assume that this state is invariant under either the full translation group, X, or some crystallographic group, Y, and that, like the quantum field $\hat{q}$, it is invariant under time reversals. In view of this latter invariance and the stationarity of equilibrium states, it follows from Eq. (7.5.2) that

$$W_2^{(L)}(f, g; t, 0) = W_2^{(L)}(f, g; -t, 0) = W_2^{(L)}(f, g; 0, t) \quad \forall f, g \in \mathbf{S}^{(n+1)}(X), \ t \in \mathbf{R}.$$

Consequently, by assumption (W.1) and Proposition 7.5.1, the stochastic process $\mathcal{M}_{\text{fluct}}$ satisfies the condition

$$E(\tilde{q}_t(f)\tilde{q}(g)) = E(\tilde{q}(f)\tilde{q}_t(g)),$$

and therefore, by Eq. (7.5.9),

$$E(\tilde{q}(f)D\tilde{q}(g)) = E(\tilde{q}(g)D\tilde{q}(f)) \quad \forall f, g \in \mathbf{S}^{(n+1)}(X). \qquad (7.6.1)$$

This is simply the condition of *detailed balance*.

Furthermore, as we prove in Appendix C, the assumption of standard clustering properties of pure phases implies that the static two-point function is given by the following counterpart, for infinite systems, of Einstein's fluctuation formula (7.1.2),

$$E(\tilde{q}(f)\tilde{q}(g)) = (f, Bg) \quad \forall f, g \in \mathbf{S}^{(n+1)}(X), \qquad (7.6.2)$$

where

$$(f, g) \equiv \sum_{k=0}^{n} \int_X dx f_k(x) g_k(x), \qquad (7.6.3)$$

and the matrix $B = [B_{kl}]$ is given by

$$B = -s''(q)^{-1}. \qquad (7.6.4)$$

We take the detailed balance condition (7.6.1) and the thermodynamical fluctuation formula (7.6.2) to constitute the *macroscopic equilibrium conditions* for the system. These conditions are evidently weaker than the KMS condition, which governs the microscopic structure of the state ρ_θ.

Onsager Relations

In fact, the above macroscopic equilibrium conditions are independent of the specific models $\mathcal{M}$ and $\mathcal{M}_{\text{fluct}}$ of Sections 7.3–7.5. In order to obtain the explicit form of D for those models, we note that the perturbative variable δq_t, of Eq. (7.3.10) is just the deviation of q_t from its uniform equilibrium value. Hence, by Eq. (7.3.11), the linear operator $\mathcal{L}_t$ reduces to the form

$$\mathcal{L}_t = K(\theta) s''(q) \Delta, \qquad (7.6.5)$$

and consequently, by Eq. (7.5.11),

$$D\tilde{q}_t(f) = \tilde{q}_t \left(s''(q) K^\star(\theta) \Delta f \right) \quad \forall f \in \mathbf{S}^{(n+1)}(X), \ t \in \mathbf{R}_+. \qquad (7.6.6)$$

It follows from this formula and Eqs. (7.6.2)–(7.6.4) that

$$E(\tilde{q}(f) D\tilde{q}(g)) = -\left(f, K^\star(\theta) \Delta g \right)$$

$$\equiv -\sum_{k,l=0}^{n} K_{lk}(\theta) \int_X dx \nabla f_k(x) . \nabla g_l(x) \quad \forall f, g \in \mathbf{S}^{(n+1)}(X). \qquad (7.6.7)$$

Hence, the detailed balance condition (7.6.1) reduces to the form

$$\sum_{k,l=0}^{n} [K_{kl}(\theta) - K_{lk}(\theta)] \int_X dx \nabla f_k(x) . \nabla g_l(x) = 0 \quad \forall f, g \in \mathbf{S}^{(n+1)}(X),$$

which implies the Onsager relations

$$K_{kl}(\theta) = K_{lk}(\theta) \quad \forall k, l \in [0, n]. \qquad (7.6.8)$$

Thus, we have established the following proposition.

Proposition 7.6.1 *Under the above assumptions, the macroscopic system $\mathcal{M}$ satisfies the Onsager relations (7.6.8) in the linear regime.*

7.7 THE NONLINEAR REGIME: LOCAL EQUILIBRIUM AND GENERALISED ONSAGER RELATIONS

The dynamics of $\mathcal{M}$, as represented by the equations of motion (7.3.8), is generally nonlinear. Correspondingly, the field $q_t(x)$ is nonuniform with respect to both position and time, and the process $\mathcal{M}_{\text{fluct}}$ is nonstationary. Our aim now is to extend the linear theory of the previous section to this general situation. Since this entails the variation of the thermodynamical field $\theta_t(x)$ with both x and t, we base our treatment on a localised version of the global macroscopic equilibrium conditions (7.6.1) and (7.6.2) that were applicable in the linear case. We assume that the range of the variable $\theta_t(x)$ here lies in a single phase region of the control space Θ for which the equilibrium states satisfy the conditions imposed on ρ_θ at the beginning of Section 7.6.

Local Macroscopic Equilibrium Conditions

In order to formulate the thermodynamical state in the neighborhood of a space-time point (x_0, t_0), we introduce the transformation $f \to f_{x_0, \epsilon}$ of $S^{(n+1)}(X)$, defined by the formula

$$f_{x_0, \epsilon}(x) = \epsilon^{-d/2} f\left(\frac{x - x_0}{\epsilon}\right) \quad \forall x, x_0 \in X, \ \epsilon \in \mathbf{R}_+ \backslash \{0\}. \qquad (7.7.1)$$

Correspondingly, we define the rescaled forward time-derivative

$$D_\epsilon = \epsilon^2 D. \qquad (7.7.2)$$

We then note that, in the *linear* regime of Section 7.6, it follows from Eqs. (7.6.2)–(7.6.4), (7.6.6), (7.7.1) and (7.7.2), together with the uniformity of the fields q and θ and the stationarity of the process $\mathcal{M}_{\text{fluct}}$, that the two-point functions $E(\tilde{q}(f)\tilde{q}(g))$ and $E(\tilde{q}(f)D\tilde{q}(g))$ are invariant under the transformations $\tilde{q} \to \tilde{q}_{t_0}$, $f \to f_{x_0, \epsilon}$, $g \to g_{x_0, \epsilon}$ and $D \to D_\epsilon$; and therefore the equilibrium conditions (7.6.1) and (7.6.2), taken in reverse order, are equivalent to the following.

$$E\left(\tilde{q}_{t_0}(f_{x_0, \epsilon})\tilde{q}_{t_0}(g_{x_0, \epsilon})\right) = -\left(f, s''(q_{t_0}(x_0))^{-1} g\right)$$

and

$$E\left(\tilde{q}_{t_0}(f_{x_0, \epsilon})D_\epsilon \tilde{q}_{t_0}(g_{x_0, \epsilon})\right) = E\left(\tilde{q}_{t_0}(g_{x_0, \epsilon})D_\epsilon \tilde{q}_{t_0}(f_{x_0, \epsilon})\right) = -\left(f, K^\star(\theta)\Delta g\right).$$

These equations evidently survive the passage to the limit $\epsilon \to 0$, which serves to localise the process $\mathcal{M}_{\text{fluct}}$ at the space-time point (x_0, t_0). In the *linear* regime, this implies that

$$\lim_{\epsilon \to 0} E\left(\tilde{q}_{t_0}(f_{x_0,\epsilon})\tilde{q}_{t_0}(g_{x_0,\epsilon})\right) = -\left(f, s''(q_{t_0}(x_0))^{-1}g\right)$$

$$\forall f, g \in \mathbf{S}^{(n+1)}(X), \ x_0 \in X, \ t_0 \in \mathbf{R}_+ \tag{7.7.3}$$

and

$$\lim_{\epsilon \to 0} E\left(\tilde{q}_{t_0}(f_{x_0,\epsilon})D_\epsilon\tilde{q}_{t_0}(g_{x_0,\epsilon})\right) = \lim_{\epsilon \to 0} E\left(\tilde{q}_{t_0}(g_{x_0,\epsilon})D_\epsilon\tilde{q}_{t_0}(f_{x_0,\epsilon})\right)$$

$$\forall f, g \in \mathbf{S}^{(n+1)}(X), \ x_0 \in X, \ t_0 \in \mathbf{R}_+, \tag{7.7.4}$$

the limits here being finite.

We now assume that Eqs. (7.7.3) and (7.7.4) are also valid in the nonlinear regime, and we term them the *local macroscopic equilibrium conditions*.

Onsager Relations

On the basis of the above local equilibrium assumption, we are able to establish the following proposition, which represents a nonlinear generalisation of the Onsager theory.

Proposition 7.7.1 *Under the above assumptions, the nonlinear system* $\mathcal{M}$ *satifies the Onsager relations*

$$K_{kl}(\theta_t(x)) = K_{lk}(\theta_t(x)) \quad \forall x \in X, \ t \in \mathbf{R}_+. \tag{7.7.5}$$

Comment This result establishes that the Onsager relations for the linear and nonlinear regimes are consistent with one another.

Proof of Proposition 7.7.1. Let $D^{(0)}$ be the counterpart of D obtained by the replacement of q and θ by $q_{t_0}(x_0)$ and $\theta_{t_0}(x_0)$, respectively, in Eq. (7.6.6), that is,

$$D^{(0)}\tilde{q}_{t_0}(f) = \tilde{q}_{t_0}\left[s''(q_{t_0}(x_0))K^\star(\theta_{t_0}(x_0))\Delta f\right]$$

$$\forall f \in \mathbf{S}^{(n+1)}(X), \ x_0 \in X, \ t_0 \in \mathbf{R}_+. \tag{7.7.6}$$

Then, defining

$$D_\epsilon^{(0)} = \epsilon^2 D^{(0)}, \tag{7.7.7}$$

it follows from Eqs. (7.3.19), (7.5.11), (7.7.1), (7.7.2), (7.7.6) and (7.7.7), together with Proposition 7.5.1 and the continuity condition contained in assumption (W.1) of Section 7.5, that

$$\lim_{\epsilon \to 0} E\Big(\tilde{q}_{t_0}(f_{x_0,\epsilon})[D_\epsilon - D_\epsilon^{(0)}]\tilde{q}_{t_0}(f_{x_0,\epsilon})\Big) = 0. \tag{7.7.8}$$

Consequently, we may replace D_ϵ by $D_\epsilon^{(0)}$ in the local detailed balance condition (7.7.4), and thereby obtain the formula

$$\lim_{\epsilon \to 0} E\Big(\tilde{q}_{t_0}(f_{x_0,\epsilon})D_\epsilon^{(0)}\tilde{q}_{t_0}(g_{x_0,\epsilon})\Big) = \lim_{\epsilon \to 0} E\Big(\tilde{q}_{t_0}(g_{x_0,\epsilon})D_\epsilon^{(0)}\tilde{q}_{t_0}(f_{x_0,\epsilon})\Big). \tag{7.7.9}$$

Further, by Eqs. (7.7.1), (7.7.6) and (7.7.7),

$$D_\epsilon^{(0)}\tilde{q}_{t_0}(f_{x_0,\epsilon}) = \tilde{q}\Big[s''(q_{t_0}(x_0))K^\star(\theta_{t_0}(x_0))(\Delta f)_{x_0,\epsilon}\Big]$$

and hence, by Eqs. (7.7.3),

$$E\Big(\tilde{q}_{t_0}(g_{x_0,\epsilon})D_\epsilon^{(0)}\tilde{q}_{t_0}(f_{x_0,\epsilon})\Big) = -\Big(g, K^\star(\theta_{t_0}(x_0))f\Big).$$

Consequently, Eq. (7.7.9) is equivalent to the formula

$$\Big(f, K^\star(\theta_{t_0}(x_0))g\Big) = \Big(g, K^\star(\theta_{t_0}(x_0))f\Big) \quad \forall f, g \in S^{(n+1)}(X), \ x_0 \in X, \ t_0 \in \mathbf{R}_+,$$

that is,

$$K^\star(\theta_{t_0}(x_0)) = K(\theta_{t_0}(x_0)),$$

which is equivalent to the Onsager relations (7.7.5). □

7.8 FURTHER CONSIDERATIONS: TOWARDS A GENERALISATION OF THE THEORY TO GALILEAN CONTINUUM MECHANICS

The theory of Sections 7.2–7.7 is limited in the following respects by the structure of the phenomenological diffusive law (7.1.7).

(a) The quantum field $\hat{q}$ is invariant under the time-reversal antiautomorphism, τ.

(b) The phenomenological dynamics executed by the corresponding classical field q is not Galilean covariant.

(c) That dynamics is covariant with respect to the scale transformations $x \to \lambda x, \ t \to \lambda^2 t$.

Consequently, the above treatment is not applicable to macroscopic laws such as those of hydrodynamics, which satisfy none of these conditions. For there, the field q has a component, namely the momentum density, that changes sign under time reversals; and, furthermore, the equations of fluid mechanics are both Galilean covariant and scale dependent. To see their scale dependence, it suffices to reduce them to the local mass conservation law and the simplified form of the Navier–Stokes equation in which thermal conductivity is ignored and the bulk and shear viscosities are taken to be equal. The fluid equations are then

$$\frac{\partial \rho}{\partial t} + \nabla.(\rho u) = 0 \tag{7.8.1}$$

and

$$\frac{\partial u}{\partial t} + u.\nabla u + \rho^{-1}\nabla p = \eta \Delta u, \tag{7.8.2}$$

where u, ρ, p and η are the drift velocity, mass density, pressure and viscosity, respectively. Now, in the *special cases* of inviscid flow, where $\eta = 0$, and incompressible flow, where ρ is constant, these equations are scale invariant. Specifically, in the former case they are invariant under the transformations

$$x \longrightarrow \lambda x, \quad t \longrightarrow \lambda t, \quad u \longrightarrow u, \quad \rho \longrightarrow \rho, \quad p \longrightarrow p,$$

and in the latter case, they are invariant under the very different transformations

$$x \longrightarrow \lambda x, \quad t \longrightarrow \lambda^2 t, \quad u \longrightarrow \lambda^{-1} u, \quad p \longrightarrow \lambda^{-2} p.$$

In general, however, the equations (7.8.1) and (7.8.2) are not invariant under any scale transformations $x \longrightarrow \lambda x$, $t \longrightarrow \lambda^c t$, with c a constant, because of the mismatch between the scaling properties of the pressure gradient and viscosity terms.

Thus, granted that the above conditions (a)–(c) are all violated in important cases such as fluid mechanics, it is natural to ask whether the macrostatistical picture we have presented could survive their removal. In fact, Casimir [Cas] liberated the original Onsager theory from the restriction (a) by extending it to cases where each component $\hat{q}_k$ of $\hat{q}$ has either even or odd parity with respect to time-reversals, that is, where

$$\tau \hat{q}_k(f) = T_k \hat{q}_k(f) \quad \text{with } T_k = \pm 1, \ k = 0, ..., n. \tag{7.8.3}$$

For such cases, Casimir's main result corresponds to the modification of the linearised Onsager relations (7.6.8) to the form (cf. [Ca, deGM])

$$K_{kl}(\theta) = T_k T_l K_{lk}(\theta^T),$$

where

$$\theta^T = (T_0 \theta_0, ..., T_n \theta_n).$$

In fact, the generalisation of this result to the nonlinear regime follows from a simple adaptation of the theory of Sections 7.6 and 7.7 that leads to the replacement of Eq. (7.7.5) by the relations

$$K_{kl}(\theta_t(x)) = T_k T_l K_{kl}(\theta_t^T(x)),$$

(7.8.4)

where

$$\theta_t^T(x) = (T_0 \theta_{t,1}(x), ..., T_n \theta_{t,n}(x)).$$

(7.8.5)

Thus, the removal of condition (a) does not present a problem.

On the other hand, the same cannot be said about conditions (b) and (c), since our formulation of the relationship of both the stochastic process $\mathcal{M}_{\text{fluct}}$ and the macroscopic model $\mathcal{M}$ to the quantum field $\hat{q}$ depended on the scale invariance of the assumed phenomenological law, as viewed in the chosen inertial frame. For this reason, it is clear that any statistical mechanical derivation of scale dependent macroscopic laws, such as those of Navier–Stokes hydrodynamics, must be based on a much more subtle analysis, of a multiscale nature. Nonetheless, in view of the overwhelming empirical support for such Galilean covariant phenomenological laws, it is surely important to have some guidelines about the strictures imposed on them by statistical mechanics. The conjecture we make, pending the advent of the relevant multiscale generalisation of the macrostatistics of this chapter, is that *the relationship between the macroscopic system $\mathcal{M}$ and the stochastic process $\mathcal{M}_{\text{fluct}}$, as represented by Eqs. (7.1.3)–(7.1.6), still prevails.*

Under this assumption, together with that of microscopic reversibility, it is a straightforward matter to recast the nonequilibrium thermodynamics of continuous media, as formulated by De Groot and Mazur [deGM], within the present macrostatistical framework and thereby to obtain the general Onsager relations assumed[10] by the latter authors.

On a similar basis, the theory could be naturally extended to phases whose order structure is represented by intensive, nonconserved variables, discussed in Section 6.5. The resultant scheme could then be applicable to phenomena such as superfluid hydrodynamics and antiferromagnetic relaxation.

APPENDIX A: TEMPERED DISTRIBUTIONS

The theory of distributions was devised by L. Schwartz [Schw] to provide a mathematical framework for objects such as the so-called Dirac δ function that were not amenable to the standard treatments of differentiable and integral calculus. There, the problem was that, according to the Lebesgue theory of integration, if ϕ is a function on $\mathbf{R}$, say, that is zero everywhere but at one point, then $\int_{\mathbf{R}} \phi(x)dx = 0$. Hence, that theory was not applicable to the δ

---

[10] An essential assumption of [deGM] was that the canonical couterparts of the Onsager relations obtained for the linear evolution of a finite set of macroscopic variables are valid for the nonlinear field equations of continuum mechanics.

function, whose total integral is supposed to be unity, despite the fact that it is nonzero at one point only. To overcome this problem, Schwartz formulated δ not as a function on $\mathbf{R}$ but as a functional on a suitable space of smooth functions, f, on $\mathbf{R}$, such that $\delta(f) = f(0)$. Thus, the action of the functional δ on f yields the same result as the formal integration of $f(x)$ against the improperly defined Dirac function $\delta(x)$. The theory of distributions is designed to generalise the theory of functions in this way. Its basic ingredients are a space of smooth test functions and a dual space of functionals on these.

To sketch the rudiments of the theory, we suppose that X is a d-dimensional Euclidean space, and that its points, x, are represented by Cartesian coordinates $(x_1, ..., x_d)$. The Schwartz space, $\mathbf{S}(X)$, of test functions is defined to comprise the infinitely differentiable functions, f, on X, such that f and its derivatives of all orders tend to zero faster than any inverse power of $|x|$ as $|x| \to \infty$. Thus, $\mathbf{S}(X)$ is a vector space: it is sometimes termed the space of *smooth, fast decreasing* functions on X. Its topology is defined so that the convergence of a sequence $\{f_n\}$ to f in $\mathbf{S}(X)$ signifies that the product of any polynomial function on X and any derivative of $(f_n - f)$ tends uniformly to zero as n tends to infinity. The space $\mathbf{S}(X)$ is complete with respect to this topology [Schw]. We denote by $\mathbf{S}_R(X)$ the subspace of $\mathbf{S}(X)$ consisting of its real-valued elements.

The space, $\mathbf{S}'(X)$, dual to $\mathbf{S}(X)$, is defined to comprise the set of continuous linear functionals on the latter space. Thus, the elements of $\mathbf{S}'(X)$ are the linear mappings, T, of $\mathbf{S}(X)$ into $\mathbf{C}$, such that, if f_n converges to f in $\mathbf{S}(X)$, then $T(f_n)$ converges to $T(f)$. $\mathbf{S}'(X)$ is therefore a vector space. Its topology is defined so that the convergence of a sequence $\{T_n\}$ to T in $\mathbf{S}'(X)$ signifies that $T_n(f)$ converges to $T(f)$ as n tends to infinity, for all $f \in \mathbf{S}(X)$. $\mathbf{S}'(X)$ is complete with respect to this topology [Schw] and is termed the space of *tempered distributions* in X.

Note It follows from this definition that $\mathbf{S}(X)$ contains the following classes of functionals.

(a) The functionals F, that correspond to measurable, polynomial bounded[11] functions, also denoted by F, on X, according to the formula

$$F(f) = \int_X F(x)f(x). \qquad (\text{A.1})$$

(b) The functionals, δ_a, defined, for any a in X, by the formula

$$\delta_a(f) = f(a). \qquad (\text{A.2})$$

This corresponds to the formal expression

---

[11] The term "polynomial bounded" signifies that there are positive constants C and n such that $|F(x)| < C(1 + |x|^n)$ for all $x \in X$.

$$\delta_a(f) = \int_X \delta(x-a)f(x).$$

Thus, $S'(X)$ contains the functionals corresponding not only to "normal" functions, but also to the Dirac δ. For this reason, the tempered distributions are sometimes termed *generalised functions*.

Notational Convention

It is sometimes convenient to denote a tempered distribution, T, as though it were a "normal" function on X and to write $T(f)$ as $\int_X dx T(x)f(x)$. This is evidently a harmless procedure as long as it does not carry the assumption that $T(x)$ is measurable.

Spaces $S^{(m)}(X)$ and $S^{(m)\prime}(X)$

$\,$), the mth topological power of $S(X)$, is the vector space whose elements $= (f^{(1)}, ..., f^{(m)}) \mid f^{(k)} \in S(X)$ for $k = 1, ..., m\}$, with addition defined by le that $(f+g)^{(k)} := f^{(k)} + g^{(k)}$ for $k = 1, ..., m$. The topology of this is defined so that the convergence of f_n to f in $S^{(m)}(X)$ signifies that of f the components $f_n^{(k)}$ to $f^{(k)}$ in $S(X)$.

$'(X)$ is the vector space of continuous linear functionals on $S^{(m)}(X)$. it comprises the linear mappings, T, of $S^{(m)}(X)$ into C, such that, if f_n ...ges to f in $S^{(m)}(X)$, then $T(f_n) \to T(f)$. The topology of $S^{(m)\prime}(X)$ is defined so that the convergence of T_n to T in this space signifies that of $T_n(f)$ to $T(f)$ for all $f \in S^{(m)}(X)$. It follows from these specifications that $S^{(m)\prime}(X)$ is the mth topological power of $S'(X)$. Thus, each of its elements, T, consists of an m-tuplet, $(T^{(1)}, ..., T^{(m)})$, of elements of $S'(X)$, and its its action on $S^{(m)}(X)$ is given by

$$T(f^{(1)}, ..., f^{(m)}) = \sum_{k=1}^{m} T^{(k)}(f^{(k)}).$$

Action of Space Translations

We define the transformation $f \to f_a$ of $S(X)$, corresponding to the space translation $x \to x + a$ in X, by the formula

$$f_a(x) = f(x-a) \quad \forall a \in X. \tag{A.3}$$

We then define the corresponding transformation $T \to T_a$ of $S'(X)$ by the equation

$$T_a(f) = T(f_a) \quad \forall a \in X, \, f \in S(X). \tag{A.4}$$

Thus, if T is represented as a generalised function, $T(x)$, of x, then $T_a(x) = T(x + a)$.

Derivatives of Tempered Distributions

The first derivatives, $\partial T/\partial x_j$, of a tempered distribution, T, are defined to be the functionals on $\mathbf{S}(X)$ given by the formula

$$\frac{\partial T}{\partial x_j}(f) = -T(\frac{\partial f}{\partial x_j}) \quad \forall f \in \mathbf{S}(X). \tag{A.5}$$

Note
  (a) In the case where T corresponds to a differentiable function F, and so is given by Eq. (A.1), its derivative $\partial T/\partial x_j$ corresponds likewise to $\partial F/\partial x_j$.

  (b) It follows easily from the above specifications that $\partial T/\partial x_j$ lies in $\mathbf{S}'(X)$. Hence, by induction, the definition (A.5) may be extended to derivatives of all orders of T according to the formula

$$\frac{\partial^{k_1+\cdots+k_d} T}{\partial x_1^{k_1} \cdots \partial x_d^{k_d}}(f) = (-1)^{k_1+\cdots+k_d} T\left(\frac{\partial^{k_1+\cdots+k_d} f}{\partial x_1^{k_1} \cdots \partial x_d^{k_d}}\right). \tag{A.6}$$

Support of a Tempered Distribution

We recall that the support of a function, f, on X is defined to be the complement of the largest open subset of X on which f vanishes, and is generally denoted by $\mathrm{supp}(f)$. Correspondingly, the support, $\mathrm{supp}(T)$, of a tempered distribution T ($\in \mathbf{S}'(X)$) is defined to be the complement of the largest open subset $\mathcal{O}$ of X, such that $T(f)$ vanishes if f is any element of $\mathbf{S}(X)$ with support in $\mathcal{O}$.

The following proposition serves to characterise the tempered distributions with support at a single point in terms of the Dirac δ.

Proposition A.1 [Schw] *The elements, T, of $\mathbf{S}(X)$ with support at a single point a are precisely the finite linear combinations of the Dirac distribution δ_a and its derivatives.*

Tempered Distributions of Zero Order

Let $C_0(X)$ be the abelian $C^\star$-algebra of continuous functions on X with compact support, equipped with supremum norm. A tempered distribution, T, of class $\mathbf{S}'(X)$ is termed to be of *zero order* if it extends by continuity to an element of the dual of this algebra, that is, if it corresponds to a measure ν on X according to the formula

$$T(f) = \int_X d\nu(x) f(x) \quad \forall f \in C_0(X).$$

This condition is indeed restrictive, since it excludes derivatives of the δ-distribution, as these do not act continuously on $C_0(X)$. Thus, the zero order distributions contain no singularities stronger than those of the Dirac δ.

The Spaces $S(X^m)$ and $S'(X^m)$

These are defined by replacing X by X^m in the above definitions of $S(X)$ and $S'(X)$. Likewise, the action of space translations on the former spaces is defined by precise analogy with those on the latter spaces. Thus, the transformations $F \rightarrow F_a$ of $S(X^m)$ and $W \rightarrow W_a$ of $S'(X^m)$, corresponding to the space translation $x \rightarrow x + a$, are defined by the formulae

$$F_a(x_1, ..., x_m) = F(x_1 - a, ..., x_m - a) \tag{A.7}$$

and

$$W_a(F) = W(F_a). \tag{A.8}$$

We term an element, W, of $S'(X^m)$ *translationally invariant* if $W_a = W$ $\forall a \in X$.

Proposition A.2 [StWi] *A translationally invariant tempered distribution, W, in X^n corresponds to a unique element, $\mathcal{W}$, of $S'(X^{m-1})$, according to the rule that, if $\mathcal{F}$ and F are elements of $S(X^{m-1})$ and $S(X^m)$ that are related by the formula*

$$F(x_1, ..., x_m) = \mathcal{F}(x_1 - x_2, ..., x_m - x_{m-1}),$$

then

$$\mathcal{W}(\mathcal{F}) = W(F).$$

Comment   This proposition corresponds to the assertion that a translationally invariant generalised function $W(x_1, ..., x_m)$ may be expressed as a function $\mathcal{W}$ of the variables $(x_1 - x_2, ..., x_{m-1} - x_m)$.

Tensor Products and Multilinear Functional.

The tensor product of m elements, $f_1, ..., f_m$, of $S(X)$ is defined to be the element $f_1 \otimes f_2 \otimes \cdots \otimes f_m$ of $S(X^m)$, whose value at the point $(x_1, ..., x_m)$ of X^m is the product $f_1(x_1) \cdots f_m(x_m)$.

On the other hand, a mapping $(f_1, ..., f_m) \rightarrow V(f_1, ..., f_m)$ of $S(X)^m$ into $\mathbf{C}$ is termed linear and continous in each argument if

$$V(f_1, .., af_l + bg_l, ..., f_m) = aV(f_1, .., f_l, ..., f_m) + bV(f_1, .., g_l, ..., f_m)$$

$$\forall a, b \in \mathbf{C}, \ l = 1, \dots, m$$

and

$$\lim_{n \to \infty} V(f_1, \dots, f_l^{(n)}, \dots, f_m) = V(f_1, \dots, f_l, \dots, f_m) \quad \text{if} \ \lim_{n \to \infty} f_l^{(n)} = f_l.$$

The following proposition, known as Schwartz's nuclear theorem, signifies that any such functional corresponds to a unique tempered distribution in X^m.

Proposition A.3 [Schw] *A functional V on $\mathbf{S}(X)^m$ that is separately linear and continuous in each argument corresponds to a unique element, W, of $\mathbf{S}(X^m)$ according to the formula*

$$W(f_1 \otimes f_2 \otimes \cdots \otimes f_m) = V(f_1, \dots, f_m).$$

This proposition permits the definition of the tensor product, $T_1 \otimes T_2 \otimes \cdots \otimes T_m$, of m elements, $T_1, \dots, T_m$, of $\mathbf{S}'(X)$ according to the prescription that it is the element T of $\mathbf{S}'(X^m)$ for which

$$T(f_1 \otimes f_2 \otimes \cdots \otimes f_m) = \prod_{j=1}^{m} T_j(f_j) \quad \forall f_1, \dots, f_m \in \mathbf{S}(X).$$

The Spaces $\mathbf{S}_V$ and $\mathbf{S}_V'$

We denote by $\mathbf{S}_V(X)$ and $\mathbf{S}_V'(X)$ the canonical counterparts of $\mathbf{S}(X)$ and $\mathbf{S}'(X)$, respectively, corresponding to vector fields in X. Thus, $\mathbf{S}_V(X)$ is the space of infinitely differentiable fast decreasing vector fields in X, $\mathbf{S}_V'(X)$ is its dual, and the topologies of this spaces are defined analogously with those of $\mathbf{S}(X)$ and $\mathbf{S}'(X)$. It is a straightforward matter to check that the whole of the above theory for the $\mathbf{S}$ and $\mathbf{S}'$ spaces is viable also for $\mathbf{S}_V$ and $\mathbf{S}_V'$.

Gaussian Probability Measures on $\mathbf{S}'$ spaces

The following proposition is important for the theory of stochastic processes (cf. Appendix B).

Proposition A.4 [GV] *Let F be a continuous, real valued, positive semi-definite, bilinear form on the real subspace, $\mathbf{S}_R$, of a Schwartz space $\mathbf{S}$, which may be be any of the spaces $\mathbf{S}(X)$, $\mathbf{S}^m(X)$ or $\mathbf{S}_V(X)$. Then there is a unique probability measure, μ, on the dual space, $\mathbf{S}'$, such that*

$$\int d\mu(\phi) \exp(i\phi(f)) = C(f) := \exp\left(-\frac{1}{2} F(f, f)\right) \quad \forall f \in \mathbf{S}_R.$$

μ is then termed a Gaussian measure and C, its Fourier transform, is termed its characteristic function.

Comment   This proposition is a generalisation to S'-spaces of the fact that, if c is the function on **R** given by

$$c(x) = \exp\left(-\frac{1}{2}ax^2\right),$$

with $a \geq 0$, then there is a unique probability measure, m, on **R**, such that

$$c(x) = \int_{\mathbf{R}} dm(y)\exp(ixy).$$

To be precise, if $a > 0$, then m is the measure corresponding to the probability density, p, given by the Fourier transform of c, that is, $p(x) = (2\pi a)^{-1/2}\exp(-x^2/2a)$; and if $a = 0$, then m is just the Dirac measure, δ_0, represented by the tempered distribution denoted by the same symbol.

APPENDIX B: CLASSICAL STOCHASTIC PROCESSES AND THE CONSTRUCTION OF $\mathcal{M}_{\text{fluct}}$ AS A CLASSICAL MARKOV FIELD

Stochastic processes, whether classical or quantal, may naturally be formulated within an operator algebraic framework [AFL]. In the classical case, where the relevant algebras are abelian, the Gelfand isomorphism[12] ensures that the resultant algebraic picture is equivalent to the standard probabilistic one, as presented in the treatises of Doob [Do], Gihman and Skorohod [GS] and Van Kampen [VK1].

B.1 Algebraic Description of Classical Stochastic Processes

We consider a classical stochastic process, $\mathcal{P}$, over a temporal range $\tilde{\mathbf{R}}$, which may be either **R** or $\mathbf{R}_+$. In the algebraic picture, $\mathcal{P}$ consists of

(a) an abelian *-algebra $\mathcal{B}$, generated by a family of subalgebras, $\{\mathcal{B}_t \mid t \in \tilde{\mathbf{R}}\}$;

(b) a two parameter family of isomorphisms, $\{J_{st} : \mathcal{B}_s \to \mathcal{B}_t \mid s,t \in \tilde{\mathbf{R}}\}$, with $J_{st}J_{tu} = J_{su}$ and $J_{tt} = I$; and

(c) a state, that is, an expectation functional, E on $\mathcal{B}$. Thus, $\mathcal{B}$ and E correspond, via the Gelfand isomorphism, to an algebra of functions and a probability measure, μ, respectively, on a space χ.

In this model, the elements of $\mathcal{B}$ are the *random variables* and their statis-

---

[12] Cf. item 35 of Section 2.4.5.

tical properties are governed by E. The subalgebra, $\mathcal{B}_t$, of $\mathcal{B}$ is that of the observables of the process at time t, and the evolute, at time $u (\geq t)$, of an element B_t of this algebra is $B_u = J_{tu} B_u$. We define $\mathcal{B}_t^{\geq}$ and $\mathcal{B}_t^{\leq}$ to be the subalgebras of $\mathcal{B}$ generated by $\{\mathcal{B}_{t'} \mid t' \geq t\}$ and $\{\mathcal{B}_{t'} \mid t' \leq t\}$, respectively, for each $t \in \tilde{\mathbf{R}}$.

The properties of the model are represented by its multitime correlation functions $E(B_{t_1}^{(1)} \cdots B_{t_m}^{(m)})$, with $B_{t_k} \in \mathcal{B}_{t_k}$.

Conditional Expectations

The conditional expectations for the process $\mathcal{P}$ are defined by the general prescription of Section 4.2.3, as simplified by the abelian property of $\mathcal{B}$. Thus, if C is a *-subalgebra of $\mathcal{B}$, then a conditional expectation (CE) of $\mathcal{B}$ with respect to C, compatible with the state E, is a positive,[13] linear mapping, $P : \mathcal{B} \to C$ that satisfies the conditions

$$P(C) = C \quad \forall C \in C, \tag{B.1.1}$$

$$P(BC) = (PB)C \quad \forall B \in \mathcal{B}, \ C \in C \tag{B.1.2}$$

and

$$E(PB) = E(B) \quad \forall B \in \mathcal{B}. \tag{B.1.3}$$

In fact, these conditions serve to define P uniquely. For, if P' is also a positive linear mapping of $\mathcal{B}$ onto C that satisfies Eqs. (B.1.1)–(B.1.3), then

$$E(C(P - P')B) = 0 \quad \forall B \in \mathcal{B}, \ B \in C,$$

and hence, putting $C = (P - P')B^\star$,

$$E\left(\mid (P - P')B \mid^2 \right) = 0 \quad \forall B \in \mathcal{B},$$

which implies that P is essentially unique, that is, that $P'B = PB$ except possibly at a set of points of zero measure in the probability space (χ, μ) corresponding to $(\mathcal{B}, E)$ via the Gelfand isomorphism.

Moreover, the requirement that P be positive would also be ensured, subject to the conditions (B.1.1)–(B.1.3), by the weaker demand that it be real, that is, that

$$(PB)^\star = PB^\star, \tag{B.1.4}$$

since it follows from Eqs. (B.1.2)–(B.1.4) that, if B and C are positive elements of $\mathcal{B}$ and C, respectively, then

---

[13] Since $\mathcal{B}$ is abelian, there is no distinction here between positive and completely positive maps.

$$E(CPB) = E(P(CB)) = E(CB) \geq 0,$$

which implies that $PB \geq 0$ except possibly on set of zero measure in the probability space (χ, μ).

Markov Conditions

The process $\mathcal{P}$ is Markovian if it is memory-free, in the sense that, for $B \in \mathcal{B}_t^{\geq}$ and $t \in \tilde{\mathbf{R}}$, the conditional expectation of B with respect to $\mathcal{B}_t^{\leq}$ is just its CE with respect to $\mathcal{B}_t$ and therefore does not depend on the observables of the process at times before t. We formulate this condition as signifying that $\mathcal{P}$ is equipped with a family of linear mappings $\{P_t : \mathcal{B}_t^{\geq} \to \mathcal{B}_t \mid t \in \tilde{\mathbf{R}}\}$, that satisfy the equations

$$P_t B = B \quad \forall B \in \mathcal{B}_t, \tag{B.1.5}$$

$$P_t B^{\star} = (P_t B)^{\star} \quad \forall B \in \mathcal{B}_t^{\geq} \tag{B.1.6}$$

and

$$E(B' P_t B) = E(B' B) \quad \forall B \in \mathcal{B}_t^{\geq}, \ B' \in \mathcal{B}_t^{\leq}. \tag{B.1.7}$$

In view of the above discussion of CEs, we see that these specifications imply that P_t is the E-compatible conditional expectation of the former algebra with respect to *both* $\mathcal{B}_t$ and $\mathcal{B}_t^{\leq}$.

B.2 CLASSICAL GAUSSIAN FIELDS

Suppose now that the basic random variables of the process $\mathcal{P}$ are given by a real, time-dependent field, ϕ_t, over a Schwartz space $\mathbf{S}$, which may be any of the spaces $\mathbf{S}(X)$ or $\mathbf{S}^{(m)}(X)$ or $\mathbf{S}_V(X)$. In this case, $\mathcal{B}_t$ is the *-algebra of polynomials in $\{\phi_t(f) \mid f \in \mathbf{S}\}$, where ϕ_t is the version of ϕ at time t. Thus, ϕ is a real field, indexed by $\tilde{\mathbf{R}}$ and $\mathbf{S}$. Its reality is represented by the condition that

$$\phi_t(f)^{\star} = \phi_t(\bar{f}) \quad \forall t \in \tilde{\mathbf{R}}, \ f \in \mathbf{S}. \tag{B.2.1}$$

We define the m-point correlation function, W_m, for the field ϕ by the formula

$$W_m(f^{(1)}, ..., f^{(m)}; t_1, ..., t_m) = E\left(\phi_{t_1}(f^{(1)}) ... \phi_{t_m}(f^{(m)})\right)$$

$$\forall t_1, ..., t_m \in \tilde{\mathbf{R}}; \ f^{(1)}, ..., f^{(m)} \in \mathbf{S}. \tag{B.2.2}$$

In view of the reality property (B.2.1), these functions satisfy the condition

$$\overline{W_m(f^{(1)}, ..., f^{(m)}; t_1, ..., t_m)} = W_m(\overline{f}^{(1)}, ..., \overline{f}^{(m)}; t_1, ..., t_m)$$

$$\forall t_1, ..., t_m \in \tilde{\mathbf{R}}; \ f^{(1)}, ..., f^{(m)} \in \mathbf{S}. \qquad (\text{B.2.3})$$

Gaussian Fields

The field ϕ is termed Gaussian, with zero expectation value, if the W_m are all determined by the two-point function, W_2, according to the formulae

$$W_m(f^{(1)}, ..., f^{(m)}; t_1, ..., t_m) = 0 \quad \text{for } m \text{ odd} \qquad (\text{B.2.4a})$$

and

$$W_m(f^{(1)}, ..., f^{(m)}; t_1, ..., t_m) = \sum_P \prod_{(k,l) \in P} W_2(f^{(k)}, f^{(l)}; t_k, t_l) \quad \text{for } m \text{ even}$$

$$(\text{B.2.4b})$$

where P denotes the set of partitions of $(1, 2, ..., m)$ into pairs (j, k).

It is convenient to express the process in terms of its *generating function*, C, which for each set of different times $t_1, ..., t_m$, takes the form

$$C_{t_1, ..., t_m}(f^{(1)}, ..., f^{(m)}) = E\left(\exp(i \sum_{k=1}^{m} \phi_{t_j}(f^{(j)}))\right)$$

$$\equiv \sum_{n=0}^{\infty} \frac{1}{n!} E\left[\left(\sum_{k=1}^{m} \phi_{t_j}(f^{(j)})\right)^n\right] \quad \forall f^{(1)}, ..., f^{(m)} \in \mathbf{S}_R \qquad (\text{B.2.5})$$

where $\mathbf{S}_R$ is the real subspace of $\mathbf{S}$. It follows then from Eqs. (B.2.1)–(B.2.5) that the form of C is determined by that of W_2 according to the equation

$$C_{t_1, ..., t_m}(f^{(1)}, ..., f^{(m)}) = \exp\left[-\frac{1}{2} \sum_{k,l=1}^{m} W_2(f^{(k)}, f^{(l)}; t_k, t_l)\right]. \qquad (\text{B.2.6})$$

Equivalently, defining the *covariance function* $F_{t_1, ..., t_m}$ to be the bilinear form on $\mathbf{S}^m$ given by the formula

$$F_{t_1, ..., t_m}(f^{(1)}, ..., f^{(m)}; g^{(1)}, ..., g^{(m)}) = E\left(\sum_{k=1}^{m} \phi_{t_k}(f^{(k)}) \sum_{l=1}^{m} \phi_{t_l}(g^{(l)})\right)$$

$$(\text{B.2.7a})$$

$$= \sum_{k,l=1}^{m} W_2(f^{(k)}, g^{(l)}; t_k, t_l) \qquad (\text{B.2.7b})$$

we may express Eq. (B.2.6) in the form

$$C_{t_1, ..., t_m}(f^{(1)}, ..., f^{(m)}) = \exp\left[-\frac{1}{2} F_{t_1, ..., t_m}(f^{(1)}, ..., f^{(m)}; f^{(1)}, ..., f^{(m)})\right]. \qquad (\text{B.2.8})$$

We now note that, by Eq. (B.2.6), $C_{t_1, ..., t_m}(\lambda_1 f^{(1)}, ..., \lambda_m f^{(m)})$ is an infinitely differentiable function of the real variables $\lambda_1, \ldots, \lambda_m$. Hence, as the elements of the algebra $\mathcal{B}_t$ are the polynomials in $\{\phi_t(f) \mid f \in S\}$, it follows from Eq. (B.2.5) that the correlations functions, $E(B_{t_1}^{(1)} \cdots B_{t_m}^{(m)})$, of the process $\mathcal{P}$ may be generated by simple algebraic operations on the derivatives of $C_{t_1,\ldots,t_m}(\lambda_1 f^{(1)}, \ldots, \lambda_m f^{(m)})$ with respect to the λ's at the point $(\lambda_1, \ldots, \lambda_m) = 0$.

Thus, the stochastic properties of $\mathcal{P}$ are determined by the form of the generating function $C_{t_1,\ldots,t_m}$, which in turn is determined by the function $F_{t_1,\ldots,t_m}$ according to Eq. (B.2.8). We note that, by Eqs. (B.2.1)–(B.2.5) and (B.2.7a), the key functions F and C possess the following properties:

(I) For fixed $t_1, \ldots, t_m$, $F_{t_1,\ldots,t_m}$ is a positive semi-definite bilinear form on $\mathbf{S}_R^m$, in that the left-hand side of Eq. (B.2.7) is nonnegative if $f^{(k)} = g^{(k)}$ for $k = 1, \ldots, m$.

(II) $C_{t_1,\ldots,t_m}(f^{(1)}, \ldots, f^{(m)})$ is invariant under the permutations $(f^{(k)}, t_k) \rightleftharpoons (f^{(l)}, t_l)$.

(III) This function also satisfies the consistency condition that

$$C_{t_1,\ldots,t_{m-1},t_m}(f^{(1)}, \ldots, f^{(m-1)}, 0) \equiv C_{t_1,\ldots,t_{m-1}}(f^{(1)}, \ldots, f^{(m-1)}). \quad \text{(B.2.9)}$$

Reconstruction of $\mathcal{P}$ from the Generating Function

We may summarise the above results by saying that the covariance and generating functions, F and C, of a real classical, Gaussian, stochastic field ϕ are related by Eq. (B.2.8) and possess the properties (I)–(III). The converse, which is a particular example of Kolmogorov's [Ko2] reconstruction theorem, is that the conditions (I)–(III) and Eq. (B.2.8) uniquely determine a real, classical Gaussian, random field ϕ for which C is the generating function.

To prove this, we infer from Proposition A.4 of Appendix A, together with Eq. (B.2.8) and condition (I), that, for each set of different times $t_1, \ldots, t_m$, $C_{t_1,\ldots,t_m}$ is the characteristic function of a unique probability measure $\mu_{t_1,\ldots,t_m}$ on $\mathbf{S}^m$, that is,

$$C_{t_1,\ldots,t_m}(f^{(1)}, \ldots, f^{(m)})$$

$$= \int d\mu_{t_1,\ldots,t_m}(\phi_{t_1}, \ldots, \phi_{t_m}) \exp\left(i \sum_{k=1}^{m} \phi_{t_k}(f^{(k)})\right) \forall f^{(1)}, \ldots, f^{(m)} \in \mathbf{S}_R. \quad \text{(B.2.10)}$$

Thus, $\mu_{t_1,\ldots,t_m}$ is a probability measure for a set of time-dependent fields $\phi_{t_1}, \ldots, \phi_{t_m}$. Further, by Eq. (B.2.9) and conditions (II) and (III), μ is invariant under the permutations $(f^{(k)}, t_k) \rightleftharpoons (f^{(l)}, t_l)$ and satisfies the consistency condition that

$$\int d\mu_{t_1,\ldots,t_m}(\phi_{t_1},\ldots,\phi_{t_m})\exp\left(i\sum_{k=1}^{m-1}\phi_{t_k}(f^{(k)})\right)$$

$$=\int d\mu_{t_1,\ldots,t_{m-1}}(\phi_{t_1},\ldots,\phi_{t_m})\exp\left(i\sum_{k=1}^{m-1}\phi_{t_k}(f^{(k)})\right). \qquad \text{(B.2.11)}$$

Consequently, the set of probability measures $\{\mu_{t_1,\ldots,t_m}\}$ corresponds to an expectation functional E on the polynomials in $\{\exp(i\phi_t(f))\}$, and Eq. (B.2.10) reduces to the form

$$E\left[\exp\left(i\sum_{k=1}^{m}\phi_{t_k}(f^{(k)})\right)\right]=C_{t_1,\ldots,t_m}(f^{(1)},\ldots,f^{(m)}). \qquad \text{(B.2.12)}$$

This formula, together with Eqs. (B.2.6), (B.2.7b) and (B.2.8), signifies that the field ϕ has been reconstructed from the functions W_2 and C, subject to the conditions (I)–(III).

Sufficient Conditions for the Markov Property

Suppose that $\{V_{t,t'} \mid t,t'(\leq t) \in \tilde{\mathbf{R}}\}$ is a two-parameter family of linear transformations of $\mathbf{S}$, that satisfy the conditions analogous to those of Eqs. (7.3.15) and (7.3.16), namely

$$V_{t,t''} = V_{t',t''}V_{t,t'}, \quad V_{t,t} = I \qquad \text{(B.2.13)}$$

and

$$\overline{V_{t,t'}f} = V_{t,t'}\bar{f}. \qquad \text{(B.2.14)}$$

Suppose also that the $\mathcal{B}_t^{\geq}$ and $\mathcal{B}_t^{\leq}$ are defined in terms of the algebras, $\mathcal{B}_t$, of polynomials in the smeared fields $\phi_t(f)$ according to the prescription of Section B.1, so that (cf. Eq. (7.5.8)),

$$E(\phi_u(f)\phi_t(g)) = E(\phi_t(V_{u,t}f)\phi_t(g)) \quad \forall u \geq t. \qquad \text{(B.2.15)}$$

Proposition B.2.1 *Under the above assumptions, the process $\mathcal{P}$ is Markovian.*

Proof. In view of our general specifications of Markov processes in Section B.1 and of the explicit form of $\mathcal{B}$ and its subalgebras in Section B.2, it suffices for us to *construct* a family of mappings $\{P_t : \mathcal{B}_t^{\geq} \to \mathcal{B}_t \mid t \in \mathbf{R}_+\}$ that satisfies the conditions (B.1.5)–(B1.7) when

$$B = \exp\left(\sum_{k=1}^{m}\phi_{u_k}(f^{(k)})\right) \quad \text{with } f_k \in \mathbf{S}, \; u_k \geq t, \; k = 1,\ldots,m \qquad \text{(B.2.16)}$$

and

$$B' = \exp\left(\sum_{j=1}^{l}\phi_{s_j}(e^{(j)})\right) \quad \text{with } e_j \in \mathbf{S}, \; s_j \leq t, \; j = 1,\ldots,m. \qquad \text{(B.2.17)}$$

To this end, we define P_t by the formula

$$P_t \exp\left(i\sum_{k=1}^{m}\phi_{u_k}(f^{(k)})\right) = \frac{C_{u_1,\ldots,u_m}(f^{(1)},\ldots,f^{(m)})}{C_t\left(\sum_{k=1}^{m}V_{u_k,t}f^{(k)}\right)} \exp\left(i\sum_{k=1}^{m}\phi_t(V_{u_k,t}f^{(k)})\right)$$

$$\forall f^{(1)},\ldots,f^{(m)} \in S, \; u_1,\ldots,u_m \geq t,$$

$$(B.2.18)$$

and we then check that it satisfies the conditions (B.1.5)–(B.1.7) when B and B' are given by Eqs. (B.2.16) and (B.2.17).

Now since $V(t,t) = I$ (by Eq. (B.2.13)), it follows from Eqs. (B.2.16), with all the u equal to t, and Eq. (B.2.18) that the condition (B.1.5) is fulfilled. Further, Eq. (B.1.6) is a simple consequence of Eqs. (B.2.1) and (B.2.14). Thus, it remains for us to verify that condition (B.1.7) is satisfied, that is, by Eqs. (B.2.16) and (B.2.17), that

$$E\left[\exp i\left(\sum_{j=1}^{l}\phi_{s_j}(e^{(j)}) + \sum_{k=1}^{m}\phi_{u_k}(f^{(k)})\right)\right]$$

$$= E\left[\exp\left(i\sum_{j=1}^{l}\phi_{s_j}(e^{(j)})\right)P_t \exp\left(i\sum_{k=1}^{m}\phi_{u_k}(f^{(k)})\right)\right]. \quad (B.2.19)$$

Now, by Eqs. (B.2.5), (B.2.6) and (B.2.18), the left- and right-hand sides of this equation are equal to

$$C_{s_1,\ldots,s_l}(e^{(1)},\ldots,e^{(l)})C_{u_1,\ldots,u_n}(f^{(1)},\ldots,f^{(m)})$$

$$\times \exp-\left[\sum_{j=1}^{l}\sum_{k=1}^{m}E\left(\phi_{u_k}(f^{(k)})\phi_{s_j}(e^{(j)})\right)\right]$$

and

$$C_{s_1,\ldots,s_l}(e^{(1)},\ldots,e^{(l)})C_{u_1,\ldots,u_n}(f^{(1)},\ldots,f^{(m)})$$

$$\times \exp-\left[\sum_{j=1}^{l}\sum_{k=1}^{m}E\left(\phi_t(V_{u_k,t}f^{(k)})\phi_{s_j}(e^{(j)})\right)\right]$$

respectively. Therefore since, by Eqs. (B.2.13) and (B.2.15),

$$E\left(\phi_{u_k}(f^{(k)})\phi_{s_j}(e^{(j)})\right) = E\left(\phi_{s_j}(V_{u_k,s_j}f^{(k)})\phi_{s_j}(e^{(j)})\right)$$

$$= E\left(\phi_{s_j}(V_{t,s_j}V_{u_k,t}f^{(k)})\phi_{s_j}(e^{(j)})\right)$$

$$= E\left(\phi_t(V_{t,s_j}f^{(k)})\phi_{s_j}(e^{(j)})\right),$$

the two sides of Eq. (B.2.19) are indeed equal. This completes the proof of the proposition. $\square$

B.3 Proof of Propositions 7.5.1 and 7.5.2

In order to relate the theory of the previous two sections to that of Section 7.5, we assume that $\tilde{\mathbf{R}}$ is $\mathbf{R}_+$, that the Schwartz space $\mathbf{S}$ is $\mathbf{S}^{(n+1)}(X)$, that the correlation functions W_m are just those of Section 7.5, and that the conditions (W.1)–(W.3) of that section are fulfilled.

Proof of Proposition 7.5.1. Under the above assumptions, it follows from (W.2) that the functions W_m possess the Gaussian property (B.2.4). Further, defining the functions F and C in terms of W_2 by Eqs. (B.2.7b) and (B.2.8), we infer from (W.1)–(W.3) that the above conditions (II) and (III) of Section B.2 are fulfilled, while, by Eqs. (7.5.2), (7.5.4) and (B.2.7b),

$$F_{t_1,\ldots,t_m}(f^{(1)},\ldots,f^{(m)};g^{(1)},\ldots,g^{(m)}) = \lim_{L\to\infty}\Big\langle \rho^{(L)}; \sum_{k=1}^{m}\tilde{q}_{t_k}(f^{(k)})\sum_{l=1}^{m}\tilde{q}_{t_l}(g^{(l)})\Big\rangle$$

(B.3.1)

from which it follows that F satisfies the positivity condition (I).

Thus, the functions $\{W_m\}$ possess the Gaussian property (B.2.4) and satisfy the conditions (I)–(III), and therefore are the correlation functions of a classical, Gaussian stochastic field, $\tilde{q}$.   □

Proof of Proposition 7.5.2. This proposition is just the particular case of Proposition B.2.1 where ϕ, $\mathbf{S}$, $V_{t,s}$ and $\tilde{\mathbf{R}}$ are identified with $\tilde{q}$, $\mathbf{S}^{(n+1)}(X)$, $T_{t,s}^{\star}$ and $\mathbf{R}_+$, respectively.

APPENDIX C: EQUILIBRIUM CORRELATIONS AND THE STATIC TWO-POINT FUNCTION

This appendix is devoted to a derivation of the formulae (7.6.2) and (7.6.4) for the static two-point function for the macroscopic fluctuation field $\tilde{q}$ in a state ρ_θ. It is assumed that this state corresponds to a pure phase, that is, that θ lies in the interior of a single phase region represented by an open domain Δ of the thermodynamical control space Θ. We again assume, as in Section 6.4, that the pure equilibrium phases are invariant under either the full translation group X or some crystallographic subgroup thereof. For simplicity, we base the main part of our treatment, carried out in Sections C.1–C.4, on the assumption that the equilibrium states of the single phase region Δ are fully translationally invariant, and then extend our results to the case of crystallographic symmetry in Section C.5. The key formulae (7.6.2) and (7.6.4) are established in Propositions C.1.1 and C.3.5.

We start in Section C.1 by proving Eq. (7.6.2) for *some* constant matrix $B = [B_{kl}]$, not neccessarily that given by Eq. (7.6.4), subject to an assumption on the spatial decay of the two-point correlation function of the field $\hat{q}$. In order to relate B to the Hessian $s''(q)$, we first show, in Section C.2, that $\pi''(\theta)$ is equal to $-s''(q)^{-1}$ and is given by the derivative with respect to θ of $\hat{q}(\rho_\theta)$. We then proceed in Section C.3 to express this derivative in terms of a linked cluster expansion for the truncated correlation functions of the field $\hat{q}$ for the state ρ_θ, and deduce therefrom the formula (7.6.4), subject to two further assumptions on the clustering properties of that state. Section C.4 is devoted to the detailed proof, for lattice systems with finite range interactions, of the linked cluster expansion formulae of Section C.3; our employment of corresponding formulae for continuous systems represents supplementary assumptions for the algebra of the unbounded smeared field $\hat{q}(f)$. Finally, in Section C.5, we extend the results obtained here to the case where ρ_θ has a crystallographic, rather than full translational, symmetry.

C.1 The Truncated Static Two-Point Function

At the microscopic level, the truncated static two-point function representing the correlations of fluctuations of $\hat{q}$ in the state ρ_θ is the function, C_2, on $\mathbf{S}^{(n+1)}(X)^2$ defined by the formula

$$C_2(f,g) := \langle \rho_\theta; \hat{q}(f)\hat{q}(g) \rangle - \langle \rho_\theta; \hat{q}(f) \rangle\langle \rho_\theta; \hat{q}(g) \rangle \quad \forall f, g \in \mathbf{S}^{(n+1)}(X). \quad \text{(C.1.1)}$$

Hence, by Eqs. (7.4.1), (7.5.1), (7.5.2), (7.5.4) and (7.5.7), C_2 is related to the static two-point function for the macroscopic fluctuation field $\tilde{q}$ by the equation

$$E(\tilde{q}(f)\tilde{q}(g)) = \lim_{L \to \infty} L^d C_2(f^{(L)}, g^{(L)}) \quad \forall f, g \in \mathbf{S}^{(n+1)}(X). \quad \text{(C.1.2)}$$

By Eq. (C.1.1) and assumption (q̂.5) of Section 7.2, C_2 is a tempered distribution, and thus corresponds to a generalised function C_2 on X^2 according to the formula

$$C_2(f,g) = \int_{X^2} dx\, dy\, C_2(x,y) f(x) g(y). \quad \text{(C.1.3)}$$

We assume that ρ_θ possesses the following cluster property, which signifies essentially that $C_2(x,y)$ tends to zero, faster than $|x-y|^{-d}$, as $|x-y|$ tends to infinity.

(Cluster 1) Given $b > 0$, there are positive constants c and η, such that, for any pair of elements, f and g, of $\mathbf{S}^{(n+1)}(X)$ whose supports are separated by a distance greater than b,

$$|C_2(f,g)| < c \int_{X^2} dx\, dy\, |f(x)|\, |g(y)|\, |x-y|^{-d-\eta}, \quad \text{(C.1.4)}$$

where, for $f = (f_0, ..., f_n)$

$$| f(x) | = \sum_{r=0}^{n} | f_r(x) | .$$ (C.1.5)

The following proposition serves to validate Eq. (7.6.2) for *some* constant $(n + 1) - by - (n + 1)$ matrix B.

Proposition C.1.1 *Under the above assumption (Cluster 1), together with those of Section 7.5, there is a constant $(n + 1)$-by-$(n + 1)$ matrix $B = [B_{kl}]$, such that*

$$E(\tilde{q}(f)\tilde{q}(g)) = (f, Bg) \equiv \sum_{k,l=0}^{n} B_{kl} \int_X dx f_k(x) g_l(x) \quad \forall f, g \in S^{(n+1)}(X).$$

(C.1.6)

Proof. Let

$$T(f, g) = E(\tilde{q}(f)\tilde{q}(g)) \quad \forall f, g \in S^{(n+1)}(X).$$ (C.1.7)

Then it follows from assumption (W.1) of Section 7.5 and Proposition 7.5.1 that T is a tempered distribution, corresponding to an $(n + 1)$-by-$(n + 1)$ matrix, whose elements T_{kl} are of class $S'(X^2)$. Further, by translational invariance, T may be identified with an $(n + 1)$-by-$(n + 1)$ matrix-valued, $S'(X)$-class tempered distribution $\tilde{T}$, according to the formal prescription

$$\tilde{T}(x - y) = T(x, y).$$ (C.1.8)

Suppose now that v and w are elements of $S^{(n+1)}(X)$ with mutually disjoint supports. Then it follows from Eqs. (7.4.2), (C.1.4) and (C.1.5) that

$$L^d | C_2(v^{(L)}, w^{(L)}) | < CL^{-d} \int_{X^2} dx dy | v(x/L) || w(y/L) || x - y |^{-d-\eta}$$

$$\equiv cL^{-\eta} \int_{X^2} dx dy | v(x) || w(y) || x - y |^{-d-\eta} .$$

Since the disjointness of the supports of v and w ensures that the last integral in this equation is finite, it follows from this inequality, together with Eqs. (C.1.2) and (C.1.7), that $T(v, w) = 0$.

Thus, $T(v, w)$ vanishes if v and w have disjoint supports, and therefore, by Eq. (C.1.8), $\tilde{T}$ has support at the origin. This implies that $\tilde{T}$ is a linear combination of the Dirac δ distribution, with support at the origin, and a finite number of its derivatives, that is,

$$\tilde{T}(x) = B\delta(x) + \sum_{k_1, ..., k_d} D(k_1, ..., k_d) \frac{\partial^{k_1 + ... + k_d}}{\partial x_1^{k_1} ... \partial x_d^{k_d}} \delta(x),$$ (C.1.9)

where B and the D's are constant $(n + 1)$-by-$(n + 1)$ matrices. Moreover, by

Eqs. (7.4.1)', (7.4.2), (C.1.2), (C.1.7) and (C.1.8), $\tilde{T}$ satisfies the scaling relation

$$\tilde{T}(\lambda x) = \lambda^{-d}\tilde{T}(x) \quad \forall \lambda \in \mathbf{R}_+ \backslash 0. \tag{C.1.10}$$

Therefore, since δ scales as λ^{-d} under the transformation $x \to \lambda x$, whereas the derivatives of δ scale as lower powers of λ, the only form of Eq. (C.1.9) that satisfies Eq. (C.1.10) is

$$\tilde{T} = B\delta.$$

The required result (C.1.6) follows immediately from this equation and Eqs. (C.1.7) and (C.1.8).

C.2 Quantum Statistical Formulation of $s''(q)$

By Eqs. (6.4.14) and (6.4.15),

$$s'(q) = \theta \tag{C.2.1}$$

and

$$\pi'(\theta) = -q, \tag{C.2.2}$$

and consequently the Hessians $s''(q)$ and $\pi''(\theta)$ are related by the formula

$$\pi''(\theta)s''(q) = s''(q)\pi''(\theta) = -I,$$

which signifies that

$$\pi''(\theta) = -s''(q)^{-1}. \tag{C.2.3}$$

Furthermore, since q is the expectation value of the functional $\hat{q}$ for the equilibrium state ρ_θ, it follows from Eq. (C.2.2) that

$$\frac{\partial \pi(\theta)}{\partial \theta_k} = -\hat{q}_k(\rho_\theta),$$

that is, in view of the translational invariance of ρ_θ,

$$\frac{\partial \pi(\theta)}{\partial \theta_k} = -\langle \rho_\theta; \hat{q}_k(u) \rangle, \tag{C.2.4}$$

where u is any element of $S(X)$ that satisfies the condition

$$\int_X dx u(x) = 1. \tag{C.2.5}$$

For our later convenience, we choose u to have compact support. We note here that Eq. (C.2.4) may be equivalently expressed in the form

$$\frac{\partial \pi(\theta)}{\partial \theta_k} = -\langle \rho_\theta; \hat{q}(u^{(k)}) \rangle, \tag{C.2.4$'$}$$

where $u^{(k)}$ is the element of $\mathbf{S}^{m+1}(X)$ whose kth component is u and whose other components are all zero, that is,

$$(u^{(k)})_l = u \delta_{kl}. \tag{C.2.6}$$

It follows immediately from Eq. (C.2.4)$'$ that

$$\frac{\partial^2 \pi(\theta)}{\partial \theta_k \partial \theta_l} = -\frac{\partial}{\partial \theta_l} \langle \rho_\theta; \hat{q}(u^{(k)}) \rangle. \tag{C.2.7}$$

Our aim now is to prove that the right-hand side of this equation is equal to B_{kl} and thus, using Eqs. (7.6.2) and (C.1.6), to establish the validity of Eq. (7.6.4).

C.3 Formulation of π'' via Perturbations of ρ_θ

In the single phase region, ρ_θ is the unique translationally invariant state characterised by the KMS condition (6.4.10), namely,

$$\langle \rho_\theta; [\alpha_\theta(t)A]B \rangle = \langle \rho_\theta; B\alpha_\theta(t + i)A \rangle, \tag{C.3.1}$$

and the generator of $\alpha_\theta(\mathbf{R})$ is of the form[13]

$$\delta_\theta = \sum_{k=0}^{n} \theta_k \delta_k, \tag{C.3.2}$$

where, formally, $\delta_k = i \int_X [\hat{q}_k(x), \cdot]$. Thus, defining

$$\hat{q}_\theta(f, t) = \alpha_\theta(t)[\hat{q}(f)], \tag{C.3.3}$$

it follows from Eqs. (7.2.3) and (7.2.4) that $\hat{q}$ satisfies a local conservation law

$$\frac{\partial}{\partial t} \hat{q}_\theta(f, t) = \hat{j}_\theta(\nabla f, t), \tag{C.3.4}$$

where the current $\hat{j}_\theta$ is of the form

$$\hat{j}_\theta = \sum_{k=0}^{n} \hat{j}_\theta^{(k)}, \tag{C.3.5}$$

and the kth term of this sum stems from δ_k.

Correlation Functions for the Smeared Field

We express below the perturbations of the state ρ_θ in terms of analytic continuations of its correlation functions for the time-dependent smeared field $\hat{q}(f, t)$. The m-point function, G_m, for this field is defined by the formula

---

[13] Cf. assumption ($\hat{Q}$.6) of Section 6.4.

$$G_m(f^{(1)}, ..., f^{(m)}; t_1, ..., t_m)$$

$$= \left\langle \rho_\theta; \hat{q}_\theta(f^{(1)}, t_1)...\hat{q}_\theta(f^{(m)}, t_m) \right\rangle \quad \forall f^{(1)}, ..., f^{(m)} \in \mathbf{S}^{(n+1)}(X), \ t_1, ..., t_m \in \mathbf{R}.$$

$$(C.3.6)$$

We assume that G_m is continuous in all its arguments.

The truncated m-point functions, G_m^T, which represent the correlations of fluctuations of the field $\hat{q}$, are constructed by the following standard prescription (cf. [Haa1]).

We define S to be the set of finite increasing sequences, s, of positive integers, and P_s to be the set of all partitions of the sequence s ($\in S$) into its subsequences, denoting the set of those obtained from a partition π by $\{s_\pi\}$. For $s = (s_1, ..., s_r)$ and $f^{(s_1)}, ..., f^{(s_r)} \in \mathbf{S}^{(n+1)}(X)$, we define $|s|$ to be r and F_s to be $(f^{(s_1)}, ..., f^{(s_r)})$. We then define the truncated m-point functions $G_{\theta,m}^T$ recursively in terms of the G_m's by the formula

$$G_{|s|}(F_s) = \sum_{\pi \in P_s} \Pi_{s_\pi} G_{|s_\pi|}^T(F_{s_\pi}). \quad (C.3.7)$$

Thus, in view of the α_θ-invariance of ρ_θ, it follows from this prescription that

$$G_1^T(f; t) = \langle \rho_\theta; \hat{q}(f) \rangle, \quad (C.3.8)$$

$$G_2^T(f, g; s, t) = \langle \rho_\theta; \hat{q}_\theta(f, s)\hat{q}_\theta(g, t) \rangle - \langle \rho_\theta; \hat{q}(f) \rangle \langle \rho_\theta; \hat{q}(f) \rangle, \quad (C.3.9)$$

etc.

Perturbations of the state ρ_θ

Suppose now that ψ is an element of $\mathbf{S}_R^{(n+1)}(X)$ with compact support, and that $\hat{q}(\psi)$ is the perturbation of the local effective Hamiltonians $\theta.Q_\Lambda$, for supp(ψ) $\subset \Lambda$. Then α_θ will be perturbed to the automorphism group, α_θ^ψ, whose generator is

$$\delta_\theta^\psi = \delta_\theta + i[\hat{q}(\psi), \cdot]. \quad (C.3.10)$$

Correspondingly, the state ρ_θ will be perturbed to a state, ρ_θ^ψ, characterised by the KMS property

$$\langle \rho_\theta^\psi; [\alpha_\theta^\psi(t)A]B \rangle = \langle \rho_\theta^\psi; B\alpha_\theta^\psi(t + i)A \rangle. \quad (C.3.11)$$

We assume the general validity of the following two propositions, even though our proofs of them, in Section C.4, are limited to lattice systems with finite range interactions. Corresponding proofs for continuous systems would involve a suitable topology for the algebra of unbounded observables comprising the polynomials of their smeared fields $\hat{q}(f)$.

Proposition C.3.1 *The expectation value of $\hat{q}(f)$ for the perturbed state ρ_θ is given by the linked cluster expansion*[14]

$$\langle \rho_\theta^\psi ; \hat{q}(f) \rangle = \sum_{m=0}^{\infty} \frac{(-1)^m}{m!} \overline{\mathcal{T}} \int_0^1 dt_1 \cdots \int_0^1 dt_m G_{m+1}^T(f, \psi, ..., \psi; 0, it_1, ..., it_m),$$

(C.3.12)

where $\overline{\mathcal{T}}$ is the antichronological operator that serves to arrange $t_1, ..., t_m$ in ascending order.

Proposition C.3.2 *Let h be an $\mathbf{S}_R(X)$-class function with compact support, that takes the value unity in the unit ball, $|x| \leq 1$, and let*

$$h_R(x) = h(R^{-1}x),$$ (C.3.13)

where R is a real-valued, positive parameter. Then, assuming that ϕ is a perturbation of the control variable θ that does not lead out of the single phase region Δ,

$$\lim_{R \to \infty} \langle \rho_\theta^{\phi h_R} ; \hat{q}(f) \rangle = \langle \rho_{\theta + \phi} ; \hat{q}(f) \rangle \quad \forall f \in \mathbf{S}^{(n+1)}(X).$$ (C.3.14)

Comments

(1) This latter proposition signifies that the change in the expectation value of $\hat{q}(f)$ due to the *global* change of state, corresponding to the increment ϕ in θ, is the same as that due to the local Hamiltonian perturbation $\hat{q}(\phi h_R)$, in the limit $R \to \infty$.

(2) In view of Eq. (C.2.4)', the following corollary is an immediate consequence of Propositions C.3.1 and C.3.2.

Corollary C.3.3 *Assuming again that the perturbation ϕ of θ does not lead out of the single phase region Δ,*

$$\frac{\partial \pi(\theta + \phi)}{\partial \theta_k}$$

$$= \lim_{R \to \infty} \langle \rho_\theta^{\phi h_R} ; \hat{q}(u^k) \rangle$$

$$= \lim_{R \to \infty} \sum_{m=0}^{\infty} \frac{(-1)^{m+1}}{m!} \overline{\mathcal{T}} \int_0^1 dt_1 \cdots \int_0^1 dt_m G_{m+1}^T(u^{(k)}, \phi h_R, ..., \phi h_R; 0, it_1, ..., it_m)$$

$$\forall f^{(1)}, ..., f^{(m)} \in \mathbf{S}^{n+1}(X).$$ (C.3.15)

---

[14] The expansion is so called because it is expressed in terms of truncated correlation functions, which may be represented by connected Feynman diagrams, and is commonly employed in both quantum field theory and statistical mechanics (cf. [Haa1]).

The next proposition depends on the further standard assumption of analyticity properties of both the reduced pressure and the locally perturbed state in the single phase region.

Proposition C.3.4 *Assuming further that, in the same situation, both* $\pi(\theta + \phi)$ *and* $\rho_\theta^{\phi h_R}(\hat{q}(f))$ *are real analytic functions of* ϕ,

$$\frac{\partial^2 \pi(\theta)}{\partial \theta_k \partial \theta_l} = \lim_{R \to \infty} \int_0^1 dt [\langle \rho_\theta; \hat{q}_k(u)\hat{q}_l(h_R, it)\rangle - \langle \rho_\theta; \hat{q}_k(u)\rangle\langle\rho_\theta; \hat{q}_l(h_R)\rangle]. \quad \text{(C.3.16)}$$

Proof. In view of the analyticity assumptions, the application of Vitali's theorem to Eq. (C.3.15) yields the formula

$$\frac{\partial^2 \pi(\theta)}{\partial \theta_k \partial \theta_l} = \lim_{R \to \infty} \frac{\partial}{\partial \phi_l} \langle \rho_\theta^{\phi h_R}; \hat{q}(u^k)\rangle_{\phi=0} = \lim_{R \to \infty} \int_0^1 dt G_2^T(u^k, h_R^l; 0, it), \quad \text{(C.3.17)}$$

where h_R^l is the element of $\mathbf{S}^{(n+1)}(X)$ whose lth component is h_R and whose others are all zero.

Furthermore, it follows from Eqs. (7.2.1), (C.2.4)', (C.2.6) and (C.3.9), together with the invariance of ρ_θ under the dynamical group α_θ^*, that the right-hand side of this last equation is precisely that of Eq. (C.3.16). Thus, Eq. (C.3.17) reduces to the required result. $\square$

Derivation of Eq. (7.6.4)

In order to pass from Proposition C.3.4 to Eq. (7.6.4), we need to show that the right-hand side of Eq. (C.3.16) is unchanged if the variable it in the integrand is replaced by 0. To this end, we invoke the local conservation law given by Eq. (C.3.4) and infer from it that

$$\int_0^1 dt \langle \rho_\theta; \hat{q}_k(u)\hat{q}_l(h_R, it)\rangle = \langle \rho_\theta; \hat{q}_k(u)\hat{q}_l(h_R)\rangle + iJ_{kl}(u, \nabla h_R), \quad \text{(C.3.18)}$$

where

$$J_{kl}(v, w) := \int_0^1 dt \int_0^t ds \langle \rho_\theta; \hat{q}_k(v)\hat{j}_{\theta,l}(w, is)\rangle \quad \forall v \in \mathbf{S}(X), w \in \mathbf{S}_V(X). \text{(C.3.19)}$$

We assume that J_{kl} is continuous in its two arguments and that it enjoys the following clustering property, which parallels that of the two-point function C_2 given by (Cluster 1) in Section C.1.

(Cluster 2) Given $c > 0$, there are positive constants D and γ, such that, if v and w are arbitrary $\mathbf{S}(X)$ and $\mathbf{S}_V(X)$ class test functions, whose supports are separated by the distance c, then

$$| J(v, w) | < D \int_{X^2} dx \, dy \, | v(x) || w(y) || x - y |^{-d-\gamma}. \quad \text{(C.3.20)}$$

Furthermore, we strengthen assumption (Cluster 1) with one concerning the components $C_{2,kl}$ of the matrix-valued tempered distribution C_2. These components are defined by the requirement that, if f and g are elements of $S(X)$ whose only nonzero components are the kth and lth, respectively, then $C_2(f, g) = C_{2,kl}(f_k, g_l)$; and, in view of the translational invariance of ρ_θ, they are canonically related to $S'(X)$-class distributions $\tilde{C}_{2,kl}$ by the generalised functional formula

$$C_{2,kl}(x, y) = \tilde{C}_{2,kl}(x - y). \tag{C.3.21}$$

Our strengthened version of (Cluster 1) is the following.

(Cluster 3) The distributions $\tilde{C}_{2,kl}$ are of zero order, that is, they correspond to continuous linear functionals on the abelian $C^\star$-algebra $C_0(X)$.

The following Proposition establishes Eq. (7.6.4).

Proposition C.3.5   *Under the further assumption of (Cluster 2) and (Cluster 3), the matrix B of Proposition C.1.1 is given by Eq. (7.6.4), that is,*

$$B = \pi''(\theta) = -[s''(q)]^{-1}. \tag{C.3.22}$$

Proof. It follows from the specifications of h_R in the statement of Proposition C.3.2 that the support of ∇h_R lies at least a distance R from the origin and that $|\nabla h_R(x)|$ is majorised by $R^{-1}\|\nabla h\|$, where $\|\cdot\|$ is the supremum norm. Therefore, as we stipulated, immediately after Eq. (C.2.5), that u has compact support, it follows from (Cluster 2) that $J_{kl}(u, \nabla h_R)$ tends to zero as $R \rightarrow \infty$. Consequently, by Eq. (C.3.18), Eq. (C.3.16) remains valid if the variable it in its integrand of is replaced by 0, that is,

$$\frac{\partial^2 \pi(\theta)}{\partial \theta_k \partial \theta_l} = \lim_{R \rightarrow \infty} [\langle \rho_\theta; \hat{q}_k(u)\hat{q}_l(h_R)\rangle - \langle \rho_\theta; \hat{q}_k(u)\rangle\langle\rho_\theta; \hat{q}_l(h_R)\rangle].$$

Hence, in view of the definition of $C_{2,kl}$, following Eq. (C.3.20),

$$\frac{\partial^2 \pi(\theta)}{\partial \theta_k \partial \theta_l} = \lim_{R \rightarrow \infty} C_{2;kl}(u, h_R). \tag{C.3.23}$$

To evaluate the right-hand side of this equation, we note that, by Eq. (C.3.21),

$$C_{2,kl}(u, h_R) = \int_{X^2} dx dy \tilde{C}_{2,kl}(x - y)u(x)h(R^{-1}y)$$

$$\equiv \int_{X^2} dx dy \tilde{C}_{2,kl}(y)u(x)h(R^{-1}y + R^{-1}x),$$

that is,

$$C_{2,kl}(u, h_R) = \int_{X^2} dxdy \tilde{C}_{2,kl}(y)u(x)h(R^{-1}y) + \int_X dy \tilde{C}_{2,kl}(y)g_R(y), \quad \text{(C.3.24)}$$

where

$$g_R(y) = \int_X dxu(x)\Big[h(R^{-1}y + R^{-1}x) - h(R^{-1}y)\Big]. \quad \text{(C.3.25)}$$

In order to estimate the second integral in Eq. (C.3.24), we note that, by Eq. (C.3.25),

$$| g_R(y) | \leq R^{-1}\|\nabla h\| \int_X dx|xu(x)|,$$

where again $\|\cdot\|$ is the supremum norm. Consequently, as both h and u are $S(X)$-class functions, the coefficient of R^{-1} in the last inequality is a finite R-independent constant, and therefore, by assumption (Cluster 3), the second integral of Eq. (C.3.24) tends to zero as $R \to \infty$. Further, by Eq. (C.2.5), the first integral in Eq. (C.3.24) is $\int_X dy \tilde{C}_{2,kl}(y)h(R^{-1}y)$, and consequently Eq. (C.3.23) reduces to the form

$$\frac{\partial^2 \pi(\theta)}{\partial\theta_k \partial\theta_l} = \lim_{R\to\infty} \int_X dy \tilde{C}_{2,kl}(y)h(R^{-1}y),$$

that is, defining $\tilde{C}_{2,kl}^{(R)}$ to be the tempered distribution given by the formula

$$\tilde{C}_{2,kl}^{(R)}(y) = R^d \tilde{C}_{2,kl}(Ry), \quad \text{(C.3.26)}$$

$$\frac{\partial^2 \pi(\theta)}{\partial\theta_k \partial\theta_l} = \lim_{R\to\infty} \tilde{C}_{2,kl}^{(R)}(h). \quad \text{(C.3.27)}$$

We now note that, by Eqs. (C.1.1) and (C.1.2), Proposition C.1.1 implies that $C_{2,kl}^{(R)}(x, y)$ converges to $B_{kl}\delta(x - y)$, that is, that $\tilde{C}_{2,kl}(y)$ converges to $B_{kl}\delta(y)$, as $R \to \infty$. Hence, by Eq. (C.3.27) and our stipulation that h takes the value unity in the unit ball,

$$\frac{\partial^2 \pi(\theta)}{\partial\theta_k \partial\theta_l} = B_{kl}.$$

This equation, coupled with Eq. (C.2.3), constitutes the required result. $\quad\square$

C.4 PROOF OF PROPOSITIONS C.3.1 AND C.3.2 FOR LATTICE SYSTEMS WITH FINITE RANGE INTERACTIONS

Analyticity Properties

Araki [Ar4,5] has proved that the KMS condition (C.3.1) implies that

(a) the multitime correlation function $\langle \rho_\theta; A_{1,t_1} \cdots A_{m,t_m} \rangle$, with $A_{j,t} \equiv \alpha_\theta(t)A_j$, has an analytic continuation to the complex domain in which the $\text{Im}(t_j)$ increases with j and $\text{Im}(t_m - t_1) \leq 1$;

(b) the vector $\pi_\theta(A_{1,t_1} \cdots A_{k,t_k})\Phi_\theta$ in the GNS representation, $(\mathcal{H}_\theta, \pi_\theta, \Phi_\theta)$, of ρ_θ has an analytic continuation to the domain $\{t_1, ..., t_k \in \mathbf{C} \mid \text{Im}(t_{j+1} - t_j) \geq 0; \text{Im}(t_k - t_1) \leq 1/2\}$, and

(c)
$$\|\pi_\theta(BA_{1,it_1}...A_{k,it_k})\Phi_\theta\| \leq \|B\|\Pi_{j=1}^{k}\|A_j\| \qquad \text{(C.4.1)}$$

$$\forall B, A_1, ..., A_k \in \mathcal{A}, \ 0 \leq t_1 \leq t_2 \leq \cdots \leq t_k \leq 1/2.$$

It follows from these analyticity properties and the KMS condition (C.3.1) that

$$\langle \rho_\theta; BA_{1,it_1} \cdots A_{m,it_m} \rangle = \left(\pi_\theta(A_{m,i(1-t_m)} \cdots A_{k+1,i(1-t_{k+1})})\Phi_\theta, \ \pi_\theta(BA_{1,it_1} \cdots A_{k,it_k})\Phi_\theta \right)$$

$$\forall B, A_1, ..., A_m \in \mathcal{A}, \ t_1, ..., t_m \in \mathbf{R}_+,$$

$$0 \leq t_1 \leq \cdots \leq t_k \leq 1/2 < t_{k+1} \leq t_{k+2} \leq \cdots \leq t_m \leq 1. \qquad \text{(C.4.2)}$$

Proof of Proposition C.3.1. Araki [Ar5] has proved that, in the case of lattice systems with finite range interactions,

$$\langle \rho_\theta^\psi; A \rangle = \frac{\left(\Phi_\theta^\psi, \pi_\theta(A)\Phi_\theta^\psi \right)}{(\Phi_\theta^\psi, \Phi_\theta^\psi)} \qquad \forall A \in \mathcal{A}, \qquad \text{(C.4.3)}$$

where Φ_θ^ψ is a cyclic vector given by the formula

$$\Phi_\theta^\psi = \sum_{m=0}^{\infty} \frac{(-1)^m}{m!} \overline{\mathcal{T}} \int_0^{1/2} dt_1 \cdots \int_0^{1/2} dt_m \, \pi_\theta(\hat{q}(\psi, it_1) \cdots \hat{q}(\psi, it_m))\Phi_\theta, \quad \text{(C.4.4)}$$

the absolute convergence of this sum being guaranteed by the estimate (C.4.1). By Eqs. (C.3.6), (C.4.1) and (C.4.4),

$$\left(\Phi_\theta^\psi, \pi_\theta(\hat{q}(f))\Phi_\theta^\psi \right) = \sum_{k=0}^{\infty}\sum_{l=0}^{\infty} \frac{(-1)^{k+l}}{k!l!}$$

$$\times \overline{\mathcal{T}} \int_0^{1/2} dt_1 \cdots \int_0^{1/2} dt_k \int_{1/2}^{1} dt_{k+1} \cdots \int_{1/2}^{1} dt_{k+l}$$

$$\times G_{k+l+1}(f, \psi, ..., \psi, 0, it_1, ..., it_{k+l})$$

$$= \sum_{m=0}^{\infty} \frac{(-1)^m}{m!} \sum_{k=0}^{m} {}^mC_k \overline{\mathcal{T}} \int_0^{1/2} dt_1 \cdots \int_0^{1/2} dt_k \int_{1/2}^{1} dt_{k+1} \cdots \int_{1/2}^{1} dt_m$$

$$\times G_{m+1}(f, \psi, ..., \psi, 0, it_1, ..., it_m) \qquad \text{(C.4.5)}$$

$$\equiv \sum_{m=0}^{\infty} \frac{(-1)^m}{m!} \overline{T} \int_0^1 dt_1 \cdots \int_0^1 dt_m G_{m+1}(f, \psi, ..., \psi; 0, it_1, ..., it_m).$$

Similarly,

$$\left(\Phi_\theta^\psi, \Phi_\theta^\psi\right) = \sum_{m=0}^{\infty} \frac{1}{m!} \overline{T} \int_0^1 dt_1 \cdots \int_0^1 dt_m G_m(\psi, ..., \psi; it_1, ..., it_m). \quad \text{(C.4.6)}$$

It follows now from Eq. (C.4.3) that in order to prove the proposition, it suffices to show that the right-hand side of Eq. (C.4.5) is equal to the product of those of Eqs. (C.3.12) and (C.4.6), that is, that

$$\overline{T} \int_0^1 dt_1 \cdots \int_0^1 dt_m [G_{m+1}(f, \psi, ..., \psi; 0, it_1, ..., it_m)$$

$$- \sum_{k=0}^{m} G_{k+1}^T(f, \psi, ..., \psi; 0, it_1, ..., it_k) G_{m-k}(\psi, ..., \psi; it_{k+1}, ..., it_m)] = 0. \quad \text{(C.4.7)}$$

It is now a simple exercise in combinatorics to infer from Eq. (C.3.7) that the integrand of this last equation vanishes and therefore that the proposition is valid. □

Proof of Proposition C.3.2. For lattice systems with finite range interactions, the local effective Hamiltonians are bounded and all elements of the local algebra $\mathcal{A}_L$ lie in the domain of the generator, δ_θ, of the automorphism group α_θ. In this situation, the KMS condition (C.3.1) for ρ_θ is equivalent to the following correlation inequalities [Se7, FV], which essentially represent local thermodynamical stability.

$$\langle \rho_\theta; A^\star \delta_\theta A \rangle \geq \rho_\theta(A^\star A) \ln\left[\rho_\theta(A^\star A)/\rho_\theta(AA^\star)\right] \quad \forall A \in \mathcal{A}_L. \quad \text{(C.4.8)}$$

Correspondingly, by Eq. (C.3.10), the KMS condition for the perturbed state $\rho_\theta^{\phi h_R}$ is equivalent to the inequalities

$$\langle \rho_\theta^{\phi h_R}; A^\star (\delta_\theta A + i[\hat{q}(\phi h_R), A]\rangle$$

$$\geq \rho_\theta^{\phi h_R}(A^\star A) \ln\left[\rho_\theta^{\phi h_R}(A^\star A)/\rho_\theta^{\phi h_R}(AA^\star)\right] \quad \forall A \in \mathcal{A}_L \quad \text{(C.4.9)}$$

By the compactness of the state space, $\rho_\theta^{\phi h_R}$ converges over some subsequence to a $w^\star$ limit, ω, as $R \to \infty$. Furthermore, it follows from Eq. (C.3.2) and our specifications of the function h_R in the statement of Proposition C.3.2 that, for any local observable A, $i[\hat{q}(\phi h_R), A]$ reduces to $\delta_\phi A$ for sufficiently large R. Hence, as δ_θ is linear in θ, the limiting form of Eq. (C.4.9), as $R \to \infty$, is

$$\langle \omega; A^\star \delta_{\theta+\phi} A \rangle \geq \omega(A^\star A) \ln\left[\omega(A^\star A)/\omega(AA^\star)\right] \quad \forall A \in \mathcal{A}_L, \quad \text{(C.4.10)}$$

which means that ω is an equilibrium state corresponding to the value $\theta + \phi$ of the control variable. Therefore, since $\theta + \phi$ lies in the single phase region Δ, $\omega = \rho_{\theta+\phi}$. The uniqueness of this state implies that the above compactness argument, which shows that it is a subsequential limit of $\rho^{\phi h_R}$, implies the stronger result that

$$w^\star : \lim_{R \to \infty} \rho_\theta^{\phi h_R} = \rho_{\theta+\phi}. \tag{C.4.11}$$

Since, for lattice systems, $\hat{q}(f)$ lies in the $C^\star$-algebra of quasi-local bounded observables $\mathcal{A}$, this last formula implies Eq. (C.3.14) and thus establishes the proposition. $\square$

C.5 Pure Crystalline Phases

We now seek to generalise Propositions C.1.1 and C.3.5, which imply the key formulae (7.6.2) and (7.6.4), to crystalline states. Thus, we assume here that $\Delta \, (\subset \Theta)$ is a single phase region and that, if the state ρ_θ is a pure equilibrium phase corresponding to $\theta \, (\in \Delta)$, then the translational invariance of ρ_θ is restricted to a crystallographic subgroup,[15] Y_θ, of X. However, since the ratio of the Y_θ-lattice spacing to L vanishes when L becomes infinite, this lattice reduces to the continuum X in the large scale limit employed in the definition of the distributions W_m (cf. Eqs. (7.5.2) and (7.5.4)). Accordingly, we assume that, as in the case of systems of particles on a lattice, the Y_θ-translational invariance of ρ_θ, at the microscopic level, leads to the full translational invariance of these distributions and hence, by Eq. (7.5.7), of the process $\tilde{q}$. Proposition C.1.1 then ensues from this assumption and (Cluster 1), the proof of Section C.1 being applicable.

It remains for us to extend Proposition C.3.5, under the assumptions (Cluster 1)–(Cluster 3) of Sections C.1 and C.3, to the present situation. To this end, we note that the proof of Proposition C.3.1 did not require the full translational invariance of ρ_θ, and is still applicable here. On the other hand, the translational invariance of the equilibrium state $\rho_{\theta+\phi}$ was invoked in the proof of Proposition C.3.2 for the purpose of establishing both that that state is unique and that the limit on the right-hand side of Eq. (C.3.14) is fully sequential. Nevertheless, on transferring the argument used in the proof of Proposition C.3.2 to the present situation, we infer that it still leads to Eq. (C.3.4), but with the interpretation that

(a) the limit on the right-hand side of the equation is subsequential, and

(b) $\rho_{\theta+\phi}$ is *an* equilibrium state corresponding to the point $(\theta + \phi)$ of the control space, Θ. Here we note that the set of all such states comprises

---

[15] In view of phenomena such as thermal expansion, the spacing of the lattice Y_θ is expected to depend on the control variable θ.

the pure crystalline phases and the mixtures thereof. Moreover, the pure phases are simply space translates of one another over a cell of the lattice $Y_{\theta+\phi}$.

In order to free the theory from the restrictions imposed by (a) and (b), we note that, since the functional $\hat{q}$ is the global space average of the field $\hat{q}(x)$ and since the pure phases corresponding to $(\theta + \phi)$ are space translates of one another over a cell of the lattice $Y_{\theta+\phi}$, all these states and their mixtures yield the same value of $\hat{q}$. Thus, $\hat{q}(\rho_{\theta+\phi})$, unlike the state $\rho_{\theta+\phi}$, is unique. Moreover, by Eq. (C.2.2),

$$\hat{q}(\rho_{\theta+\phi}) = -\pi'(\theta + \phi). \tag{C.5.1}$$

Furthermore, it follows from the definition of $\hat{q}$ that the left-hand side of this formula may be expressed in the form

$$\hat{q}(\rho_{\theta+\phi}) = \lim_{N\to\infty} \langle \rho_{\theta+\phi}; \hat{q}(f_N) \rangle, \tag{C.5.2}$$

where $\{f_N\}$ is a sequence of positive $S^{n+1}(X)$-class functions, such that

(i)
$$\int_X dx f_N(x) = 1; \tag{C.5.3}$$

(ii) f_N is positive and constant for $|x| \le N$; and

(iii) $f_N(x)$ decreases to zero as $|x|$ increases from N to $N + 1$ and is zero when $|x| > N + 1$.

Hence, by Eqs. (C.3.14), (C.5.2) and (C.5.3),

$$\frac{\partial \pi(\theta + \phi)}{\partial \theta_k} = -\lim_{N\to\infty} \lim_{R\to\infty} \langle \rho_\theta^{\phi h_R}; \hat{q}_k(f_N) \rangle, \tag{C.5.5}$$

where, by the above remark (a), the first limit might be merely subsequential. However, since this formula stems from the compactnes argument of Section C.4, the uniqueness of its left-hand side ensures that the limit is, in fact, fully sequential.

The treatment of this last formula according to the procedure of Section C.3 then leads to Propositions C.3.4 and C.3.5, the latter being the required result.

Part III

<hr>
<hr>

Superconductive electrodynamics as a consequence of off-diagonal long range order, gauge covariance and thermodynamical stability: prospectus

Part III is devoted to a general, model-independent derivation of superconductive electrodynamics from very general assumptions, namely those of off-diagonal long range order, gauge covariance and thermodynamical stability.

We start in Chapter 8 with a brief critical historical survey of the principal macroscopic and microscopic theories of the superconductive phase, and explain the need for a treatment of the electromagnetic properties of this phase on the basis of the above specified general principles. In Chapter 9, we proceed to the derivation of the Meissner effect, flux quantisation and persistent currents from these principles by methods that do not involve any many-body theoretic combinatorics.

Chapter 8

Brief historical survey of theories of superconductivity

The object of this chapter is to provide a brief critical survey of the development of the current picture of the superconductive phase. This is based on

(a) the macroscopic quantum phenomena of persistent currents, the Meissner and Josephson effects, and flux quantisation;

(b) the macroscopic theories of the Londons [Lon, Lo], Pippard [Pip] and Ginzburg and Landau (GL) [GL, Gi];

(c) Fröhlich's [Fr2, Fr3] electron-phonon model and the ensuing Bardeen–Cooper–Schrieffer (BCS) [BCS] microscopic theory of superconductivity; and

(d) Yang's [Ya] characterisation of the superconductive phase by off-diagonal long range order (ODLRO).

This survey contains a critique of the theories of superconductive electrodynamics based on the original forms of the BCS theory, and a motivation for the alternative derivation of this electrodynamics provided in Chapter 9. Our discussion here is centered on theories of metallic superconductivity, supplemented by some brief comments at the end of the chapter concerning the modifications required for their extension to ceramic superconductors.

The Phenomenological Picture [Scho, Lo]

A large class of metals and alloys undergo second order phase transitions at critical temperatures, T_c, of the order of $1°K$, with the following key features. At temperatures $T > T_c$, their electromagnetic properties are just those of normal metals, whereas for $T < T_c$ they are superconductors, as characterised by zero resistivity and perfect diamagnetism. The former superconductive property ensures that the materials support persistent currents[1] in circuits free of any electromotive force; while the latter property manifests itself in the form of the Meissner effect, whereby magnetic flux is expelled from the interior of a body in this phase. Most importantly, this effect *must* stem from

---

[1] These currents may be initially induced by magnetic fields.

an underlying quantum mechanism, since a classical system of charged particles has zero magnetic susceptibibility (cf. [Lo, Section 25]). Moreover, it
cannot be a consequence of zero resistivity alone, as is evident from the fact
that the ideal charged Fermi gas model is a perfect conductor, but nevertheless
has only a very weak diamagnetic susceptibility [La2].

It follows from these considerations that the primary objective of any theory
of the superconductive state is to provide a model that exhibits the properties
of zero resistivity and perfect static diamagnetism. Such theories were
achieved at the macroscopic level by the London brothers [Lon] and by
Ginzburg and Landau [GL, Gi].

The London Theory [Lon, Lo]

This is based on the following assumptions for the charge-carrying fluid
formed by the electrons.

(1) The electronic fluid behaves as a mixture of two components, the first of
which behaves as an Ohmic conductor and the second as an inviscid
fluid. These are generally termed the "normal fluid" and "superfluid"
components, and are indicated by the suffixes n and s, respectively.
Thus, the densities, ρ and j, of the electron particle number and electric
current, respectively, take the form

$$\rho = \rho_n + \rho_s, \quad j = j_n + j_s. \tag{8.1}$$

The ratio ρ_s/ρ is temperature dependent, decreasing monotonically from
1 to 0 as the temperature T increases from absolute zero to the transition
point, T_c. This two-fluid model, with the additional imput that the superfluid component carries no entropy, has provided an accurate thermodynamical picture of the superconductive phase.

(2) The Ohmic behavior of the normal fluid is represented by the constitutive equation

$$j_n = \sigma E, \tag{8.2}$$

where σ is a constant.

(3) The superfluid is a charged inviscid fluid that evolves according to
classical hydrodynamics and electromagnetism. Thus, the Euler equation of motion for its drift velocity, v_s, is

$$\frac{\partial v_s}{\partial t} + v_s.\nabla v_s = \frac{-1}{m}\nabla\mu - \frac{e}{m}E - \frac{e}{mc}v_s \times B, \tag{8.3}$$

where (E, B) is the electromagnetic field, $-e$ and m are the electron
charge and mass, respectively, and μ is the chemical potential. The
Londons observed that this equation implies a simple generalisation

of Kelvin's theorem concerning the preservation of irrotational flow by classical, inviscid, electrically neutral fluids. Specifically, it implies the analogous result obtained by the replacement of v_s by the *drift momentum*, $p_s = mv_s - eA/c$, A being the magnetic vector potential, that is, $B = \nabla \times A$. Thus, if $\nabla \times p_s$ is initially zero throughout the charged inviscid fluid, then it subsequently remains so. The key assumption of the London theory, which has been amply justified *a posteriori*, is that *superfluid flow is characterised by the irrotationality of the drift momentum p_s*. In other words, of all possible inviscid flows compatible with the Euler-cum-Maxwell equations, the superfluid one is selected by the special condition that

$$\nabla \times p_s = 0, \tag{8.4}$$

that is,

$$\nabla \times v_s = \frac{e}{mc} B. \tag{8.5}$$

This property was interpreted [Lo] as a *rigidity* of the superconductive state against rotations of the superfluid drift momentum vector, and is sometimes termed "London stiffness".

The superfluid constitutive equations may now be obtained from Eqs. (8.3) and (8.5), subject to the assumption that the superfluid density ρ_s and the chemical potential μ are uniform. Thus, since $j_s = -e\rho_s v_s$, it follows from this assumption and Eq. (8.5) that

$$\nabla \times j_s = -\frac{e^2 \rho_s}{mc} B \tag{8.6a}$$

and also that the linearised form of Eq. (8.3) is

$$\frac{\partial j_s}{\partial t} = \frac{e^2 \rho_s}{m} E. \tag{8.6b}$$

Eqs. (8.6) are the celebrated London equations. We see now that both Eq. (8.6b) and the nonlinear Euler equation (8.3) from which it stems do not carry any frictional term, and also that they are invariant under time reversals $t \to -t$, $v_s \to -v_s$, $\mu \to \mu$, $E \to E$. Hence the superfluid does indeed have zero resistance. Furthermore, it follows from Eqs. (8.1), (8.2) and (8.6a), together with the Maxwell equations

$$\nabla \times B = \frac{j}{c}, \quad \nabla . B = 0 \quad \text{and} \quad \nabla \times E = -c^{-1} \frac{\partial B}{\partial t},$$

that

$$\Delta B = -\nabla \times (\nabla \times B) = -c^{-1} \nabla \times j_s - c^{-1} \nabla \times j_n = \frac{e^2 \rho_s}{mc^2} B + \frac{\sigma}{c} \frac{\partial B}{\partial t},$$

and hence that, in equilibrium states,

$$\Delta B = \frac{e^2 \rho_s}{mc^2} B. \tag{8.7}$$

It follows from this equation and the standard boundary conditions of electro-magnetic theory that any magnetic field entering the body decays exponen-tially within a distance $\lambda = (mc^2/e^2 \rho_s)^{1/2}$ of its surface. Since this distance is typically of the order of 10^{-5} cm, this signifies that the magnetic field is expelled from the deep interior of a macroscopic specimen. Furthermore, by Eq. (8.5), the penetration of the magnetic field to the surface "shell" of thickness λ implies that electric currents prevail in this region; and, in fact, these currents are the source of the magnetic field that serves to screen out any externally applied ones from the interior of the body.

Thus, the macroscopic London model exhibits both zero resistivity and perfect diamagnetism. Most importantly, the connection between these prop-erties is that the persistent supercurrents in a ring of superconducting material are screening currents that serve to implement the Meissner effect.

Furthermore, F. London [Lo] combined the theory of the macroscopic model with a particular quantum mechanical interpretation of London stiffness in such a way as to predict the phenomenon of *flux quantisation*, whereby the magnetic flux trapped in any hollow tunnel of a multiply-connected super-conductor is an integral multiple of $2\pi c\hbar/e$.

The Pippard Theory

Pippard [Pip] proposed a nonlocal modification of the London theory on the basis of experiments on the penetration of electromagnetic fields of short wavelengths into superconductors. This modification consisted of the replace-ment of the field vectors $(E(x), B(x))$ of Eqs. (8.3), (8.5) and (8.6) by

$$\int dy K(x-y)\big(E(y), B(y)\big)$$

respectively, where $K(x)$ is a decaying function of $|x|$ of finite range ξ and total integral equal to unity. ξ is generally termed the Pippard *coherence length* and interpreted as a range of cooperative interaction between the electrons. The Pippard model also exhibits perfect conductivity and the Meissner effect, and reduces to that of the Londons in the long wavelength limit. As regards the magnitude of the coherence length ξ, this can be either greater or less than the London penetration length λ. A rough rule is that ξ tends to exceed λ for pure metals, but not for highly impure alloys.

The Ginzburg–Landau Theory [GL, Gi]

This is based on Landau's theory of phase transitions of the second kind [LL2], wherein the ordered phase is characterised by an order parameter that varies continuously with temperature and vanishes at the critical point. In the GL theory, this parameter is assumed to be a complex, possibly position-dependent, quantity, ϕ, whose coupling to the electromagnetic field is of the same form as that of a quantum mechanical wave function. Specifically, it is assumed that the free energy of a superconductor in a magnetic field $B = \nabla \times A$, is given formally by

$$F(\phi, A) = \int dx \left[f(|\phi|^2) + \frac{1}{2m} |i\hbar \nabla \phi - eA\phi/c|^2 + \frac{1}{2c^2} (\nabla \times A)^2 \right], \quad (8.8)$$

where the first term in the integrand is the free energy density when ϕ is uniform and $B = 0$, the second term represents the coupling of the magnetic field to the gradient of the order field, ϕ, and the third term is the magnetic energy density. Thus, F satisfies the gauge invariance condition

$$F(\phi, A) = F(\phi \exp(-ie\chi/\hbar c), A + \nabla \chi). \quad (8.9)$$

As regards the intrinsic free energy density f, it is assumed that this attains its absolute minimum at a non-zero value, ϕ_0, of ϕ, such that[2]

$$f'(|\phi_0|^2) = 0 \quad \text{and} \quad f''(|\phi_0|^2) > 0, \quad (8.10)$$

The equilibrium condition is given by the minimisation of F, as represented by the equations

$$\frac{\delta F}{\delta \phi(x)} = 0, \quad \frac{\delta F}{\delta A(x)} = 0,$$

that is,

$$\frac{1}{2m} (i\hbar \nabla - eA/c)^2 \phi + f'(|\phi|^2)\phi = 0 \quad (8.11)$$

and

$$\nabla \times B \equiv \nabla \times (\nabla \times A) = \frac{j}{c}, \quad (8.12)$$

where

$$j = -\frac{ie\hbar}{2m} (\phi \nabla \overline{\phi} - \overline{\phi} \nabla \phi) - \frac{e^2}{mc^2} |\phi|^2 A. \quad (8.13)$$

---

[2] Thus, a typical ansatz for f is given by $f(|\phi|^2) = -a|\phi|^2 + b|\phi|^4$, where a, b are temperature-dependent parameters such that $b > 0$ and $a(= 2b|\phi_0|^2)$ decreases from a positive value, a_0 to zero as T increases from 0 to the transition point T_c.

Since Eq. (8.12) is Maxwell's static $B-j$ equation, it follows that j is the electric current density.

We note here that if an electromagnetic field, due to an external current source j_{ext}, is applied to the system, then the formula (8.8) for F becomes modified by the addition of the term, $-c^{-1} \int j_{ext}.A dx$, on the right hand side. However, this does not affect Eqs. (8.11)-(8.13) for ϕ and A in the interior of the superconductor: it just leads to a Maxwell equation of the same form as Eq. (8.12), but with j replaced by j_{ext}, for the exterior region. On the other hand, in the case where no external fields are applied, it follows immediately from Eqs. (8.8) and (8.10) that F is minimised, and therefore equilibrium prevails, when $A = 0$ and ϕ is a constant, ϕ_0, which is determined up to a phase factor by Eq. (8.10).

As for the time-dependent GL theory, this is assumed to take the form of the nonlinear Schrödinger equation

$$i\hbar \frac{\partial \phi}{\partial t} = \frac{\delta F}{\delta \phi} = (-i\hbar \nabla + eA/c)^2 \phi + f'(|\phi|^2)\phi. \tag{8.14}$$

Hence, defining

$$\rho = |\phi|^2, \tag{8.15}$$

it follows immediately from Eqs. (8.13)–(8.15) that ρ satisfies the local conservation law

$$\frac{\partial \rho}{\partial t} - e^{-1}\nabla.j = 0,$$

and may therefore be naturally interpreted as the number density of the charge-carrying fluid.

The GL and London theories are, in fact, very closely related, as we see from the following argument, due to Fröhlich [Fr4]. On expressing ϕ in the form

$$\phi = \rho^{1/2} \exp(i\theta/\hbar), \tag{8.16}$$

Eq. (8.13) reduces to

$$j = -e\rho(\nabla\theta + eA/c), \tag{8.17}$$

which signifies that the drift velocity, u, and momentum, p, are given by the formulae

$$u = m^{-1}(\nabla\theta + eA/c) \tag{8.18}$$

and

$$p := mu - eA/c = -m\nabla\theta.$$

Hence p satisfies the London stiffness condition that (cf. Eq. (8.4))

$$\nabla \times p = 0,$$

or equivalently, u satisfies the constitutive equation analogous to Eq. (8.5), that is,

$$\nabla \times u = \frac{eB}{mc}. \tag{8.19}$$

Furthermore, on employing the form (8.16) for ϕ in Eq. (8.14), and equating the real parts of coefficients of $\exp(i\theta/\hbar)$ on the left- and right-hand sides of the resultant formula, the following equation of motion for θ ensues:

$$\frac{\partial \theta}{\partial t} = \frac{\hbar^2}{2m} \frac{\Delta \rho^{1/2}}{\rho^{1/2}} - \frac{1}{2m} (\nabla \theta + eA/c)^2 - f'(\rho).$$

The gradient of this equation, together with the formula $E = -c^{-1} \partial A/\partial t$ and Eqs. (8.18) and (8.19), then yields the Euler equation

$$\frac{\partial u}{\partial t} + u.\nabla u = -\nabla \zeta - \frac{e}{m} E - \frac{e}{mc} u \times B, \tag{8.20}$$

where

$$\zeta = f'(\rho) - \frac{\hbar^2}{2m^2} \frac{\Delta \rho^{1/2}}{\rho^{1/2}}. \tag{8.21}$$

Thus, Eq. (8.20) has precisely the same form as the Euler equation (8.3) of the London theory, with u and ζ corresponding to v_s and μ, respectively. However, the GL theory does have the advantage of being a self-contained model, which requires no ad hoc assumptions such as those concerning the constancy of ρ_s and μ that were employed in the passage from the stiffness condition (8.4) to the London equations (8.6).

The electrical resistance of the GL model is manifestly zero, since its Euler equation (8.20) contains no frictional term. To show that it exhibits the Meissner effect for sufficiently weak magnetic fields, consider the situation where ϕ and ρ are perturbed from their equilibrium values, ϕ_0 and $\rho_0 = |\phi_0|^2$, by the action of a 'small' vector potential A_1. Then it follows from Eqs. (8.11) and (8.15) that the resultant linearised increments, ϕ_1 and ρ_1, in ϕ and ρ satisfy the equations

$$\rho_1 = \overline{\phi}_0 \phi_1 + c.c.$$

and

$$\Delta \rho_1 = l^{-2} \rho_1,$$

where

$$l = \left[\frac{(\hbar^2}{4m\rho_0 f''(\rho_0)} \right]^{1/2},$$

which is typically of the order of 10^{-4} or 10^{-5} cm [DeG]. It follows that the perturbation ρ_1 of ρ decays exponentially within a distance l of the surface of the body, and hence that ρ may be equated to ρ_0 in its interior. Hence, for weak fields, ρ may be replaced by ρ_0 in Eq. (8.17) and consequently

$$\nabla \times j = -\frac{e^2 \rho_0}{mc^2} B.$$

Since this is precisely analogous to the London equation (8.6a), it signifies that the GL model exhibits the Meissner effect.

For strong magnetic fields, the Ginzburg–Landau theory has proved to be extraordinarily fruitful in predicting a great wealth of experimentally confirmed nonlinear phenomena [Abr, DeG], that lie beyond the scope of the linear London equations (8.6).

The Microscopic Theory: Fröhlich's Electron-Phonon Model

Since it was known from the quantum theory of metals that their electrical resistance arose from the coupling of electrons to the vibrations of the ionic lattice [Bl1], it was generally assumed in the early days of the quantum theory of condensed matter that the phenomenon of superconductivity stemmed from some mechanism that overrode this coupling. Accordingly, the early attempts by Heisenberg [Hei] and Born and Cheng [BC] to provide a microscopic theory of the superconductive phase were based on models of interacting electrons that moved through a rigid ionic lattice. However, such attempts proved to be abortive.

A major breakthrough towards a theory of superconductivity was provided by Fröhlich [Fr2, Fr3], who proposed that the mechanism behind this phenomenon arose *because of, not despite*, the interaction between conduction electrons and the phonon field generated by the ionic lattice vibrations. Fröhlich's theory was based on a model, whose formal Hamiltonian was of the form

$$H_F = H_e + H_p + H_{ep}, \tag{8.22}$$

where H_e is the kinetic energy of a system of free electrons, H_p is the Hamiltonian of a free phonon field, which is bosonic, and H_{ep} represents the electron–phonon interaction. Most importantly, this interaction leads to attractive interelectronic forces, just as, in nuclear physics, the coupling between nucleons and mesons leads to attractive nuclear forces. In fact, Fröhlich showed that the above Hamiltonian H_F leads, in Born approximation, to an effective electronic Hamiltonian, H'_F, of the form

$$H'_F = H_e + H_{\text{int}}, \tag{8.23}$$

where the contribution H_{int} corresponds to two-body interactions. Their principle qualitative features are that they are attractive, highly momentum dependent and concentrated predominantly on the electrons in a thin shell of momentum space close to the Fermi surface.

Fröhlich's essential idea was that the phonon-mediated interelectronic attraction could lead to some condensation phenomenon, corresponding to the superconductive phase. Although the specific *ansatz* he proposed was shown by Schafroth [Scha2] to be defective, in that it did not lead to the Meissner effect, his idea that the electron-phonon coupling was responsible for superconductivity was vindicated by the experimental discovery of the isotope effect, whereby the transition temperature was inversely proportional to the square root of the ionic mass. This amounts to a direct proof of the involvement of the ionic vibrations, that is, the phonons.

Schafroth's Suggestion of Bound Electron Pairs

In his critique of Fröhlich's theory, Schafroth [Scha2] formulated the response of a charged ideal Bose gas to an external magnetic field, and thereby deduced that it exhibits the Meissner effect below the Bose–Einstein condensation temperature. This led him to suggest in a subsequent work [Scha1] that superconductivity might arise as a result of the formation of bosons comprising electron pairs bound together by the Fröhlich interaction H_{int}.

The Cooper Pairing Model

Cooper [Co] treated the problem of the binding of electron pairs in a relatively simple model, in which the states of a Fermi sphere are occupied and a further two electrons of opposite spin interact with one another via attractive momentum-dependent forces confined to a thin shell around the Fermi surface. In this model, the electrons inside the Fermi sphere are passive and serve only to exclude the remaining two electrons from that sphere. Cooper solved the model exactly, obtaining the remarkable result that the latter two electrons had precisely one bound state of zero total momentum.

The Bardeen–Cooper–Schrieffer (BCS) Theory [BCS]

This is based on an *ansatz* in which the electrons of the Fröhlich model form Cooper-like pairs and yields an accurate description of the thermal properties of superconductors, including their phase transition of the second kind into the normal metallic state.

Furthermore, the pairing hypothesis, with its implication that the effective charge of the current carriers is $-2e$, rather than $-e$, has been experimentally substantiated by two important phenomena, which we now discuss. The first of these is the value of the trapped flux quantum in multiply-connected superconductors, which F. London [Lo] calculated to be ch/e ($\equiv 2\pi c\hbar/e$), on the basis of his macroscopic theory, together with a particular interpretation of

London stiffness.[3] Since that calculation was based on the assumption that the charge carrier was $-e$, the BCS pairing hypothesis leads to a replacement of e by $2e$ in the above expression, that is, to the prediction that the flux quantum is $ch/2e$. This is precisely the value that has been observed experimentally [DF].

The second phenomenon carrying direct confirmation of the pairing hypothesis is the Josephson effect [Jo], whereby a potential difference of ΔV between two coupled superconductors, separated by a thin insulating film, leads to an alternating tunnelling current of frequency $2e\Delta V/\hbar$. This is precisely the value predicted by the Ginzburg–Landau theory, based on the assumption of charge carriers $-2e$ [Jo].

Thus, the BCS ansatz and pairing hypothesis have the great merit of yielding both the thermodynamical properties of the superconductive phase and the correct values of both the trapped flux quantum and the Josephson frequency. Here, we note that the calculations of these latter values were based on the *assumption* of the London and Ginzburg–Landau theories, coupled with the pairing hypothesis that the effective charge carrier was $-2e$. They did *not* entail any derivation of the macroscopic equations of superconductive electrodynamics.

In fact, the BCS ansatz does not provide a natural basis for the derivation of that electrodynamics, since it violates the principle of local gauge covariance [Scha3, Fr5] and consequently does not even admit a precise definition of the local current density. Its violation of the gauge principle arises from the fact that the BCS theory is based on a *truncation* of a fully gauge invariant model (Fröhlich's electron-phonon system), that retains only the interactions that give rise to the electron pairing. The results of attempts to remedy the situation by taking some account of the residual interactions [An1, Ri, Na] are inconclusive, since they are all based on approximative schemes that are virtually impossible to check.[4]

Thus, there is a need for a conceptually clear quantum theory of superconductive electrodynamics.

Off-diagonal Long Range Order (ODLRO) and Superconductive Electrodynamics

Yang [Ya] observed that the BCS ansatz possessed the property of *off-diagonal long range order* (ODLRO), which was defined for fermion systems in Section 3.3 by the formula

---

[3] Cf. the above sketch of the London theory.

[4] For example, the treatments of Anderson [An1] and Rickayzen [Ri] are based on random phase approximations, while that of Nambu [Na] is based on a generalised Hartree–Fock approximation.

$$\lim_{|y|\to\infty}\left[\langle\rho;\Psi(x_1,x_2)\Psi^\star(x_1'+y,x_2'+y)\rangle-\Phi(x_1,x_2)\overline{\Phi}(x_1'+y,x_2'+y)\right]=0,$$

$$(8.24)$$

where now ρ is the microstate, Ψ is the pair wave operator given by the equation

$$\Psi(x_1,x_2):=\psi_\uparrow(x_1)\psi_\downarrow(x_2) \qquad (8.25)$$

and Φ is a classical two point field, termed the macroscopic wave function, that does not tend to zero at infinity. Furthermore, Yang proposed that the ODLRO property provided a characterisation of the superconductive phase, and Gorkov [Go] argued, on the basis of a mean field theoretic argument, that the macroscopic wave function Φ was related to the Ginzburg–Landau wave function ϕ by the formula

$$\phi(x)=\Phi(x,x).$$

In fact, Yang's proposal that ODLRO characterised the superconductive phase was brought to fruition by the present author [Se11, Se1], who proved, by model-independent arguments, that the assumptions of ODLRO, local gauge covariance and thermodynamical stability implied the phenomena of the Meissner effect, persistent currents and flux quantisation, with the correct value for the flux quantum. Chapter 9 is devoted to an updated review of this work.

Note on Ceramic Superconductors

These are remarkable for their relatively high critical temperatures, typically of the order of 100 K [BM], and for the fact that their "normal" phase is quite complex and certainly different from that of a metal [An2]. Their macroscopic electrodynamical properties, however, appear to be the same as those of metallic superconductors.

As regards the quantum mechanisms underlying their behavior, the picture appears to be much less developed than that of their metallic counterparts, despite the proposal of various interesting ideas (cf. [An2, KRS, AR, AM, Lau]). Since some of the proposed ansätze appear to carry the ODLRO property (cf. [An2, AR, AM]), we conjecture that our derivation of superconductive electrodynamics is also applicable to ceramic superconductors.

Chapter 9

Off-diagonal long range order and superconductive electrodynamics

9.1 INTRODUCTION

The electromagnetic properties of the superconductive phase, as manifested by the phenomena of persistent electric currents, the Meissner effect and the quantisation of trapped magnetic flux, are remarkably simple, and qualitatively different from those of other phases of matter. Since there is such a wide variety of superconducting materials, this suggests that a theory of their electrodynamics should be derivable from arguments based only on very general features, rather than detailed microscopic properties, of the superconductive phase. Moreover, such a theory is eminently desirable, in view of the radical problems, discussed in Chapter 8, that arise in the derivation of the Meissner effect from detailed microscopic ansätze.

The present chapter is devoted to a general, model independent, quantum statistical derivation of superconductive electrodynamics from the assumptions of off-diagonal long range order (ODLRO), as formulated in Section 3.3.4, local gauge covariance and thermodynamic stability. On this basis, we obtain the Meissner effect, the quantisation of trapped magnetic flux, and the metastability of supercurrents. The key to these results, as shown in Section 9.3, is that it follows from the above assumptions that the macroscopic wave function, specified by the ODLRO condition, enjoys the London stiffness property discussed in Chapter 8.

The presentation of the theory is organised as follows. We start in Section 9.2 with a general formulation of our gauge-covariant model of matter in interaction with a classical magnetic field. This formulation covers the cases of both continuous and lattice systems. A key result here is that the dynamics of the model, in the presence of a uniform magnetic induction B, is covariant with respect to the *regauged space translations* [Za, Se11] whereby the transformation of the quantised electronic field ψ corresponding to the space translation $x \rightarrow x + a$ is given by

$$\psi(x) \rightarrow \psi(x + a) \exp\left(\frac{-ie(B \times a).x}{2\hbar c}\right), \qquad (9.1.1)$$

where $-e$ is the electronic charge. Most importantly, the sinusoidal factor in this formula arises from a regauging of the magnetic vector potential to compensate for its change under the above space translation.

In Section 9.3, we provide a general formulation of ODLRO and prove, in Proposition 9.3.2, that it is incompatible with a nonzero uniform induction in any state that is invariant with respect to the regauged space translations. Hence, the question of whether ODLRO or a uniform magnetic induction prevails in a translationally invariant equilibrium state is a thermodynamical one.

In Section 9.4, we employ the key result of Proposition 9.3.2 to derive the Meissner effect from the assumptions of ODLRO, translational invariance and thermodynamical stability. Here it will be seen that the Meissner effect stems from a certain rigidity[1] of the macroscopic wave function in the face of a sufficiently weak applied field. Remarkably, the principle of local gauge covariance, which was an obstacle to the theories based on the BCS ansatz, is an essential ingredient of our argument leading to the Meissner effect!

In Section 9.5, we extend that argument, along the lines of the treatment by Nieh, Su and Zhao [NSZ], to derive the quantisation of trapped magnetic flux in multiply-connected systems.

In Section 9.6, we formulate the theory of persistent currents in a multiply-connected body, such as a ring. Here, the supercurrents are none other than the currents that implement the Meissner effect by screening the trapped magnetic flux from the interior of the body [Lo], and the essential problem is that of the stabilisation of both the trapped flux and its screening current. This is evidently a problem of *metastability*, since the current-carrying state has higher free energy than that of true thermal equilibrium at the same temperature. Thus, adopting our characterisation of metastable states[2] by thermodynamical stability against local perturbations, we formulate the condition for the persistent currents in terms of a variational principle, that is subject to the constraint of flux quantisation. In this way, we characterise the phenomenon of superconductivity itself, that is, of persistent currents, by a relatively simple thermodynamical condition. Furthermore, we show that this phenomenon is intimately related to a superselection rule that forbids locally induced transitions between states with different flux quantum numbers.

The theory of Sections 9.3–9.6 is centered on translationally invariant states and is therefore inapplicable to the vortex phase of type II superconductors, which supports periodic magnetic induction and vorticity profiles (cf. [Abr, DeG]). In Section 9.7, we extend the above theory to this phase and prove that

---

[1] To be precise, it is a macroscopic version of the London stiffness, which was originally conceived as a rigidity of the *microscopic* many-particle wave function of the ground state of the system (cf. [Lo, Section 26]).

[2] Cf. Chapter 5.4 and, for more detailed discussions, [Se8] and [Se2, Chapter 8].

the magnetic flux through a plaquette is an integral multiple of $ch/2e$. Thus it is quantised according to the same rule as that governing the flux through a tunnel in a multiply-connected superconductor.

We conclude, in Section 9.8, with a very brief discussion of open problems of the theory of superconductivity.

In the Appendix we establish a useful technical point, namely that a magnetic induction with compact support may be represented by a vector potential whose support is also compact.

9.2 THE GENERAL MODEL

The model is that of an infinitely extended quantum dynamical system, Σ, consisting of electrons and possibly a second species of particles, for example, phonons or ions. We assume that the system occupies a three-dimensional space, X, which may be either a Euclidean continuum or a simple cubic lattice. For simplicity, we generally employ a notation appropriate to the continuum case. This may easily be translated into the corresponding notation for lattice systems by standard procedures employed in gauge theories [Sei].

Our reason for assuming that X is three-dimensional is that ODLRO, like the Bose–Einstein condensation of an ideal Bose gas, cannot be sustained in spaces of lower dimensionality [GWM]. At a more heuristic level, we note that, although some of the proposed theories of ceramic superconductivity are based on two-dimensional models (cf. [ZA, KRS, CWWH, Lau]), there are persuasive arguments [An2] to the effect that this phenomenon is essentially three-dimensional by virtue of the crucial role played by the interactions between different planar layers.

The model is constructed so that its dynamics is covariant with respect to gauge transformations of both the first and second kind, space translations and time reversals. These covariance properties are demanded by the requirements of quantum mechanics and electromagnetism, and so should be realised by any satisfactory microscopic model of the electrodynamical properties of matter. We note here that they are indeed realised by the models of Fröhlich [Fr2, Fr3] and Hubbard [Hu], which are the bases of theories of metallic and ceramic superconductivity, respectively.

We denote the electronic component of Σ by Σ_{el}, and its other component, if any, by Σ'. We assume that the algebra of observables, $\mathcal{A}$, of Σ, has the local structure described in Section 2.5.1. For the purposes of this chapter, it is unimportant whether $\mathcal{A}$ is a $C^\star$ or a $W^\star$ algebra. We assume that $\mathcal{A}$ is generated by the algebras of observables of Σ_{el} and Σ', which we denote by $\mathcal{A}_{el}$ and $\mathcal{A}'$, respectively. In particular, we take $\mathcal{A}_{el}$ to be the globally gauge invariant algebra of the fermionic electron field

$$\psi = \begin{pmatrix} \psi_1 \\ \psi_{-1} \end{pmatrix} \equiv \begin{pmatrix} \psi_\uparrow \\ \psi_\downarrow \end{pmatrix},$$

as formulated according to the prescription of Section 2.5.3. Thus, ψ is quantised according to the canonical anticommutation relations

$$[\psi_s(x), \psi_{s'}^\star(x')]_+ = \delta_{ss'}\delta(x-x'), \quad [\psi_s(x), \psi_{s'}(x')]_+ = 0 \quad \forall s, s' = \pm 1,$$

$$(9.2.1)$$

and $\mathcal{A}_{\text{el}}$ is generated algebraically by monomials of the form

$$\psi_{s_1}^\star(x_1)\cdots\psi_{s_n}^\star(x_n)\psi_{s_{n+1}}(x_{n+1})\cdots\psi_{s_{2n}}(x_{2n}),$$

or, more precisely, by smeared versions thereof.

We are concerned with the properties of the system in the presence of a classical static magnetic field B, represented by a vector potential A. Thus

$$B = \text{curl}\, A. \qquad (9.2.2)$$

We assume that the dynamics of the model is covariant with respect to gauge transformations of the second kind, as given by the formula

$$A \to A + \nabla\chi, \quad \psi \to \psi\exp\left(-\frac{ie\chi}{\hbar c}\right), \quad \mathcal{O}' \to \mathcal{O}', \qquad (9.2.3)$$

where χ is an arbitrary continuously differentiable function of position and $\mathcal{O}'$ is any observable of Σ'.

We assume that the electronic observables, $q(x)$, $j_A(x)$ and $m(x)$, representing the position-dependent densities of charge, electric current and magnetic polarisation are given by the standard formulae

$$q(x) = -e\psi^\star(x)\psi(x), \qquad (9.2.4a)$$

$$j_A(x) = \frac{ie\hbar}{2}\left(\psi^\star(x)\nabla\psi(x) - \nabla\psi^\star(x)\psi(x)\right) - \frac{e^2}{c}A(x)\psi^\star(x)\psi(x), \qquad (9.2.4b)$$

and

$$m(x) = \psi^\star(x)\sigma\psi(x) \qquad (9.2.4c)$$

respectively, where $-e$ (< 0) is the electronic charge and σ is the spin vector, whose components $(\sigma_x, \sigma_y, \sigma_x)$ are the Pauli matrices, as given by Eq. (2.2.8). Evidently, q, j_A and m are all gauge invariant.

We assume that the formal Hamiltonian that governs the dynamics of the model takes the form

$$H_A = \int\left[\frac{1}{2}(\hbar\nabla\psi^\star - \frac{ieA}{c}\psi^\star).(\hbar\nabla\psi + \frac{ieA}{c}\psi) - B.m\right]dx + V_{\text{el}} + H_{\text{int}} + H',$$

$$(9.2.5)$$

where V_{el} is the potential energy of the interelectronic interactions, H' is the Hamiltonian for Σ' and H_{int} is the energy of interaction between Σ_{el} and Σ'. Here we stipulate that the magnetic interactions between the electrons are not incorporated into either V_{el} or H_{int} but are represented by the dependence of the field B on the sources m and j; specifically, B is the resultant of the magnetic field due to external sources and the expectation value of that due to the above internal ones. Furthermore, we assume that V_{el}, H' and H_{int} are all invariant under the gauge transformations (9.2.3), and thus, by Eq. (9.2.5), that H_A is fully gauge invariant.

Covariance of the Dynamics with respect to Gauge Transformations, Space Translations and Time Reversals

We represent the dynamics of the model by the one-parameter group of auto-morphisms, $\alpha_A(\mathbf{R})$, of $\mathcal{A}$, whose generator is $i[H_A, \cdot]/\hbar$. In order to formulate the gauge covariance of this dynamics of the system in algebraic terms, we define $C(X)$ to be the additive group of all continuously differentiable real-valued functions on X and γ to be the representation of $C(X)$ in the auto-morphisms of $\mathcal{A}$ defined by the formula

$$\gamma(\chi)\psi(x) = \psi(x)\exp(-ie\chi(x)/\hbar c) \quad \text{and} \quad \gamma(\chi)_{|\mathcal{A}'} = I \quad \forall \chi \in C(X). \; (9.2.6)$$

Thus, $\gamma(\chi)$ is the automorphism corresponding to the transformation of the quantum observables given by Eq. (9.2.3), and the assumption of gauge covariance is just that of the invariance of $\alpha_A(t)$ under the joint action of the automorphism $\gamma(\chi)$ and the transformation $A \to A + \nabla\chi$. To obtain the action of $\gamma(\chi)$ on $\alpha_A(t)$, we note that the automorphism that sends the transform, by $\gamma(\chi)$, of an observable Q to that of its evolute, $\alpha_A(t)Q$, is just

$$\tilde{\alpha}_A(t) := \gamma(\chi)\alpha_A(t)\gamma(-\chi).$$

Hence, the action of $\gamma(\chi)$ on $\mathcal{A}$ induces the transformation of the dyna-mical automorphism $\alpha_A(t)$ to $\tilde{\alpha}_A(t)$, and therefore the condition for gauge covariance is that

$$\tilde{\alpha}_{A+\nabla\chi}(t) = \alpha_A(t),$$

that is, that

$$\alpha_{A+\nabla\chi}(t) = \gamma(-\chi)\alpha_A(t)\gamma(\chi). \tag{9.2.7}$$

Since the generator of $\alpha_A(\mathbf{R})$ is $i/\hbar[H_A, \cdot]$, it may readily be seen that this condition is equivalent to the invariance of H_A under the transformation (9.2.3).

We formulate the covariance of the dynamics with respect to space transla-tions in a similar way. Thus, we first assume that these translations correspond to a representation, ξ, of the additive group X in the automorphisms of $\mathcal{A}$, such that the local subalgebras of $\mathcal{A}$ transform covariantly with respect to $\xi(X)$, as in Section 2.5.1, and, in particular,

$$\xi(a)\psi(x) = \psi(x+a) \quad \forall \ x, \ a \ \in X. \tag{9.2.8}$$

The space translational covariance condition is then that $\alpha_A(t)$ is invariant under the combined effect of the automorphism $\xi(a)$ and the transformation $A \rightarrow A_a$, with

$$A_a(x) = A(x+a). \tag{9.2.9}$$

Hence, by repeating the argument leading to the gauge covariance condition (9.2.7), we see that the condition for space translational covariance is given by the formula

$$\alpha_{A_a}(t) = \xi(-a)\alpha_A(t)\xi(a). \tag{9.2.10}$$

This is equivalent to the condition that H_A is invariant under the combined action of the automorphism $\xi(a)$ and the transformation $A \rightarrow A_a$.

We assume that $\mathcal{A}$ is equipped with a time-reversal antiautomorphism τ, which, in the absence of any magnetic field, is related to the dynamics represented by the formal Hamiltonian H_0, that is, by the automorphisms $\alpha_0(\mathbf{R})$, according to the specifications of Section 4.1. Thus, the action of τ on $\mathcal{A}_{\mathrm{el}}$ is given by the formula

$$\tau\psi_s(x) = \psi^{\star}_{-s}(x), \tag{9.2.11}$$

and its action on H_0 leaves this formal Hamiltonian invariant. Hence, since Eqs. (9.2.5) and (9.2.11) imply that the antiautomorphism τ sends $(H_A - H_0)$ to $(H_{-A} - H_0)$, and therefore H_A to H_{-A}, it follows that

$$\tau\alpha_A(t)\tau = \alpha_{-A}(-t), \tag{9.2.12}$$

which is the condition for covariance with respect to time reversals. It is the generalisation of the microscopic reversibility condition (4.1.7) to systems in magnetic fields.

The Regauged Space Translations

By Eqs. (9.2.9) and (9.2.10), the space and time translational automorphisms, $\xi(a)$ and $\alpha_A(t)$, do not generally intercommute when the vector potential A is nonzero. However, we now show that, in the special case of a uniform magnetic field, B, there is a modified version of the space translational automorphisms that does commute with the time translational ones and so provides the structure required for a statistical thermodynamical treatment of the model Σ. Moreover, the case where B is uniform is crucially important for superconductive electrodynamics, since the essential difference between a normal magnetism and the Meissner effect is that the former admits a uniform, nonzero magnetic induction at equilibrium and the latter does not.[3]

In order to formulate space translations when B is uniform, we first note that, by Eq. (9.2.2), this induction may be represented by the vector potential

$$A(x) = \frac{1}{2} B \times x. \tag{9.2.13}$$

Hence, by Eq. (9.2.9),

$$A_a(x) - A(x) = \frac{1}{2} B \times a,$$

and consequently, defining

$$\chi_{B,a}(x) = -\frac{1}{2}(B \times a).x, \tag{9.2.14}$$

$$A_a = A + \nabla \chi_{B,a}.$$

This signifies that the left-hand sides, and hence the right-hand sides, of Eqs. (9.2.7) and (9.2.10) are equal when $\chi = \chi_{B,a}$, and therefore that

$$\xi(a)\gamma(-\chi_{B,a})\alpha_A(t) = \alpha_A(t)\xi(a)\gamma(-\chi_{B,a}).$$

Hence, defining the automorphism

$$\xi_B(a) = \xi(a)\gamma(-\chi_{B,a}), \tag{9.2.15}$$

we see that the automorphisms $\{\xi_B(a) \mid a \in X\}$ commute with the time translational ones, $\alpha_A(\mathbf{R})$. We term them the *regauged space translations* since, by Eq. (9.2.15), $\xi_B(a)$ consists of a space translation, compensated by a gauge transformation that equalises the space translate, A_a, and the gauge transform, $A + \nabla\chi_{B,a}$, of A. Since it is $\xi_B(a)$, rather than $\xi(a)$, that commutes with the dynamical group $\alpha_A(\mathbf{R})$, we regard it as the automorphism corresponding to the space translation a in the presence of the uniform magnetic field B. The explicit form of its action on $\mathcal{A}$ is easily obtained from Eqs. (9.2.6), (9.2.8), (9.2.14) and (9.2.15), and is given by the formula

$$\xi_B(a)\psi(x) = \psi(x + a)\exp(\frac{-ie(B \times a).x)}{2\hbar c}) \quad \text{and} \quad (\xi_B(a) - \xi(a))_{|\mathcal{A}'} = 0. \tag{9.2.16}$$

Note Although ξ_B is a mapping of the translation group, X, into Aut($\mathcal{A}$), it is *not* a representation of that group. This may be seen from the fact that, by Eq. (9.2.16),

$$\xi_B(a)\xi_B(b)\psi(x) = \xi_B(a + b)\psi(x)\exp(-ieB.(a \times b)/2\hbar c),$$

---

[3] In other words, the Meissner effect corresponds to zero permeability in the long wavelength limit.

which, in general, is not equal to $\xi_B(a+b)\psi(x)$. In fact, this seemingly negative result, which stems from a mismatch between the normal space translations, $\xi(a)$, and the gauge transformations, $\gamma(\chi_{B,a})$, is crucial to superconductive electrodynamics, since, as we discuss in the following two sections, it lies at the root of our derivation of the Meissner effect.

9.3 ODLRO VERSUS MAGNETIC INDUCTION

As in Section 3.3.4, we formulate ODLRO in terms of the *pair field*

$$\Psi(x_1, x_2) = \psi_\uparrow(x_1)\psi_\downarrow(x_2). \tag{9.3.1}$$

We then say that a state ρ of Σ possesses the property of ODLRO if there is a *classical* two-point field $\Phi(x_1, x_2)$ such that

$$\lim_{|y|\to\infty}\left[\langle\rho; \Psi(x_1,x_2)\Psi^\star(x_1'+y, x_2'+y)\rangle - \Phi(x_1,x_2)\overline{\Phi}(x_1'+y, x_2'+y)\right] = 0,$$
$$\tag{9.3.2}$$

and further, $\Phi(x_1+y, x_2+y)$ does not tend to zero as $|y| \to \infty$. In this case, Φ is termed the *macroscopic wave function* of the state.

Note   The product of the Ψ and the $\Psi^\star$ in Eq. (9.3.2) is affiliated to the observable algebra $\mathcal{A}_{\text{el}}$, even though each of these fields is affiliated only to the field algebra $\mathcal{A}_{\text{el}}^{(F)}$, as defined according to the procedure of Section 2.5.3.

The following two propositions, which we prove at the end of this section, provide us with the key to the relationship between ODLRO and the electromagnetic properties of our model.

Proposition 9.3.1
  (1)   *The ODLRO condition (9.3.2) defines the macroscopic wave function Φ up to a phase factor, that is, if Φ_1 and Φ_2 are two such functions that satisfy this condition for the same state, then $\Phi_1 = e^{ik}\Phi_2$, where k is some real constant.*

  (2)   *In the case where $A = 0$ and the state ρ is invariant with respect to the time reversals, the macroscopic wave function Φ is real-valued, up to a constant phase factor.*

Proposition 9.3.2   *Assume that the magnetic induction is uniform, that the state ρ is invariant with respect to the regauged space translations, and that it possesses the property of ODLRO. Then*

  (1)   *in the case where X is a continuum, the magnetic induction B vanishes, and*

(2) *in the case where X is a lattice, B satisfies the quantisation condition that the magnetic flux through any plaquette is an integral multiple of $hc/2e \equiv \pi\hbar c/e$.*

Comments

(1) Proposition 9.3.2 tells us that, if X is a continuum, then ODLRO and uniform nonzero magnetic induction cannot coexist in translationally invariant states; whereas, if X is a lattice, ODLRO might be compatible with a nonzero quantised induction B, of the order of $\hbar c/el^2$, where l is the lattice spacing. However, for a real crystalline material, where l is typically of the order of 10^{-8} cm, such an induction is of the order of 10^9 G, which is not only many orders of magnitude larger than any known critical fields for superconductors, but also much larger than the internal fields in ferromagnets. On this basis, *we henceforth assume that, for lattice systems as well as continuous ones, ODLRO and nonzero uniform magnetic induction cannot coexist in translationally invariant states.*

(2) In view of this latter assumption, we conclude that the question of whether ODLRO or a uniform, nonzero magnetic induction prevails at equilibrium, when the system is subjected to a uniform external magnetic field, is a thermodynamical one, which we address in Section 9.4. There we establish that if it is ODLRO that prevails when the applied field is sufficiently weak, then the system exhibits the Meissner effect.

(3) Although the assumption at the end of Comment (1) above serves to dismiss the possibility of the coexistence of ODLRO and a uniform nontrivial quantised induction B in lattice models, the result given by Proposition 9.3.2 (2) is of significance for the theory of type II superconductors, as we can see in Section 9.7. There the essential point is that such superconductors support a phase in which the electronic fluid forms a lattice, $\mathcal{L}$, of vortices whose spacing is very much larger than that between the atoms of a typical crystal, and the magnetic induction has the periodicity of $\mathcal{L}$ [Abr]. In Section 9.7 we adapt the argument leading to Proposition 9.3.2 (2) to show that the flux through a plaquette of $\mathcal{L}$ is quantised in multiples of $hc/2e$. This result is a generalisation of that obtained by Abrikosov [Abr] on the basis of the Ginzburg–Landau theory.

Proof of Proposition 9.3.1. (1) Assuming that Φ_1, Φ_2 both satisfy the ODLRO condition (9.3.2), it follows from that equation that

$$\lim_{|y|\to\infty}\left[\Phi_1(x_1,x_2)\overline{\Phi}_1(x_1'+y,x_2'+y) - \Phi_2(x_1,x_2)\overline{\Phi}_2(x_1'+y,x_2'+y)\right] = 0.$$

$$(9.3.3)$$

Since this is valid for all $x_1, x_2, x'_1, x'_2 \in X$, we may replace x_1, x_2 here by arbitrary points x''_1, x''_2, thereby obtaining the equation

$$\lim_{|y| \to \infty} \left[\Phi_1(x''_1, x''_2)\overline{\Phi}_1(x'_1 + y, x'_2 + y) - \Phi_2(x''_1, x''_2)\overline{\Phi}_2(x'_1 + y, x'_2 + y) \right] = 0.$$

(9.3.4)

On multiplying (9.3.3) by $\Phi_2(x''_1, x''_2)$ and (9.3.4) by $\Phi_2(x_1, x_2)$, and then taking the difference, we see that

$$\lim_{|y| \to \infty} \overline{\Phi}_1(x'_1 + y, x'_2 + y)[\Phi_1(x_1, x_2)\Phi_2(x''_1, x''_2) - \Phi_1(x''_1, x''_2)\Phi_2(x_1, x_2)] = 0.$$

Since, by the definition of ODLRO, Φ does not tend to zero at infinity, it follows that the quantity in the square brackets of this last equation vanishes. Therefore, as Φ_1 and Φ_2 are nonzero, by the same stipulation,

$$\Phi_2(x_1, x_2) = c\Phi_1(x_1, x_2) \quad \forall x_1, x_2 \in X,$$

where c is a complex-valued constant. Consequently, as Φ_1, Φ_2 both satisfy Eq. (9.3.2), it follows immediately that c is just a constant phase factor $\exp(ik)$, with k real, as required.

(2) Assuming now that ρ is invariant under time-reversals and that $A = 0$, it follows from Eq. (9.2.11) and the antiautomorphic property of τ that, if Φ satisfies the ODLRO condition (9.3.2), then so too does its complex conjugate $\overline{\Phi}$. Hence, by Part (1) of this proposition, $\overline{\Phi} = \exp(ik)\Phi$, where k is a real-valued constant. In other words, the function $\Phi \exp(ik/2)$ is real-valued, which is the required result.   □

Proof of Proposition 9.3.2. The invariance of the state ρ under the regauged space translational automorphisms, $\xi_B(a)$, implies that

$$\langle \rho; \Psi(x_1, x_2)\Psi^\star(x'_1, x'_2) \rangle \equiv \langle \rho; [\xi_B(a)\Psi(x_1, x_2)][\xi_B(a)\Psi^\star(x'_1, x'_2)] \rangle. \quad (9.3.5)$$

Further, by Eqs. (9.2.16) and (9.3.1),

$$\xi_B(a)\Psi(x_1, x_2) = \Psi(x_1 + a, x_2 + a)\exp\left(\frac{-ie(B \times a).(x_1 + x_2)}{2\hbar c}\right). \quad (9.3.6)$$

Hence, extending the definition of $\xi_B(a)$ to the macroscopic wave functions by the analogous formula, namely

$$\xi_B(a)\Phi(x_1, x_2) = \Phi(x_1 + a, x_2 + a)\exp\left(\frac{-ie(B \times a).(x_1 + x_2)}{2\hbar c}\right), \quad (9.3.7)$$

we infer from Eqs. (9.3.5–9.3.7) that, if Φ satisfies the ODLRO condition (9.3.2) for the state ρ, then so too does $\xi_B(a)\Phi$. In view of Proposition 9.3.1 (1), this signifies that

$$\xi_B(a)\Phi = g(a)\Phi \quad \forall a \in X, \quad (9.3.8)$$

where $g(a)$ is an a-dependent complex constant of unit modulus. Consequently,

$$\xi_B(a)\xi_B(b)\Phi = \xi_B(b)\xi_B(a)\Phi = g(a)g(b)\Phi \quad \forall a,b \in X. \qquad (9.3.9)$$

We now note that equation (9.3.7) implies that $\xi_B(a)$ and $\xi_B(b)$, as applied to the macroscopic wave functions, do not intercommute, but that

$$\xi_B(a+b)[\xi_B(a)\xi_B(b) - \xi_B(b)\xi_B(a)] = 2i\sin\!\left(\frac{eB.(a \times b)}{\hbar c}\right)\!I. \qquad (9.3.10)$$

Hence, by Eqs. (9.3.9) and (9.3.10),

$$\sin\!\left(\frac{eB.(a \times b)}{\hbar c}\right)\!\Phi = 0 \quad \forall a,b \in X.$$

Since $\Phi \neq 0$, by the assumption of ODLRO, this implies that

$$\sin\!\left(\frac{eB.(a \times b)}{\hbar c}\right) = 0 \quad \forall a,b \in X,$$

which is equivalent to the required result, whether X is a lattice or a continuum.

9.4 STATISTICAL THERMODYNAMICS OF THE MODEL AND THE MEISSNER EFFECT

As noted in Comment (2) following the statement of Proposition 9.3.2, the problem of whether ODLRO or a uniform magnetic induction prevails in translationally invariant equilibrium states is a thermodynamical one. In order to address this problem, we now formulate the equilibrium states and the thermodynamical potentials of the system.

9.4.1 The Equilibrium States

The equilibrium states, $\rho_{A,\beta}$, of the system, corresponding to the vector potential A and the inverse temperature β are characterised by the KMS condition (cf. Chapter 5)

$$\langle \rho_{A,\beta}; [\alpha(t)Q_1]Q_2 \rangle = \langle \rho_{A,\beta}; Q_2\alpha_A(t + i\hbar\beta)Q_1 \rangle \quad \forall Q_1, Q_2 \in \mathcal{A}. \qquad (9.4.1)$$

We assume that *this KMS condition is satisfied by one state only*, that is, that it defines $\rho_{A,\beta}$ uniquely. This implies that the equilibrium states on the *observable algebra $\mathcal{A}$* undergo no symmetry breakdown. However, since the electronic subalgebra, $\mathcal{A}_{\text{el}}$, of $\mathcal{A}$ consists of the globally gauge invariant elements of the field algebra, $\mathcal{A}_{\text{el}}^{(F)}$, generated by ψ and $\psi^\star$, it does *not* preclude the possibility that the canonical extension of the model to that larger algebra might support equilibrium states that break the global gauge symmetry; and in fact, as pointed out at the end of Section 3.3.4, the condition of ODLRO on the observable algebra $\mathcal{A}_{el}$ is equivalent to that of global gauge symmetry break-down on the field algebra $\mathcal{A}_{el}^{(F)}$.

Now, assuming that the induction B is uniform, the regauged space translational automorphisms, $\xi_B(a)$, commute with the dynamical group $\alpha_A(\mathbf{R})$, and therefore, if $\rho_{A,\beta}$ satisfies the KMS condition (9.4.1), then so too does $\xi_B^\star(a)\rho_{A,\beta}$. Consequently, in view of our assumption of the uniqueness of the KMS state $\rho_{A,\beta}$, this state is invariant under the regauged space translations $\{\xi_B^\star(a) \mid a \in X\}$ and therefore Proposition 9.3.2 is applicable to it.

Note (i)   The assumption of uniform induction, which has led to this translational invariance, is not applicable to the vortex phase of type II superconductors [DeG, Abr], and we leave our treatment of that phase until Section 9.7.

Note (ii)   It follows from the time reversal covariance and KMS conditions, given by Eqs. (9.2.12) and (9.4.1), respectively, that the KMS property of $\rho_{A,\beta}$ with respect to the automorphisms α_A implies that of $\tau^\star \rho_{A,\beta}$ with respect to α_{-A}. Hence, in view of our assumption of the uniqueness of the KMS state corresponding to any given (A, β), $\tau^\star \rho_{A,\beta} = \rho_{-A,\beta}$.

9.4.2 Thermodynamical Potentials

To formulate the thermodynamical potentials of Σ, we first note that the magnetic induction, B, whether uniform or not, must stem from an external source, of current density J_{ext}, which we take to be classical. This is therefore the source of an external field, H_{ext}, that satisfies the Maxwell equation

$$\text{curl}\, H_{ext} = c^{-1} J_{ext}, \tag{9.4.2}$$

whereas the induction B is generated by both the external current, J_{ext}, and the internal one, $J\ (= \rho_{A,\beta}(j_A))$, that is,

$$\text{curl}\, B = c^{-1}(J_{ext} + J). \tag{9.4.3}$$

We denote the conservative system comprising Σ, the magnetic field, B, and the fixed source of current, J_{ext}, by $\tilde{\Sigma}$. We represent the states, $\tilde{\rho}$, of this latter system by pairs (ρ, B), where ρ and B run through the states of Σ and the classical magnetic fields in X, respectively.

In the above formulation of Σ, B is taken to be a given magnetic field that acts on the system; on the other hand, it may equally be regarded as a J_{ext}-dependent variable of the system $\tilde{\Sigma}$. The following thermodynamical treatment involves both of these mutually compatible views of the induction B.

We denote the free energy density of Σ, corresponding to fixed values of T and of a uniform induction B, by $f(B, T)$; this may be formulated in statistical mechanical terms by the procedure of Chapter 5.

The free energy density of $\tilde{\Sigma}$, on the other hand, comprises $f(B,T)$, together with the energy density of the magnetic field B and that of its interaction with the current source J_{ext}. Further, the energy of this latter interaction is $-c^{-1}\int_X dx A.J_{ext}$, which, in view of the Maxwell equation (9.4.2), reduces to the form $-\int_X dx H_{ext}.B$; while the energy of the magnetic field B is $\int_X dx B^2/2$. Hence, in the case where H_{ext} and B are uniform, the total free energy density of $\tilde{\Sigma}$ is

$$\tilde{f}(B,T,H_{ext}) = f(B,T) - B.H_{ext} + \frac{1}{2}B^2. \qquad (9.4.4)$$

The equilibrium value of B, corresponding to given H_{ext} and T, is then obtained by minimising $\tilde{f}$, and the resultant Gibbs free energy density of $\tilde{\Sigma}$ is

$$\tilde{g}(H_{ext}, T) = \min_B \tilde{f}(B,T,H_{ext}),$$

that is, by Eq. (9.4.4),

$$\tilde{g}(H_{ext}, T) = \min_B \left(f(B,T) - B.H_{ext} + \frac{1}{2}B^2 \right). \qquad (9.4.5)$$

In order to combine this thermodynamical scheme with Proposition 9.3.2 and the subsequent Comments (1)–(3), we now introduce the following assumptions (I)–(IV). The first of these is merely a consequence of Proposition 9.3.2, Comment (1) and the above assumption that the equilibrium states are translationally invariant.

(I)   The equilibrium states of the system Σ that possess the property of ODLRO cannot support a uniform nonzero magnetic induction.

This implies that even an infinitesimal uniform induction suffices to destroy ODLRO and so suggests that if Σ possesses this order in a certain temperature range, then it should undergo a phase transition there at $B = 0$. Accordingly, we make the following assumption, which stems from the hypothesis that the system does indeed exhibit ODLRO in a regime where $B = 0$ and the temperature is sufficiently low. This is our key hypothesis, and the subsequent theory is designed to show that it leads to the electrodynamics observed in superconductors.

(II)   (a) For $B = 0$, the system Σ undergoes a phase transition at a temperature T_c, such that, for $T < T_c$ only, it exhibits the property of ODLRO.
(b) For $B \neq 0$, on the other hand, the thermodynamic potential $f(B,T)$ is a smooth function of both its arguments and converges to a definite limit as $B \to 0$.

In view of (I) and (II), we define the free energy densities of Σ for the ODLRO and normal phases at temperatures below T_c to be

$$f_O(T) = f(0, T) \qquad (9.4.6)$$

and

$$f_{\mathcal{N}}(B, T) = f(B, T) \text{ for } B \neq 0 \qquad (9.4.7)$$

respectively. We note here that, by assumption (IIb), the potential $f_{\mathcal{N}}$ may be extended by continuity to $B = 0$ according to the formula

$$f_{\mathcal{N}}(0, T) = \lim_{B \to 0} f_{\mathcal{N}}(B, T). \qquad (9.4.8)$$

We define the Gibbs potentials, $\tilde{g}_O$ and $\tilde{g}_{\mathcal{N}}$, for the ordered and normal phases of $\tilde{\Sigma}$ to be the versions of $\tilde{g}$ obtained by replacing f by f_O and $f_{\mathcal{N}}$, respectively, in Eq. (9.4.5) and setting B equal to zero in the former phase. Thus, by Eqs. (9.4.4)–(9.4.8),

$$\tilde{g}_O(H_{\text{ext}}, T) = f_O(T) \qquad (9.4.9)$$

and

$$\tilde{g}_{\mathcal{N}}(H_{\text{ext}}, T) = \min_B \left(f_{\mathcal{N}}(B, T) + \frac{1}{2} B^2 - H_{\text{ext}}.B \right). \qquad (9.4.10)$$

The question of whether $\tilde{\Sigma}$ takes up the ordered or normal phase for given (H_{ext}, T) now reduces to that of whether $\tilde{g}_O(H_{\text{ext}}, T)$ is less or greater than $\tilde{g}_{\mathcal{N}}(H_{\text{ext}}, T)$. The following empirically based assumption [Scho, Lo] asserts that the ordered phase prevails at sufficiently low temperatures in the absence of any external magnetic field.

(III) For $T < T_c$ and $H_{\text{ext}} = 0$, the equilibrium state of $\tilde{\Sigma}$ possesses ODLRO, that is, by Eq. (9.4.9),

$$f_O(T) \equiv \tilde{g}_O(0, T) < \tilde{g}_{\mathcal{N}}(0, T) \quad \forall T < T_c. \qquad (9.4.11)$$

Our final assumption concerning $\tilde{\Sigma}$ is the following, which signifies that this system undergoes no subphase transitions within its normal phase for $T < T_c$.

(IV) The potential $\tilde{g}_{\mathcal{N}}(H_{\text{ext}}, T)$ is continuous in both its arguments.

It follows immediately from assumptions (III) and (IV) that, for $T < T_c$ and $|H_{\text{ext}}|$ sufficiently small,

$$\tilde{g}_O(H_{\text{ext}}, T) \equiv f_O(T) < \tilde{g}_{\mathcal{N}}(H_{\text{ext}}, T),$$

which signifies that the equilibrium state of $\tilde{\Sigma}$ enjoys the property of ODLRO and thus carries no magnetic induction. This establishes the following proposition.

Proposition 9.4.1   *Under the above assumptions, and for $T < T_c$ and $|H_{ext}|$ less than some critical value, $H_c(T)$, the equilibrium state of the system $\tilde{\Sigma}$ has the property of ODLRO and carries no magnetic induction. Thus, it exhibits the Meissner effect.*

Comments

(1) This result stems from the stability of ODLRO against applied magnetic fields, and corresponds to a rigidity of the macroscopic wave function similar to that envisaged by F. London [Lo] for the microstate of the system.

(2) In general, when B and H_{ext} might not be uniform, the current density $J(A; x)$ is still given by the formula

$$J(A; x) = \langle \rho_{A,\beta}; j_A(x) \rangle, \qquad (9.4.12)$$

and satisfies the gauge invariance condition

$$J(A; x) = J(A + \nabla \chi; x). \qquad (9.4.13)$$

The constitutive equation governing the relationship between the current and the magnetic field is given by the explicit form of J. That equation, together with the Maxwell equations (9.4.2) and (9.4.3), serves to determine the induction resulting from the application of the field H_{ext} and hence the wavelength dependent magnetic susceptibility of the model. Thus, it provides more detailed information than Proposition 9.4.1.

(3) According to the London theory, the essential difference between a Meissner effect and a normal diamagnetism may be simply expressed in terms of the functional $J_{lin}(A; x)$ obtained by linearising $J(A; x)$ with respect to A. The distinction is just that [Lon, Lo], in the case of the Meissner effect, $J_{lin}(A; x)$ is proportional to the transversely gauged vector potential $A_{tr}(x)$ in a large scale limit; while for normal magnetism it is proportional to $\nabla^2 A_{tr}$ ($= -\text{curl } B$) in the same limit.

(4) The relation between the critical field, $H_c(T)$, and the thermodynamic potentials of the model takes a particularly simple form in the case where the magnetic susceptibility is negligible in the normal phase, for example, in metallic type I superconductors. For, in this case, $f_{\mathcal{N}}(B, T)$ may be equated with $f_{\mathcal{N}}(0, T) \equiv f_{\mathcal{N}}(T)$. Hence, by Eqs. (9.4.9) and (9.4.10), the condition (9.4.11) for the Meissner effect reduces to the form

$$f_{\mathcal{O}}(T) < f_{\mathcal{N}}(T) - \frac{1}{2} H_{ext}^2,$$

and, consequently, the critical field, H_c is given by the standard formula [Lo]

$$H_c(T) = (f_{\mathcal{N}}(T) - f_{\mathcal{O}}(T))^{1/2}. \qquad (9.4.14)$$

9.5 FLUX QUANTISATION

We now consider the situation where a cylindrical region $\Gamma(\subset X)$ is removed from the body of Σ, so that the system occupies the complementary region $X' = X \backslash \Gamma$. We assume that both ODLRO and the Meissner effect prevail in this geometry for $T < T_c$. Here, the basis of the assumption of the latter effect is not Proposition 9.3.2, since that required the translational symmetry of X, but the hypothesis that the constitutive equation for the current, $J(A;x)$, remains unchanged, except possibly for inessential modifications due to the new boundary conditions.

Thus, under these assumptions, a sufficiently weak magnetic field in the tunnel Γ will be screened from the body of the system by currents in a "surface layer", and magnetic flux will be trapped in Γ and that layer.

We derive the quantisation rule for this trapped flux, $\mathcal{F}$, by an adaptation of the argument of Nieh, Su and Zhao [NSZ], from general properties of the macroscopic wave function Φ that stem from gauge covariance, symmetry and topology. This argument is centered exclusively on the properties of the model in the asymptotic region, far from the tunnel, where the magnetic induction can be taken to be zero and where we assume that translational covariance prevails. Thus, although we shall not keep repeating the point, the properties we invoke here are all asymptotic ones that become exact only in a limit where the regions in question are infinitely far from the tunnel.

Suppose now that C is a closed loop in X', in the deep interior of the body, that encircles the tunnel Γ. Then it follows immediately from the above specifications and Stokes's theorem that

$$\int_C A.dl = \mathcal{F}. \tag{9.5.1}$$

Thus, as in the Aharonov-Bohm effect [AB], the magnetic vector potential survives, with nontrivial circulation, even in the field-free region.

On the other hand, if C' is a loop in the asymptotic field-free region that does not encircle the tunnel, then, by Stokes's theorem,

$$\int_{C'} A.dl = 0. \tag{9.5.2}$$

We now use this formula to obtain an asymptotic covariance with respect to certain regauged space translations, similar to those given by Eq. (9.2.16). To this end, we choose C' to be the parallelogram whose vertices are $x - a, x, x + h$ and $x - a + h$. Thus, denoting by $\int_y^z A.dl$ the integral of A along the straight line leading from y to z, we see from Eq. (9.5.2) that

$$\left(\int_{x-a}^{x} + \int_{x}^{x+h} + \int_{x+h}^{x-a+h} + \int_{x-a+h}^{x-a} \right) A.dl = 0.$$

Hence, defining

$$\theta_a(x) = \int_{x-a}^{x} A.dl, \qquad (9.5.3)$$

we obtain the formula

$$\theta_a(x+h) - \theta_a(x) = \left(\int_{x}^{x+h} - \int_{x-a}^{x-a+h} \right) A.dl,$$

from which it follows that

$$\nabla \theta_a(x) = A(x) - A(x-a). \qquad (9.5.4)$$

We now invoke the covariance of the model with respect to the gauge transformation

$$A \rightarrow A - \nabla \theta_a, \quad \psi \rightarrow \psi \exp\left(\frac{ie\theta_a}{2\hbar c} \right),$$

which, in view of Eq. (9.5.4), reduces to the form

$$A(x) \rightarrow A(x-a), \quad \psi(x) \rightarrow \psi(x) \exp\left(\frac{ie\theta_a(x)}{\hbar c} \right). \qquad (9.5.5)$$

Thus, the dynamics is covariant with respect to this transformation in the
asymptotic induction-free region. We also assume that it is covariant there
with respect to space translations

$$A(x) \rightarrow A(x+a), \quad \psi(x) \rightarrow \psi(x+a). \qquad (9.5.6)$$

Hence, by Eqs. (9.5.5) and (9.5.6), the dynamics is asymptotically covariant
with respect to the *regauged space translation* (cf. Eq. (9.2.16))

$$A(x) \rightarrow A(x), \quad \psi(x) \rightarrow \psi(x+a) \exp\left(\frac{ie\theta_a(x+a)}{\hbar c} \right). \qquad (9.5.7)$$

Correspondingly, assuming uniqueness of the equilibrium state for given
(A, β), as in Section 9.4 (cf. assumption (IIa)), the expectation value of
$\Psi(x_1, x_2)\Psi^{\star}(x_1', x_2')$ becomes (asymptotically) invariant under this transformation. Hence, defining

$$\Phi_a(x_1, x_2) := \Phi(x_1 + a, x_2 + a) \exp\left(\frac{i(\theta_a(x_1 + a) + \theta_a(x_2 + a))}{\hbar c} \right), \quad (9.5.8)$$

it follows from Eqs. (9.3.1), (9.5.7) and (9.5.8) that the ODLRO condition
(9.3.2) remains valid when Φ is replaced is Φ_a. Consequently, by Proposition
9.3.1 (1), $\Phi_a = \exp(i\eta(a))\Phi$, for some a-dependent phase angle $\eta(a)$, that is,
by Eqs. (9.5.3) and (9.5.8),

$$\Phi(x_1 + a, x_2 + a) = \Phi(x_1, x_2) \exp\left(i\eta(a) - \frac{ie}{\hbar c} \left(\int_{x_1}^{x_1+a} + \int_{x_2}^{x_2+a} \right) A.dl \right). (9.5.9)$$

Now let $a_0(= 0), a_1, a_2, ..., a_n$ be a set of spatial displacements whose sum is zero and, for $j = 1, 2$ and $r = 0, ..., n$, let $P_{j,r}$ be the point whose position vector is $x_j + a_0 + \cdots + a_r$. We choose the x_j's and the a_r's in such a way that, for $j = 1, 2$, the contour C_j, formed from the segments $P_{j,0}P_{j,1}, ..., P_{j,n-1}P_{j,n}$, lies in the asymptotic region and encircles the tunnel Γ. Then by Eq. (9.5.1),

$$\int_{C_j} A.dl = \mathcal{F} \quad \text{for } j = 1, 2. \tag{9.5.10}$$

Furthermore, by Eq. (9.5.9),

$$\Phi(x_1 + a_1 + \cdots + a_r, x_2 + a_1 + \cdots + a_r)$$

$$= \Phi(x_1 + a_1 + \cdots + a_{r-1}, x_2 + a_1 + \cdots + a_{r-1})$$

$$\times \exp\left(i\eta(a_r) - \frac{ie}{\hbar c} \sum_{j=1,2} \int_{x_1+a_1+ +a_{r-1}}^{x_1+a_1+ a_r} A.dl \right)$$

for $r = 1, ..., n$.

Hence, in view of the above specifications of the contours C_1 and C_2, together with the fact that both a_0 and $\sum_{r=1}^{n} a_r$ are zero, the recursive application of this equation yields the formula

$$\Phi(x_1, x_2) = \Phi(x_1, x_2) \exp\left(i\sum_{r=1}^{n} \eta(a_r) - \frac{ie}{\hbar c} \sum_{j=1}^{2} \int_{C_j} A.dl \right),$$

that is, by Eq. (9.5.10),

$$\Phi(x_1, x_2) = \Phi(x_1, x_2) \exp\left(i\eta - 2ie \frac{\mathcal{F}}{\hbar c} \right), \tag{9.5.11}$$

where

$$\eta = \eta(a_1) + \cdots + \eta(a_n). \tag{9.5.12}$$

Hence, since the ODLRO condition demands that Φ be nonzero in the asymptotic region,

$$\exp\left(i(\eta - \frac{2e\mathcal{F}}{\hbar c}) \right) = 1. \tag{9.5.13}$$

In order to disentangle the two terms in the exponent of this equation, we now employ the same argument with x_1, x_2 replaced by points x_1', x_2', and C_1, C_2 by corresponding contours C_1', C_2' that lie in the asymptotic region *but do not encircle* Γ. It follows from this last condition that, instead of Eq. (9.5.13), we obtain the equation

$$\exp(i\eta) = 1.$$

Hence, by Eq. (9.5.13),

$$\exp(-2ie\,\mathcal{F}/\hbar c) = 1,$$

which signifies that $\mathcal{F}$ must be an integral multiple of $hc/2e$. Thus, we have established the following proposition, which accords with the experiment of Deaver and Fairbank [DF].

Proposition 9.5.1 *Under the above assumptions, the trapped flux, $\mathcal{F}$, is quantised in integral multiples of $hc/2e$, that is,*

$$\mathcal{F} = \frac{\nu hc}{2e}, \tag{9.5.14}$$

where ν is an integer.

9.6 METASTABILITY OF SUPERCURRENTS AND SUPERSELECTION RULES

The phenomenon of superconductivity itself is that of the persistence of currents induced in a ring, in the absence of any applied electric field. According to the London macroscopic theory [Lon, Lo], this phenomenon is intimately connected to the Meissner effect in that the supercurrents are the source of a magnetic field, which they themselves screen from the interior of the body. Thus, the currents and magnetic field stabilise one another. Further, it is clear that the supercurrent carrying states must be *metastable*, rather than absolutely stable, since both the currents and the magnetic field carry positive energy.

Our objective now is to provide a quantum mechanical basis for the above phenomenological picture. We base our considerations on the quantum mechanics of a system occupying a multiply-connected region, for example, one with the topology of a torus, *in the absence of any external sources*. For simplicity, we employ the model of Section 9.5.

So again we consider the state of the system in which magnetic induction is trapped in the tunnel, Γ, and a "surface region" of the matter. The supercurrents are then currents that implement the Meissner effect by screening the induction out of the deep interior of the body. Our objective now is to investigate the stability properties of the trapped magnetic flux and the associated supercurrents, corresponding to the flux quantum number ν of Proposition 9.5.1. Evidently, ν is related to B by the formula

$$\int_S B.dS = \nu hc/2e, \tag{9.6.1}$$

where S is a surface in the deep interior of the body that is orthogonal to the generators of the tunnel Γ.

We start by noting that since, by Eqs. (9.2.4), the observables representing the position-dependent densities of charge, current and polarisation are fully gauge invariant, their equilibrium expectation values may be expressed as functionals, Q, J and M, respectively, of B, as defined over the whole space X, and the position x in the material region X', according to the formulae

$$Q(B;x) = \langle \rho_{A,\beta}; q(x) \rangle, \tag{9.6.2}$$

$$J(B;x) = \langle \rho_{A,\beta}; j_A(x) \rangle \tag{9.6.3}$$

and

$$M(B;x) = \langle \rho_{A,\beta}; m(x) \rangle. \tag{9.6.4}$$

We now extend the definitions of Q, J and M to the whole space X by the specification that they vanish in the tunnel, Γ.

Since we assume here that there are no external sources, the magnetic field $H(x) = B(x) - M(B;x)$ and induction B satisfy the Maxwell equations

$$\text{curl}\, H(x) = c^{-1} J(B;x) \quad \text{and} \quad \text{div}\, B = 0,$$

that is,

$$\text{curl}\, B(x) = \text{curl}\, M(B;x) + c^{-1} J(B;x) \quad \text{and} \quad \text{div}\, B(x) = 0 \quad \forall x \in X. \tag{9.6.5}$$

We now make the following assumptions concerning the solution of these equations for B.

(I)   There is a unique solution, B_ν, of the equations (9.6.5) for B, subject to the constraint (9.6.1) and the standard conditions on the boundary of Γ and infinitely far from this cylinder, where B vanishes, by Meissner effect.

(II)   Given the vector potential A_ν representing the induction B_ν, there is a unique equilibrium state, $\rho_{A_\nu,\beta}$, of Σ that satisfies the KMS equilibrium condition (9.4.1). We denote by $\tilde{\rho}_{A_\nu,\beta} \equiv (\rho_{A_\nu,\beta}, B_\nu)$ the corresponding state of $\tilde{\Sigma}$.[4]

It follows from our specifications that both $J(B;\cdot)$ and B are nonzero in a surface region of the matter when $\tilde{\Sigma}$ is in the state $\tilde{\rho}_{A_\nu,\beta}$. Therefore, the question of the persistence of the supercurrent and trapped magnetic field reduces to that of the metastability of this state. Again, we characterise metastability[5] by thermodynamical stability against localised perturbations of state.

---

[4]   Recall that a state of $\tilde{\Sigma}$, as defined in the paragraph following Eq. (9.4.3), comprises a state, ρ, of Σ and a magnetic induction B.

[5]   Cf. section 5.4 and, for more detailed discussion, [Se8] and [Se2, Chapter 6].

Thus, we reduce the problem of the persistence of supercurrents to that of the thermodynamical stability of the state $\tilde{\rho}_{A_\nu,\beta}$ of $\tilde{\Sigma}$ against local perturbations. To formulate this problem, we define the local perturbations of the induction, B_ν, to comprise the set, $\mathcal{B}$, of divergence-free vector fields, b, with compact support. Since b vanishes outside some bounded region and div $b = 0$, these perturbations do not change the flux defined by the left-hand side of Eq. (9.6.1). Further, as we prove in the appendix, any divergence-free vector field b with compact support may be represented by a vector potential a, whose support is also compact. In the following argument, we represent the elements b of $\mathcal{B}$ exclusively by vector potentials a of compact support.

We employ the following procedure to formulate the incremental free energies, $\phi_\nu(b)$ and $\tilde{\phi}_\nu(b)$, of Σ and $\tilde{\Sigma}$, respectively, due to the perturbation of B_ν by $b(\in \mathcal{B})$. By Eqs. (9.2.4) and (9.2.5), a perturbation λb of B_ν, with $\lambda \in \mathbf{R}$, leads to a change in the Hamiltonian of Σ given by

$$K(\lambda) = -\int_{X'} (\lambda c^{-1} j_{A_\nu}.a + \frac{1}{2} e\lambda^2 c^{-2} qa^2 + \lambda m.b)dx. \qquad (9.6.6)$$

Evidently, this is a local observable of the system, since b and a have bounded support. We now note that, by classical thermodynamics, the increment in free energy of a system, due to an infinitesimal change ΔH in its Hamiltonian, is given by the expectation value of ΔH for the unperturbed equilibrium state. In the present situation, this implies that the change in free energy of Σ due to an infinitesimal increment $d\lambda$ in the parameter λ of Eq. (9.6.6), is $\langle \rho_{A_\nu + \lambda a, \beta}; K'(\lambda) \rangle d\lambda$, where $K'(\lambda)$ is the derivative of $K(\lambda)$ with respect to λ. Hence, denoting by $\phi_\nu(b)$ the increment in the free energy of Σ due to the change of induction from B_ν to $B_\nu + b$,

$$\frac{\partial}{\partial\lambda}\phi_\nu(\lambda b) = \langle \rho_{A_\nu + \lambda a, \beta}; K'(\lambda) \rangle. \qquad (9.6.7)$$

Further, by Eqs. (9.2.4) and (9.6.6),

$$K'(\lambda) = -\int_{X'}(c^{-1}j_{A_\nu}.a + e\lambda c^{-2}qa^2 + m.b)dx \equiv -\int_{X'}(c^{-1}j_{A_\nu + \lambda a}.a + m.b)dx,$$

and therefore, by Eqs. (9.6.2)–(9.6.4) and (9.6.7),

$$\frac{\partial}{\partial\lambda}\phi_\nu(\lambda b) = -\int_{X'} dx\Big[c^{-1}J(B_\nu + \lambda b, x).a(x) + \mathcal{M}(B_\nu + \lambda b, x).b(x)\Big].$$

Here the domain of integration may equivalently be taken to be X, since both the current J and the polarisation $\mathcal{M}$ vanish in the tunnel Γ, and consequently we may re-express the last equation as

$$\frac{\partial}{\partial\lambda}\phi_\nu(\lambda b) = -\int_{X} dx\Big[c^{-1}J(B_\nu + \lambda b, x).a(x) + \mathcal{M}(B_\nu + \lambda b, x).b(x)\Big]. \qquad (9.6.8)$$

Moreover, since $\phi_\nu(0) = 0$, by definition of incremental free energy, this formula implies that

$$\phi_\nu(b) = -\int_0^1 d\lambda \int_X dx \Big[c^{-1} J(B_\nu + \lambda b, x).a(x) + \mathcal{M}(B_\nu + \lambda b, x).b(x) \Big].(9.6.9)$$

The corresponding incremental free energy of $\tilde{\Sigma}$ is

$$\tilde{\phi}_\nu(b) = \phi_\nu(b) + \int_X dx \Big(b(x).B_\nu(x) + \frac{1}{2} b(x)^2 \Big), \qquad (9.6.10)$$

the integral being the contribution due to the magnetic field energy.

We now recall that, for fixed A, the KMS condition (9.4.1) ensures that the free energy of Σ cannot be reduced by any localised modification of its equilibrium state (cf. Section 5.3.2). Hence, the state $\tilde{\rho}_{A_\nu,\beta}(= (\rho_{A_\nu,\beta}, B_\nu))$ of $\tilde{\Sigma}$ will be thermodynamically stable against all local perturbations if $\tilde{\phi}_\nu$ is minimised at $b = 0$. The following proposition asserts that this functional is certainly stationary there.

Proposition 9.6.1 *Under the above assumptions, the functional $\tilde{\phi}_\nu$ on $\mathcal{B}$ is stationary at $b = 0$, that is,*

$$\frac{\partial}{\partial \lambda} \tilde{\phi}_\nu(\lambda b)_{|\lambda=0} = 0 \quad \forall b \in \mathcal{B}. \qquad (9.6.11)$$

Proof. By Eqs. (9.6.8) and (9.6.10), the left-hand side of Eq. (9.6.11) is equal to

$$\int_X dx \Big[B_\nu(x).b(x) - c^{-1} J(B_\nu, x).a(x) - \mathcal{M}(B_\nu, x).b(x) \Big].$$

Since $b = \text{curl } a$, and since both b and a have compact support, this expression reduces to

$$\int_X dx \Big[\text{curl } B_\nu(x) - c^{-1} J(B_\nu, x) - \text{curl } \mathcal{M}(B_\nu, x) \Big].a(x),$$

which vanishes, by Eq. (9.6.5). □

Metastability Conditions

In view of the this proposition, we have the following possibilities.

(A) $\tilde{\phi}_\nu$ is minimised at $b = 0$. In this case, the current-carrying state $\tilde{\rho}_{A_\nu,\beta}$ is stable under all local perturbations, and so is metastable in the sense defined in Section 5.4. The lifetime of the state is then infinite, and so the current persists forever.

(B) $\tilde{\phi}_\nu$ attains a local, but not absolute, minimum at $b = 0$. In this case, the state $\tilde{\rho}_{A_\nu,\beta}$ will still be metastable, in the sense proposed by Penrose and Lebowitz [PL], if its lifetime is enormously long by normal laboratory observational standards. In fact, this situation can prevail if the local minimum at $b = 0$ is surrounded by maxima in such a way that the activation energy required for escape from the state $\tilde{\rho}_{A_\nu,\beta}$ is very large by comparison with $k_{\text{Boltzmann}}T.$ [6]

(C) $\tilde{\phi}_\nu$ is not even locally minimised at the stationary point $b = 0$. In this case, the state $\tilde{\rho}_{A_\nu,\beta}$ is unstable against even against certain infinitesimal local perturbations and thus cannot be metastable.

In view of these considerations, we conclude that the conditions for metastability of the state $\tilde{\rho}_{A_\nu,\beta}$, and thus for the persistence of the associated currents, is that either (A) or (B) prevails; and that, in the latter case, the activation energy governing the escape from the state $\tilde{\rho}_{A_\nu,\beta}$ is very large by comparison with $k_{\text{Boltzmann}}T$.

Superselection Rules

We may gain another perspective of the metastability of supercurrents by relating it to a *superselection rule*, forbidding transitions between states with different flux quantum numbers. For this purpose, we note that, in the asymptotic field-free region, A_ν is the gradient of a scalar potential, χ_ν. Further, by Eq. (9.5.1) and Proposition 9.5.1, this potential is many-valued, its change over a contour passing once round the tunnel being

$$[\chi_\nu] = \nu hc/2e. \tag{9.6.12}$$

We note that, since the contour lies in the asymptotic region, this formula implies that the quantum number ν represents a *global* property of the state $\rho_{A_\nu,\beta}$. Furthermore, since a reversal of the magnetic field in the tunnel corresponds to a reversal of this quantum number,

$$\chi_{-\nu} = -\chi_\nu. \tag{9.6.13}$$

We now observe that, by the quantisation rule (9.6.12), $\exp(2ie\chi_\nu/\hbar c)$ is a single-valued function of position. Thus, by gauge covariance of the second kind, the dynamics of the model is covariant with respect to the transformation

$$A \to A + 2\nabla\chi_\nu, \quad \psi \to \psi\exp(2ie\chi_\nu/\hbar c). \tag{9.6.14}$$

Hence, by Eqs. (9.3.1) and (9.3.2), the macroscopic wave function, Φ_A, corresponding to the vector potential A, satisfies the condition

---

[6] Cf. [Se8], [Se2, Chapter 6] and [SS] for treatments of other models that support states whose metastability arises in this way.

$$\Phi_{A+2\nabla\chi_\nu}(x_1, x_2) = \Phi_A(x_1, x_2) \exp\left(\frac{2ie}{\hbar c}(\chi_\nu(x_1) + \chi_\nu(x_2))\right).$$

Consequently, choosing $A = -\nabla\chi_\nu := -A_\nu$, and denoting Φ_{A_ν} by Φ_ν, it follows from Eq. (9.6.13) that

$$\Phi_{-\nu}(x_1, x_2) = \Phi_\nu(x_1, x_2) \exp\left(\frac{2ie}{\hbar c}(\chi_\nu(x_1) + \chi_\nu(x_2))\right). \qquad (9.6.15)$$

On the other hand, as noted at the end of Section 9.4.1, $\rho_{-A,\beta} \equiv \tau^\star \rho_{A,\beta}$, while by Eqs. (9.2.1), (9.2.11) and (9.3.1), $\tau^\star \Psi(x_1, x_2) = -\Psi^\star(x_2, x_1)$. Hence, by our defintion of Φ_ν, $\overline{\Phi}_\nu(x_2, x_1)$ satisfies the ODLRO formula (9.3.2) for the macroscopic wave function $\Phi(x_1, x_2)$ when the state if $\rho_{A_{-\nu,\beta}}$; and consequently, by Proposition 9.3.1 (1),

$$\Phi_{-\nu}(x_1, x_2) = \overline{\Phi}_\nu(x_2, x_1) \exp(ik), \qquad (9.6.16)$$

where k is a real constant. On combining this equation with Eq. (9.1.15), we obtain the following result.

Proposition 9.6.2 *Assuming covariance with respect to gauge transformations and time reversals, the phase of the macroscopic wave function Φ_ν in the asymptotic field-free region is given by the formula*

$$\arg\left(\Phi_\nu(x_1, x_2)\right) + \arg\left(\Phi_\nu(x_2, x_1)\right) = \frac{2e}{\hbar c}(\chi(x_1) + \chi(x_2)) + const. \quad (9.6.17)$$

Consequence: A Superselection Rule

In view of Eq. (9.6.12) and the remark immediately after it, this proposition implies that states corresponding to the different flux quantum numbers ν are *globally, that is, macroscopically,* different from one another, and that consequently there is a *superselection rule* that forbids transitions between them via the agency of localised operations (cf. [Haa1]). Thus, the superselection sector comprising the *family* of states with fixed flux quantum number ν is stable against such operations, and the metastability conditions (A) and (B) pertain to the stability of the particular state $\tilde{\rho}_{A_\nu,\beta}$ against any local modifications.

9.7 NOTE ON TYPE II SUPERCONDUCTORS

The hallmark of a type II superconductor is that, for a certain range of the applied magnetic field strength, it supports a phase in which the electronic fluid forms a lattice, $\mathcal{L}$, of vortices and the magnetic induction has the periodicity of $\mathcal{L}$ (cf. [Abr, DeG]). Thus, in this phase, it exhibits an incomplete Meissner effect. Abrikosov [Abr] derived the theory of the profiles of the vortices and the magnetic flux in this phase from the phenomenological

Ginzburg–Landau theory, and argued, on that basis, that the flux, $\mathcal{F}_{II}$, through a plaquette of the lattice is $chle^{\star}$, where $-e^{\star}$ is the charge of a single carrier.

We now obtain a generalisation of this latter result on the basis of our quantum treatment of Sections 9.3 and 9.4. Specifically, assuming that the state of the system Σ has the translational symmetry of the Abrikosov lattice $\mathcal{L}$, we show that the flux $\mathcal{F}_{II}$ through a plaquette is given by the formula

$$\mathcal{F}_{II} = vhcle^{\star} \equiv vhc/2e, \qquad (9.7.1)$$

where v is an integer and the identification of $e^{\star}$ with $2e$ arising from the Cooper pairing. Thus $\mathcal{F}_{II}$ takes exactly the same form as the quantised flux in a tunnel cut out of material in a fully superconductive phase (cf. Proposition 9.5.1).

To obtain this result, we assume, following Abrikosov, that the state of the model Σ is invariant under the crystallographic group of a two-dimensional Bravais lattice, $\mathcal{L}$ ($\subset X$), in the plane perpendicular to the magnetic field. The sites, a, of $\mathcal{L}$ are thus integral linear combinations of basis vectors a_1 and a_2. We denote by $\mathcal{L}^{\star}$ the lattice dual to $\mathcal{L}$. Its basis vectors, k_1 and k_2, are defined by the condition that

$$a_i.k_j = 2\pi\delta_{ij} \quad \forall i,j = 1,2. \qquad (9.7.2)$$

The assumption of $\mathcal{L}$-translational invariance implies that the induction B may be expressed as a Fourier series

$$B(x) = \sum\nolimits_{k \in \mathcal{L}^{\star}} B_k \exp(ik.x). \qquad (9.7.3)$$

Thus, its zero wave-vector Fourier component, B_0, is simply the space-average of B over a plaquette.

The above formula for B may be represented by the transversely gauged vector potential

$$A(x) = \frac{1}{2}B_0 \times x + \sum_{k \neq 0} \frac{ik \times B_k}{k^2} \exp(ik.x), \qquad (9.7.4)$$

from which it follows that

$$A(x+a) = A(x) + \frac{1}{2}B_0 \times a \quad \forall a \in \mathcal{L}. \qquad (9.7.5)$$

This corresponds precisely to the formula obtained from Eq. (9.2.13) for the case of a uniform induction, but with X replaced by $\mathcal{L}$ and B by B_0. Hence, the argument leading to the proof of Proposition 9.3.2 (2) becomes applicable when the displacements a and b are restricted to $\mathcal{L}$ and B is replaced by B_0. Consequently,

$$\frac{e}{\hbar c}B_0.(a \times b) = \nu\pi \quad \forall \ a, b \in \mathcal{L},$$

where ν is an integer. This implies the following proposition.

Proposition 9.7.1 *Under the specified assumptions, the magnetic flux, $\mathcal{F}_{II}$, through a plaquette of the lattice $\mathcal{L}$ is given by the formula (9.7.1), with ν an integer.*

Comment For $\nu = 1$, this result reduces to that of Abrikosov [Abr], as given by his Eq. (34), for a superconductor with carrier charge $-2e$. This restriction of ν to the value unity stems from Abrikosov's apparently ad hoc assumption that the integer N of his Eq. (20) was equal to unity.

9.8 CONCLUDING REMARKS

The theory presented here provides a general derivation of the electro-dynamics of superconductors from assumptions of off-diagonal long range order, gauge covariance and thermodynamic stability. In our view, the outstanding general problems are the following.

(1) The substantiation of our various assumptions for concrete models, including ones of ceramic superconductors;

(2) The precise characterisation of the metastability of supercurrents, that is, the determination of which of the alternatives (A) and (B) of Section 9.6 prevails;

(3) The reformulation of the whole theory on a fully quantum basis, in which the electromagnetic field as well as the matter is treated quantum mechanically. We have provided an indication of how this might be achieved in [Se12, Section 4].

APPENDIX A: VECTOR POTENTIALS REPRESENTING MAGNETIC FIELDS WITH COMPACT SUPPORT

Here we show that any divergence-free vector field of compact support in a three-dimensional Euclidean space, X, may be represented as the curl of a vector potential whose support is also compact.

Thus, we assume that b is a divergence-free vector field with bounded support, D_b, in X, and we suppose that a_1 is a vector potential that represents b, that is, that satisfies the equation

$$b = \operatorname{curl} a_1. \tag{A.1}$$

We choose x_0 to be a fixed reference point in $D_b^{(c)} := X \backslash D_b$, and, for $x \in D^{(c)}$, we construct a curve, C, in $D_b^{(c)}$, leading from x_0 to x, and we define

$$\chi_1(x) = \int_C a_1 . dl. \tag{A.2}$$

To show that this function is independent of the choice of C, we note that, if C' is another curve in $D_b^{(c)}$ that leads from x_0 to x, then

$$\left(\int_C - \int_{C'} \right) a_1 . dl = \int_S b . dS, \tag{A.3}$$

where S is a closed surface bounded by C and C'. Since X is three-dimensional, we may choose S so that it does not intersect D_b, thereby ensuring that the right-hand side of Eq. (A.3) vanishes. Hence, the function χ_1, defined by Eq. (A.2), does not depend on the choice of the curve C. It is therefore a simple consequence of that formula that

$$a_1 = \nabla \chi_1 \quad \text{in } D_b^{(c)}. \tag{A.4}$$

We now introduce

(a) a compact subset, D, of X, that contains D_b and whose boundary does not intersect that of the latter region, and

(b) a smooth, real-valued function, g, on X that takes the value unity in D_b and whose support lies in D.

We then define the scalar field χ and the vector field a by the prescription that

$$\chi = 0 \quad \text{in } D_b, \tag{A.5a}$$

$$\chi = (1 - g)\chi_1 \quad \text{in } D_b^{(c)} \tag{A.5b}$$

and

$$a = a_1 - \nabla \chi. \tag{A.6}$$

It now follows from Eqs. (A.1) and (A.6) that

$$b = \text{curl } a, \tag{A.7}$$

while Eqs. (A.4)–(A.6), together with the above specifications (a) and (b), imply that a vanishes in the complement $D^{(c)} := X \backslash D$ of the bounded region D. Thus, we conclude that b is represented by a vector potential, a, with compact support.

Part IV

Ordered and chaotic structures far from equilibrium: prospectus

The study of ordered and chaotic structures far from thermal equilibrium is of fundamental importance throughout the natural sciences. Such structures are, in fact, ubiquitous in physics, chemistry and biology, as the following examples indicate.

- The light emitted by a laser undergoes a transition from a normal, that is, incoherent, phase to a coherent one when the intensity of the random optical pumping exceeds a certain critical value (cf. [Ha1]).

- Hydrodynamical turbulence, that is, chaotic flow, generically prevails at sufficiently high values of the Reynolds number [LL1].

- A fluid heated from below takes up a cellular structure when the temperature gradient becomes sufficiently large (cf. [Cha, GP]). This is the Bénard effect. The ordering of its cellular states constitutes a translational symmetry breakdown.

- Various chemical reactions produce patterns that are periodic in both space and time when the relative concentrations of the reactants lie within certain ranges (cf. [GP]). These patterns therefore carry order corresponding to the breakdown of both space and time translational invariance.

- Biological cells, which are randomly pumped with energy from their environment, exhibit off-diagonal long range order when the pumping is sufficiently strong [Fr6]. The mechanism is captured by Fröhlich's pumped phonon model [Fr1], whose ordered phase is that of a nonequilbrium Bose–Einstein condensation.[1]

This part of the book is concerned with the theory of phase transitions far from equilibrium in dissipative quantum systems. We start in Chapter 10 by presenting a general scheme for the theory of nonequilibrium phase structures: it can be seen that this has some parallels with its equilibrium counterparts, in accordance with Haken's Synergetics [Ha2]. In Chapter 11, we present a

---

[1] See [Se2, Section 7.4] and [Duf1,2] for rigorous treatments of this model.

rigorous treatment of a dissipative quantum model, representing a laser, that provides a realisation of the scheme of Chapter 10. To be specific, this model is a version, due to Alli and Sewell [AlSe], of that of Dicke [Dic], Hepp and Lieb [HL1], as recast in terms of quantum dynamical semigroups. We prove that it exhibits transitions between normal, coherent and chaotic radiation at critical values of a parameter representing the optical pumping strength; and further that, in all phases, the time-dependent microstate is determined by a maximum entropy principle, subject to the constraints imposed by the instantaneous values of a certain set of macroscopic variables.

Chapter 10

Schematic approach to a theory of nonequlibrium phase transitions, order and chaos

Phase transitions far from equilibrium abound in physics [GH, Ha1, Ha2, GP], chemistry [GP] and biology [Fr1,6], and are indeed crucial to fundamental questions concerning the generation of ordered and chaotic structures in the natural sciences. The quantum theory of these transitions, however, is still rather undeveloped at the level of mathematical physics, being confined to a few special models, such as those of a laser [HL1, AlSe, BS] and a cell [Duf1, Duf2].

Our objective here is to propose a scheme for a general, if rudimentary, framework for such a theory. The picture it provides of nonequilibrium phase structures has some remarkable parallels with those of equilibrium phases. This accords with Haken's Synergetics [Ha2], wherein different classes of phenomena conform to similar mathematical structures. A concrete realisation of the scheme of this chapter is provided by the laser model of Chapter 11.

The Quantum Picture

At the microscopic level, we take our generic model, Σ, to be that of an open, dissipative quantum system $(\mathcal{A}, S, T)$, where $\mathcal{A}$ is a $C^\star$ or $W^\star$-algebra of observables, S is a space of states on $\mathcal{A}$, and $T(\mathbf{R}_+)$ is a quantum dynamical semigroup[1] of transformations of $\mathcal{A}$. This model is designed to represent an infinitely extended system of particles, whose dissipativity stems from the coupling of its finite constituent parts to reservoirs. The influence of these reservoirs is then buried in the structure of $T(\mathbf{R}_+)$ (cf. Section 4.4). In general, the semigroup T depends on a set of control variables, $\theta = (\theta_1, ..., \theta_k)$ that are typically parameters, such as friction constants or energy pumping strengths, that govern the dissipative properties of the system. These are therefore quite different from the the thermodynamical control variables denoted by the same symbol in Chapter 6. We henceforth indicate the dependence of the dynamical semigroup on the control variables by a suffix, θ, to T. We assume that the

---

[1] See Section 4.2.2 for the definition of quantum dynamical semigroups.

range of accessible values of θ form a connected subspace, Θ, of a Euclidean continuum. We term Θ the control space.

The Macroscopic Picture of Σ

We assume that this is given by a classical dissipative dynamical system $\mathcal{M} = (X, \tau_\theta)$, where X, the phase space, is a connected subapace of a Euclidean continuum, and $\tau_\theta(\mathbf{R}_+)$ is a one-parameter semigroup of transformations of X, induced by the quantum semigroup T_θ. Here, the points of X representing the values of a set of variables $x = (x_1, ..., x_l)$, corresponding to macroscopic observables of Σ, that evolve according to the deterministic law whereby $\tau_\theta(t)x$ is the evolute of x at time t. In the interesting cases, the transformations $\tau_\theta(t)$ are nonlinear.

The properties of nonlinear classical dissipative systems (X, τ) have been extensively studied, independently of their quantum origins (cf. [Ru3]). In particular, the generic picture obtained by Ruelle and Takens [RT] may be described as follows.

There is a connected domain, Θ_0, of the control space, Θ, such that, for each θ in Θ_0, the dynamical system (X, τ_θ) has just one *stable attractor*, that is, a subset, A_θ, of X that is characterised by the following conditions.

(A.1) A_θ is stable under τ_θ;

(A.2) the orbit, $\{\tau_\theta(t)x \mid t \in \mathbf{R}_+\}$, of any point x in A_θ is a dense subset of A_θ; and

(A.3) the orbits emanating from points in any sufficiently small neighborhood $\mathcal{N}_\theta$ of A_θ all converge to $\mathcal{A}_\theta$ in the long time limit.

Furthermore, depending on the value of the control variable θ, the attractor A_θ is *either* a fixed point *or* a periodic orbit *or* a so-called "strange attractor", wherein the motion is chaotic, in that is is hypersensitive to changes in the initial conditions. The domain Θ_0 is then the union of the subdomains Θ_1, Θ_2 and Θ_3, for which the attractors correspond to fixed points, periodic orbits and strange attractors, respectively. In the Ruelle–Takens scheme, Θ_1 and Θ_3 are contiguous with Θ_2, but not with one another, and the change in the structure of A_θ that arises when θ crosses a boundary between these regions corresponds to a Hopf bifurcation. Specifically, the fixed points of τ_θ lose their stability and give way to stable periodic orbits as θ crosses from Θ_1 to Θ_2; while those periodic orbits become unstable and give way to strange attractors as θ crosses from Θ_2 to Θ_3.

The Attractors of Σ

In order to relate the states and dynamics of $\mathcal{M}$ to those of the underlying quantum system Σ, we introduce the following assumptions, which can be seen to be fully realised in the case of the laser model of Chapter 11.

(S.1) For each point x of the attractor A_θ there is a unique primary state, ρ_x, of Σ that maximises the entropy density of Σ, subject to the contraints on its macroscopic observables that place the phase point of $\mathcal{M}$ at x. This is evidently a macroscopic completeness condition, analogous to that given by $(\hat{Q}.8)$ for equilibrium states in Section 6.4. We define S_θ to be $\{\rho_x \mid x \in A_\theta\}$.

(S.2) If $\mathcal{M}$ is confined to the attractor A_θ, with time-dependent phase point $\tau_\theta(t)x$, then the state of Σ at time t is $\rho_{\tau_\theta(t)x}$, up to transients.

(S.3) Further, if the initial phase point x of $\mathcal{M}$ lies in a sufficiently small neighborhood of the attractor A_θ, then the time-dependent state of Σ converges to the set S_θ in the long time limit.

In view of (S.2) and (S.3), we term S_θ an attractor of Σ, although an extra assumption would be needed to ensure that the canonical analogue of condition (A.2) is satisfied.

Phase Structure

Since the domains Θ_1, Θ_2 and Θ_3 of the control space corresponds to the three qualitatively different classes of attractors A_θ, or S_θ, we take them to represent different nonequilibrium phases of the system, namely those corresponding to fixed points, periodic orbits and chaotic orbits of $\mathcal{M}$. Thus, the resolution of the stable control space, Θ_0, into these domains constitutes a phase diagram analogous to that of equilibrium states. We have, therefore, a remarkable parallel between equilibrium and nonequilibrium phase structures (cf. [Ha2]).

Symmetry Breakdown, Order and Chaos

We now assume that the quantum system Σ is equipped with a dynamical symmetry group, G. This then has a representation, Φ, in the automorphisms of $\mathcal{A}$, such that $\Phi(g)$ and $T_\theta(t)$ intercommute, for all g in G, t in $\mathbf{R}_+$ and $\theta \in \Theta$. Correspondingly, G is also a dynamical symmetry group of the classical system $\mathcal{M}$, in that Φ induces a representation, ϕ, of this group in the transformations of X, such that $\phi(g)$ and $\tau_\theta(t)$ intercommute. We assume that for each $g \in G$, $\phi(g)$ acts continuously on X.

To discuss the role of the G-symmetry and its breakdown in the generation of ordered and chaotic states, we note that it follows from the above assumptions that *the attractor A_θ is invariant, as a set, under $\phi(g)$*, for the following reasons. Since $\phi(g)$ commutes with $\tau_\theta(t)$ and acts continuously on X, $\phi(g)A_\theta$

satisfies the conditions (A.1) and (A.2), together with the version of (A.3) obtained by replacing $\mathcal{N}_\theta$ by $\phi(g)\mathcal{N}_\theta$ there. Furthermore, the group property of $\phi(G)$ and the continuity of the actions of its elements on X ensure that (A.3) is unaffected by this replacement. Hence, since A_θ is completely characterised by (A.1)–(A.3), it follows that this attractor is invariant under $\phi(g)$.

We now recall that A_θ consists of a single point, x_θ, if and only if θ lies in the domain Θ_1. Hence, it follows from the G-invariance of A_θ that if $\theta \in \Theta_1$, then the point x_θ is G-invariant and therefore so too is the corresponding state ρ_{x_θ} of Σ. On the other hand, if θ lies in Θ_2 or Θ_3, then the G-invariance of the attracting set A_θ does not imply that of its constituent points, x. Hence, there might be a G-symmetry breakdown in the corresponding states, ρ_x, of Σ.

Order and Chaos

In view of these last observations, we arrive at the following picture of the different phases of Σ in the case of G-symmetry breakdown.

(1) For $\theta \in \Theta_1$, the attractor S_θ consists of a single stationary, G-invariant state ρ_{x_θ}.

(2) For $\theta \in \Theta_2$, the attractor S_θ consists of G-asymmetric, nonstationary states, with periodic time-dependence. These are generally ordered states, by virtue of the G-symmetry breakdown.

(3) For $\theta \in \Theta_3$, the attractor S_θ consists of G-asymmetric, time-dependent states, whose evolution is chaotic, since the orbits of their macroscopic variables lie on strange attractors. Thus, from a *dynamical* point of view, these are chaotic states.

To summarise, the system Σ undergoes phase transitions when θ crosses the $\Theta_1 - \Theta_2$ or the $\Theta_2 - \Theta_3$ boundary. In the former case, the transition is from a normal, stationary state to an ordered, periodically time-dependent one. In the second case, the transition is from a state of the latter kind to a chaotically evolving one.

Parallels between Equilibrium and Nonequilibrium Phase Structures

As we have already noted, there are remarkable parallels between the equilibrium and nonequilibrium and phase structures in the picture presented here, notwithstanding the fact that they pertain to physical contexts, governed by quite different controls. We may summarise the correspondences and differences between the equilibrium and nonequilibrium structures as follows.

(1) An attractor, S_θ, in the nonequilibrium state space is the counterpart of the set of equilibrium states for the given values of the control variables. However, by contrast with the latter states, those belonging to S_θ are

nonstationary except in the case where this attractor corresponds to a fixed point of the classical system $\mathcal{M}$.

(2) The nonequilibrium transitions from unordered to ordered states, characterised by symmetry breakdown, are analogous to those at equilibrium.

(3) Similarly, the nonequilibrium phase diagrams have obvious analogies with those for equilibrium states.

(4) The dynamically chaotic states, corresponding to the region Θ_3 of the control space, are the counterparts of spatially chaotic ones that occur in disordered systems; the latter states can even be pure, as in the example discussed in Section 3.2.4.

Chapter 11

Laser model as a paradigm of nonequilibrium phase structures

11.1 INTRODUCTION

In a seminal article on quantum optics, Graham and Haken [GH] argued that a laser model, comprising matter and radiation driven by a certain array of pumps and sinks, underwent a transition from incoherent to coherent radiation when the pumping strength reached a critical value; and further that this transition bore striking analogies with that occurring under equilibrium conditions between normal and ordered phases of superconductors. The parallel between equilibrium and nonequilibrium phase structures obtained there subsequently became a cornerstone of Haken's Synergetics [Ha2].

The theory of laser phase transitions was placed on a firm mathematical footing by Hepp and Lieb (HL) in a remarkable series of works [HL1–HL3] on the Dicke model [Dic]. To be precise, they treated two versions of the model. The first consisted of a chain of two level atoms that was coupled to a finite number of oscillators, representing radiation modes [HL2, HL3]: the second [HL1, HL3] consisted of the same ingredients, together with a certain array of reservoirs that were coupled to both the atoms and the radiation. The first version of the model was studied under equilibrium conditions, where it was shown to exhibit a transition from a normal to a superradiant phase. The second version, on the other hand, was shown to undergo a reservoir driven transition, far from equilibrium, from a normal to a coherent radiative phase. Thus, there was a remarkable parallel between the equilibrium and nonequilibrium phase transitions exhibited by these models, the ordered phases of both being characterised by the off-diagonal long range order that is also the hallmark of superfluids.

In the period following the Hepp–Lieb works, there were important general developments in the theory of quantum stochastic processes [Li3, HP, AFL], whose methodology opened the way to a new approach [AlSe, BS] to the theory of the laser model that led to further advances, which fully realised the general scheme described in the preceding chapter. This approach was based on a new version, due to Alli and Sewell (AS) [AlSe], of the Dicke model, in which the composite of matter and radiation was represented as an *open, dissipative system*. Thus, whereas the Hepp–Lieb model was a conservative system, comprising matter, radiation modes, pumps and sinks, that of Alli and Sewell was a dissipative one, which consisted of matter and radiation and whose dynamical semigroup carried the phenemenological parameters (damping constants) representing the effects of its interactions with reservoirs.[1]

The realisation of the scheme of Chapter 10 by the AS model was achieved in two main stages. In the first of these, it was proved [AlSe] that the macroscopic dynamics of the model yielded a nonequilibrium phase structure, that admitted not only the normal and coherent radiation of the HL picture, but also a variety of types of chaotic radiation. In the second stage, Bagarello and Sewell [BS] proved that the microscopic dynamics is piloted by that of the macroscopic variables and that, after the decay of transients, the microstate at any instant is determined by a maximum entropy principle, which comprises a simple generalisation of the one governing the equilibrium states.

This chapter is devoted to a review of the developments summarised in the previous paragraph. In Section 11.2, we formulate the AS version of the Dicke model as a dissipative quantum dynamical system. In Section 11.3, we specify the macroscopic description of the model, and prove that its phenomenological dynamics reduces to a classical deterministic form in a limit where the number of its atoms becomes infinite. In Section 11.4, we show that this macroscopic dynamics yields a nonequilibrium phase structure, with phases that correspond to quiescent, coherent and chaotic radiation.

In Section 11.5, we pass to the microscopic dynamics of the model and prove that this is piloted by its time-dependent macroscopic variables, which in turn evolve according to the self-contained law formulated in Section 11.3. In Section 11.6 we show that, after the decay of transients, the instantaneous state of the model is determined by a nonequilibrium variational principle. Specifically, this state maximises the global entropy density, subject to the constraints imposed by the prevailing values of the macroscopic observables. Thus, the results of Sections 11.3–11.6 constitute a realisation of the scheme of Chapter 10. We conclude in Section 11.7 with a further brief discussion of these results and also of some open problems.

Appendix A to this chapter is devoted to the proofs of a lemma and a proposition, both stated in Section 11.5.

11.2 THE MODEL

The model is a dissipative quantum system, consisting of a chain of $(2N + 1)$

---

[1] A further difference between the HL and AS models is that the two-level atoms were represented by pairs of Fermi oscillators (one for each level) in the former model and by Pauli spins in the latter one.

identical two-level atoms, coupled to $(2n + 1)$ radiation modes. We are concerned with its properties in a limit where N tends to infinity but n remains fixed and finite. The problem of formulating a corresponding limit with n proportional to N is discussed in the concluding section.

We formulate the model as a dissipative quantum system, $\Sigma^{(N)} = (\mathcal{A}^{(N)}, S^{(N)}, T^{(N)})$, where $\mathcal{A}^{(N)}$ is a $W^\star$-algebra of observables, $S^{(N)}$ is a space of normal states on $\mathcal{A}^{(N)}$ and $\{T^{(N)}(t) \mid t \in \mathbf{R}_+\}$ is a quantum dynamical semigroup of transformations of $\mathcal{A}^{(N)}$. We build the model Σ from its elements by first formulating the dissipative systems representing the individual atoms and modes, when decoupled from one another, and then introducing the matter–field interactions. For simplicity, we employ units in which Planck's constant is unity.

The Single Atom

We assume that the matter consists of two-level atoms, each one of which corresponds to a dissipative $W^\star$-dynamical system $\Sigma_{at} = (\mathcal{A}_{at}, S_{at}, T_{at})$, when decoupled from both the radiation and the other atoms. We assume that $\mathcal{A}_{at}$ is the algebra of 2-by-2 matrices, that is, the algebra of operators in the Hilbert space $\mathcal{H}_{at} = \mathbf{C}^2$. Thus, it is just the linear span of the Pauli matrices $(\sigma_x, \sigma_y, \sigma_z)$ and the identity I, and its structure is given by the standard relations

$$\sigma_x^2 = \sigma_y^2 = \sigma_z^2 = I, \quad \sigma_x \sigma_y = -\sigma_y \sigma_x = i\sigma_z, \text{ etc.} \qquad (11.2.1)$$

We define the spin raising and lowering operators

$$\sigma_\pm = \frac{1}{2}(\sigma_x \pm i\sigma_y). \qquad (11.2.2)$$

We assume that S_{at} is the space of all states on $\mathcal{A}_{at}$: these are normal and so are represented by the 2-by-2 density matrices.

We assume that $T_{at}(\mathbf{R}_+)$ is a quantum dynamical semigroup of transformations of $\mathcal{A}_{at}$, with $T_{at}(t)$ continuous in t. The generator, L_{at}, of this semigroup is therefore of the standard Lindblad form, given by Eq. (4.2.6), that is,

$$L_{at} = i[H_{at}, A] + \sum_j \left(V_j^\star A V_j - \frac{1}{2}[V_j^\star V_j, A]_+ \right) \quad \forall A \in \mathcal{A}_{at}, \qquad (11.2.3)$$

where H_{at} and the V_j are elements of $\mathcal{A}_{at}$, the Hamiltonian operator, H_{at}, being self-adjoint. We shall presently provide further specifications of H_{at} and the V_j.

Note Kümmerer [Kum] has proved that the model $\Sigma_{at} = (\mathcal{A}_{at}, S_{at}, T_{at})$ is an induced dynamical system, in the sense specified in Section 4.3. In this case, the semigroup T_{at} of transformations of $\mathcal{A}_{at}$ is induced by a one-parameter group, $\hat{T}_{at}$, of automorphisms of a larger algebra, $\hat{\mathcal{A}}_{at}$, which operates in a

Hilbert space $\hat{\mathcal{H}}_{at}$ containing $\mathcal{H}_{at}$. This latter algebra may naturally be interpreted as that of a composite of the two-level atom and a reservoir, and consequently the dissipative dynamics of Σ_{at} may be regarded as that induced by the interactions of this atom with a reservoir. In particular, the form of V_j is governed by those interactions. However, for our purposes, there is no need to specify the reservoir explicitly.

We now assume that the Hamiltonian operator, H_{at}, of Eq. (11.2.3) is $\epsilon\sigma_z/2$, with ϵ a positive constant, which is therefore the difference between the two atomic energy levels. We also assume that there are just three V_j in the formula (11.2.3), namely $b_{\pm}^{1/2}\sigma_{\pm}$ and $b_z^{1/2}\sigma_z$, where $b_{\pm}$ are positive constants and b_z is a nonnegative one. We shall discuss the physical significance of these parameters after Eqs. (11.2.4). First, however, we note that, since the components of σ and the identity form a basis for $\mathcal{A}_{at}$, the generator L_{at} is completely determined by its action on these operators. Conequently, by Eqs. (11.2.1)–(11.2.3) and our specifications of H_{at} and the V_j, L_{at} is fully defined by the following equations:

$$L_{at}\sigma_z = -b_+\left(\sigma_z - \frac{I}{2}\right) - b_-\left(\sigma_z + \frac{I}{2}\right), \qquad (11.2.4a)$$

$$L_{at}\sigma_{\pm} = -\left(\frac{1}{2}(b_+ + b_-) + 2b_z \mp i\epsilon\right)\sigma_{\pm} \qquad (11.2.4b)$$

and

$$L_{at}I = 0. \qquad (11.2.4c)$$

Since the eigenvalues of σ_z are ± 1, corresponding to the atomic energy levels $\pm\epsilon/2$, the first and second terms on the right-hand side of Eq. (11.2.4a) represent processes that act so as to drive the atom to its excited and ground state, respectively; and hence, in view of the note following Eq. (11.2.3), b_+ and b_- may naturally be interpreted as strengths of coupling of the atom to a pump and to a sink, respectively. Likewise, by Eqs. (11.2.4b), b_z represents the strength of its coupling to another reservoir, which acts so as to damp down the spin components $\sigma_{\pm}$, or equivalently σ_x and σ_y.

We now re-express Eqs. (11.2.4) in the concise form

$$L_{at}\sigma_{\pm} = -(\gamma_{\perp} \mp i\epsilon)\sigma_{\pm}, \quad L_{at}\sigma_z = -\gamma_{\parallel}(\sigma_z - \eta I), \quad L_{at}I = 0, \quad (11.2.5)$$

where the damping constants γ and the parameter η are given by the formulae

$$\gamma_{\perp} = \frac{1}{2}(b_+ + b_-) + 2b_z, \quad \gamma_{\parallel} = (b_+ + b_-), \quad \eta = \frac{b_+ - b_-}{b_+ + b_-}. \quad (11.2.6)$$

It follows from these definitions that the γ and η satisfy the restrictive conditions that

$$\gamma_{\|} \leq 2\gamma_{\perp} \tag{11.2.7}$$

and

$$-1 < \eta < 1. \tag{11.2.8}$$

A state, ρ_{at}, of Σ_{at} is stationary if it is invariant under the dual of the semigroup $T_{at}^{\star}$, that is, if

$$\rho_{at}(L_{at}A) = 0 \quad \forall A \in \mathcal{A}_{at}.$$

In view of the fact that $\mathcal{A}_{at}$ is the linear span of the Pauli matrices and the identity, it follows from Eq. (11.2.5) that this stationarity condition signifies precisely that

$$\rho_{at}(\sigma_z) = \eta \quad \text{and} \quad \rho_{at}(\sigma_{\pm}) = 0. \tag{11.2.9}$$

Since this formula defines a unique state ρ_{at}, it follows that this is the only stationary state of the atom. Further, it follows from Eq. (11.2.9) that η is the difference between the occupation probabilities of the excited and ground states of the atom. Hence, the stationary state of Σ_{at} carries an *inverted population*, in that it its excited state is the more heavily populated one, if and only if $\eta > 0$.

The Matter

We assume that the matter, when not coupled to the radiation, consists of $(2N + 1)$ independent copies of Σ_{at}, that are located on the sites $\{r \in \mathbf{Z} \mid -N \leq r \leq N\}$ of the one-dimensional lattice $\mathbf{Z}$. Thus, the atom at the site r is a copy, $\Sigma_r = (\mathcal{A}_r, S_r, T_r)$, of Σ_{at} and the matter constitutes the dissipative quantum system $\Sigma_{mat}^{(N)} = (\mathcal{A}_{mat}^{(N)}, S_{mat}^{(N)}, T_{mat}^{(N)})$, where $\mathcal{A}_{mat}^{(N)}$ and $T_{mat}^{(N)}$ are the tensor products of the $\mathcal{A}_r$'s and the T_r's, respectively, and $S_{mat}^{(N)}$ is the set of all states on $\mathcal{A}_{mat}^{(N)}$. This consists of the linear transformations of the Hilbert space $\mathcal{H}_{mat}^{(N)} = \mathbf{C}^{4N+2}$.

We identify the spin component $\sigma_{u,r}$ ($u = x, y, z, \pm$), of the atom at r with the element of $\mathcal{A}_{mat}^{(N)}$ given by the tensor product of $(2N + 1)$ copies of elements of $\mathcal{A}_{at}$, of which the rth is σ_u and the others are I. It follows from these specifications and Eq. (11.2.5) that the generator, $L_{mat}^{(N)}$, of the semigroup $T_{mat}^{(N)}$ is given by the formulae

$$L_{mat}^{(N)}\sigma_{\pm,r} = -(\gamma_{\perp} \mp i\epsilon)\sigma_{\pm,r}, \quad L_{mat}^{(N)}\sigma_{z,r} = -\gamma_{\|}(\sigma_{z,r} - \eta I), \quad L_{mat}^{(N)}I = 0, \tag{11.2.10}$$

and, for any set of *different* sites $r_1, \ldots, r_k$,

$$L_{mat}^{(N)}\left(\sigma_{r_1,u_1}\cdots\sigma_{r_k,u_k}\right) = \Pi_{j=1}^{k} L_{mat}^{(N)}\sigma_{r_j,u_j}. \tag{11.2.11}$$

The Radiation

We assume that the radiation, when not coupled to the matter, consists of $(2n + 1)$ attenuated modes, each one being a version of the damped harmonic oscillator represented by the model of Chapter 4, Appendix B. Thus, as in that appendix, the damping stems from a coupling of each oscillator to its own reservoir. We assume that the radiation modes correspond to waves propagated along the line of the atoms, with wave numbers

$$k_l = \frac{l}{2n + 1} \quad \text{for } l = -n, -n + 1, ..., n. \tag{11.2.12}$$

To formulate the model of the radiation as a dissipative $W^\star$-dynamical system, $\Sigma_{\text{rad}} = (\mathcal{A}_{\text{rad}}, S_{\text{rad}}, T_{\text{rad}})$, we proceed along the lines of Chapter 4, Appendix B. Thus we start by defining the Fock–Hilbert space $\mathcal{H}_{\text{rad}}$ and the closed, densely defined operators $\{a_l^\star, a_l \mid l = -n, -n + 1, ..., n\}$ in this space by the following conditions.

(1) There is a unit vector, Φ_{rad}, in $\mathcal{H}_{\text{rad}}$, such that $a_l \Phi_{\text{rad}} = 0$ for $l = -n, ..., n$;

(2) $\mathcal{H}_{\text{rad}}$ is generated by the application to Φ_{rad} of the polynomials in the $a^\star$'s; and

(3) the a and $a^\star$ satisfy the canonical commutation relations

$$[a_l, a_m^\star] = \delta_{lm} I, \quad [a_l, a_m] = 0. \tag{11.2.13}$$

We define the Weyl map $z = (z_{-n}, ..., z_n) \rightarrow W(z)$ of $\mathbf{C}^{2n+1}$ into $\mathcal{A}_{\text{rad}}$ by the standard prescription

$$W(z) = \exp i(z.a + (z.a)^\star) \quad \text{with } z.a = \sum_{l=-n}^{n} z_l a_l. \tag{11.2.14}$$

Thus, by Eq. (11.2.13), W satisfies the Weyl algebraic relation

$$W(z)W(z') = W(z + z') \exp(i \operatorname{Im}(z, z')_n), \tag{11.2.15}$$

where $(\cdot, \cdot)_n$ is the $\mathbf{C}^{2n+1}$ inner product.

We define $\mathcal{A}_{\text{rad}}$, the algebra of bounded observables of Σ_{rad}, to be $\mathcal{B}(\mathcal{H}_{\text{rad}})$, the set of bounded linear transformations of $\mathcal{H}_{\text{rad}}$. This algebra is therefore generated by the Weyl operators $\{W(z) \mid z \in \mathbf{C}^{2n+1}\}$.

We assume that S_{rad} is the space of normal states on $\mathcal{A}_{\text{rad}}$. In particular, the vacuum state, ϕ_{rad}, is that corresponding to the Fock vacuum vector Φ_{rad}. Thus, the formula (B.4) of Chapter 4, Appendix B is now applicable with the present notation, that is,

$$\phi_{\text{rad}}[W(z)] = \exp\left(-\frac{1}{2}|z|^2\right) \quad \text{with } |z|^2 = \sum_{l=-n}^{n} |z_l|^2. \tag{11.2.16}$$

We assume that the quantum dynamical semigroup, T_{rad}, is the canonical $(2n + 1)$-oscillator generalisation of that given by Eq. (B.25) of Chapter 4, Appendix B. Thus, T_{rad} is the Vanheuverzwijn [Vh] semigroup defined by the formula

$$T_{\text{rad}}(t)[W(z)] = W(z(t)) \exp\left(-\frac{1}{2}(|z|^2 - |z(t)|^2)\right), \qquad (11.2.17)$$

where $z(t) = (z_{-n}(t), ..., z_n(t))$ and the time-dependence of $z_k(t)$ is governed by a frequency, ω_l, and a damping constant, κ_l, both positive, according to the formula

$$z_l(t) = z_l \exp((i\omega_l - \kappa_l)t). \qquad (11.2.18)$$

We assume that the modes of wave vectors $\pm k$ have the same frequencies and damping constants, that is, that

$$\omega_l = \omega_{-l} \quad \text{and} \quad \kappa_l = \kappa_{-l} \quad \text{for } l = 1, 2, ..., n. \qquad (11.2.19)$$

Note T_{rad} is not a Lindblad semigroup, since $T_{\text{rad}}(t)$ is weakly, but not norm-wise, continuous in t. However, its generator, L_{rad}, is formally of the Lindblad type, since it is given by the following equation.

$$L_{\text{rad}} = \sum_{l=-n}^{n} \left(i[\omega_l a_l^\star a_l, \cdot]_- + 2\kappa_l a_l^\star (\cdot) a_l - \kappa_l [a_l^\star a_l, \cdot]_+\right). \qquad (11.2.20)$$

This generator is therefore unbounded.

The Uncoupled Matter-Radiation System

As a preliminary to formulating the system formed by coupling $\Sigma_{\text{mat}}^{(N)}$ to Σ_{rad}, we first note some elementary features of their uncoupled composite, namely the system $\Sigma_0^{(N)} = (\mathcal{A}^{(N)}, S^{(N)}, T_0^{(N)})$, where $\mathcal{A}^{(N)}$ and $T_0^{(N)}$ are $\mathcal{A}_{\text{mat}}^{(N)} \otimes \mathcal{A}_{\text{rad}}$, and $T_{\text{mat}}^{(N)} \otimes T_{\text{rad}}$, respectively and $S^{(N)}$ is the set of normal states on $\mathcal{A}^{(N)}$. In particular, $\mathcal{A}^{(N)}$ is the algebra of bounded operators in the Hilbert space $\mathcal{H}^{(N)} := \mathcal{H}_{\text{mat}}^{(N)} \otimes \mathcal{H}_{\text{rad}}$.

We identify elements A_{mat} and A_{rad} of $\mathcal{A}_{\text{mat}}^{(N)}$ and $\mathcal{A}_{\text{rad}}$, respectively, with their canonical counterparts, $A_{\text{mat}} \otimes I_{\text{rad}}$ and $I_{\text{mat}} \otimes A_{\text{rad}}$, in $\mathcal{A}^{(N)}$. Thus, $\mathcal{A}_{\text{mat}}^{(N)}$ and $\mathcal{A}_{\text{rad}}$ become intercommuting subalgebras of $\mathcal{A}^{(N)}$.

It follows directly from these specifications that the generator of the dynamical semigroup $T_0^{(N)}$ is

$$L_0^{(N)} = L_{\text{mat}}^{(N)} + L_{\text{rad}}, \qquad (11.2.21)$$

where $L_{\text{mat}}^{(N)}$ and L_{rad} are identified with $L_{\text{mat}}^{(N)} \otimes I$ and $I \otimes L_{\text{rad}}$, respectively.

We now recall that, as discussed above, the dynamical semigroups representing both the individual atoms and the radiation mode oscillators are induced on their algebras of observables by unitarily implemented one-parameter groups of automorphisms of larger algebras. Consequently, in view of our specifications of $\Sigma_0^{(N)}$, the dynamical semigroup $T_0^{(N)}$ is induced on $\mathcal{A}^{(N)}$ by a unitarily implemented one-parameter group, $\hat{T}_{0,}^{(N)}$ of automorphisms of a larger algebra, $\hat{\mathcal{A}}^{(N)}$, which acts in a Hilbert space, $\hat{\mathcal{H}}^{(N)}$, that contains $\mathcal{H}^{(N)}$.

The Coupled Matter-Radiation System

This is the system, $\Sigma^{(N)}$, formed by coupling Σ_{mat} to Σ_{rad}. Specifically, we assume that each radiation mode is linearly coupled to a matter (or spin) wave of the same wave length, so that the interaction Hamiltonian is

$$H_{\text{int}}^{(N)} = i(2N + 1)^{-1/2} \sum_{r=-N}^{N} \sum_{l=-n}^{n} \lambda_l(\sigma_{-,r} a_l^\star \exp(-2\pi i k_l r) - h.c.), \quad (11.2.22)$$

where k_l is given by Eq. (11.2.12) and the λ_l are real-valued, N-independent coupling constants that satisfy the symmetry condition

$$\lambda_l = \lambda_{-l}. \tag{11.2.23}$$

We remark here that the formula (11.2.22) for $H_{\text{int}}^{(N)}$ may be equivalently written in the form

$$H_{\text{int}}^{(N)} = \sum_{r=-N}^{(N)} (\phi_r^{(N)} \sigma_{r,+} + \phi_r^{(N)\star} \sigma_{r,-}), \tag{11.2.24}$$

where

$$\phi_r^{(N)} = -i(2N + 1)^{-1/2} \sum_{l=-n}^{n} \lambda_l a_l \exp(2\pi i k_l r). \tag{11.2.25}$$

Thus, by Eq. (11.2.24), $\phi_r^{(N)}$ may naturally be interpreted as the radiation field at the point r.

We assume that the algebra of observables and the state space of $\Sigma^{(N)}$ are the same as for $\Sigma_0^{(N)}$, namely $\mathcal{A}^{(N)}$ and $S^{(N)}$, respectively. We represent the dynamics of the system by a quantum dynamical semigroup, $T^{(N)}(\mathbf{R}_+)$, of transformations of $\mathcal{A}^{(N)}$ which was constructed in [AlSe] by the following procedure.[2] Firstly, we modified the unitarily implemented group, $\hat{T}_0^{(N)}$, of dynamical automorphisms of the extended algebra $\hat{\mathcal{A}}^{(N)}$ by introducing the perturbation represented by the interaction Hamiltonian $H_{\text{int}}^{(N)}$. This resultant dynamics corresponded to a one-parameter group, $\hat{T}^{(N)}$, of automorphisms of $\hat{\mathcal{A}}^{(N)}$, which induced a quantum dynamical semigroup $T^{(N)}$ of transformations

---

[2] The method there was devised to overcome the problems raised by the perturbation of the unbounded generator, $L_0^{(N)}$ of $T_0^{(N)}$ by the unbounded derivation $i[H_{\text{int}}^{(N)}]$.

of $\mathcal{A}^{(N)}$. We take $T^{(N)}$ to represent the dynamics of $\Sigma^{(N)}$. Thus, our model for $\Sigma^{(N)}$ is $(\mathcal{A}^{(N)}, S^{(N)}, T^{(N)})$. The evolute of a state $\rho^{(N)}$ of the model at time t is therefore $T^{(N)\star}(t)\rho^{(N)}$, and we denote this sometimes by $\rho_t^{(N)}$.

In order to obtain an explicit formulation of the evolution of $\Sigma^{(N)}$ in states that are sufficiently regular to admit well-defined photon statistics, we introduce the following definitions.

Definition 11.2.1

(1) We define $\mathcal{B}^{(N)}$ to be the $\star$-algebra of polynomials in the operators $\sigma_{r,u}$, $W(z)$, a_l and $a_l^\star$, where r and l run through the integers in the ranges $[-N.N]$ and $[-n, n]$, respectively, and z runs through $\mathbf{C}^{2n+1}$. This algebra is therefore affiliated to $\mathcal{A}^{(N)}$ and contains both bounded and unbounded operators.

(2) We define $\mathcal{D}_0^{(N)}$ to be the set of states $\rho^{(N)}$ of $\Sigma^{(N)}$ whose canonical extensions to all $\mathcal{B}^{(N)}$ are well defined, in that their density matrices, $\hat{\rho}^{(N)}$, satisfy the condition that

$$\mathrm{Tr}(\hat{\rho}^{(N)}B^\star B) < \infty \quad \forall B \in \mathcal{B}^{(N)}.$$

(3) For $\rho^{(N)} \in \mathcal{D}_0^{(N)}$ and $m \in \mathbf{N}$, we define $M_m(\rho^{(N)})$ to be the maximum value of $\rho^{(N)}(B^\star B)$ as B runs through the monomials of degree m in the operators $\{a_l^\star, a_l \mid l \in [-n.n]\}$.

(4) We then define $\mathcal{D}^{(N)}$ to be the subset of $\mathcal{D}_0^{(N)}$ whose elements, $\rho^{(N)}$, satisfy the condition

$$\sum_{m=1}^{\infty} \frac{[M_m(\rho^{(N)})]^{1/2}v^m}{m!} < \infty \quad \forall v \in \mathbf{R}_+. \qquad (11.2.26)$$

It follows easily from this definition that $\mathcal{D}^{(N)}$ is a norm dense subset of $S^{(N)}$.

The following proposition, pertaining to the evolution of $\Sigma^{(N)}$, was proved in [AlSe].

Proposition 11.2.3 *Assuming that $\rho^{(N)} \in \mathcal{D}^{(N)}$, its evolute, $\rho_t^{(N)}$, is always confined to $\mathcal{D}_0^{(N)}$ and satisfies the equation of motion*

$$\frac{d}{dt}\rho_t^{(N)}(B) = \rho_t^{(N)}(LB) \quad \forall B \in \mathcal{B}^{(N)}, \qquad (11.2.27)$$

where

$$L^{(N)} = L_{\mathrm{mat}}^{(N)} + L_{\mathrm{rad}} + i[H_{\mathrm{int}}^{(N)}, \cdot]. \qquad (11.2.28)$$

This result provides a precise sense in which the putative generator of $T^{(N)}$, as *given by the right-hand side of Eq. (11.2.28), is indeed the effective generator of that semigroup.*

A Gauge Symmetry Group

It follows from Eqs. (11.2.1), (11.2.2), (11.2.11), (11.2.13), (11.2.20), (11.2.22) and (11.2.28) that, for any given l in $[-n, n]$ and real θ, $L^{(N)}$, and hence $T^{(N)}$, is invariant under the gauge automorphisms of $\mathcal{A}^{(N)}$ given by the transformations

$$a_l \rightarrow a_l \exp(i\theta), \quad \sigma_{r,\pm} \rightarrow \sigma_{r,\pm} \exp(\mp i\theta), \quad \sigma_{r,z} \rightarrow \sigma_{r,z}. \qquad (11.2.29)$$

These automorphisms therefore constitute a continuous dynamical symmetry group as θ runs through the range $[0, 2\pi]$.

11.3 THE MACROSCOPIC DYNAMICS

We formulate the macroscopic description of the model in terms of the global intensive observables

$$s_l^{(N)} = (2N + 1)^{-1} \sum_{r=-N}^{N} \sigma_{-,r} \exp(-2\pi i k_l r), \quad l = -n, ..., n \qquad (11.3.1)$$

and

$$p_l^{(N)} = (2N + 1)^{-1} \sum_{r=-N}^{N} \sigma_{z,r} \exp(-2\pi i k_l r), \quad l = -n, ..., n \qquad (11.3.2)$$

together with the radiation mode operators

$$\alpha_l^{(N)} = (2N + 1)^{-1/2} a_l, \quad l = -n, ..., n, \qquad (11.3.3)$$

the scaling in this last formula corresponding to that of the number operators $a_l^\star a_l$ in units of $(2N + 1)$. We note here that, as $\sigma_{z,r}$ is self-adjoint, Eq. (11.3.2) implies that

$$p_l^{(N)\star} = p_{-l}^{(N)}. \qquad (11.3.4)$$

Further, by Eqs. (11.2.1), (11.2.2), (11.3.1) and (11.3.2), the operators $s_l^{(N)}$ and $p_l^{(N)}$ are bounded, and

$$\|s_l^{(N)}\| = 1, \quad \|p_l^{(N)}\| = 1. \qquad (11.3.5)$$

We denote by $\mathcal{M}^{(N)}$ the set of operators comprising all the $s^{(N)}$'s, $p^{(N)}$'s, $\alpha^{(N)}$'s and their adjoints. Thus, by Eq. (11.3.4), this set is simply $\{s_l^{(N)}, s_l^{(N)\star}, p_l^{(N)}, \alpha_l^{(N)}, \alpha_l^{(N)\star}\}$, with l running from $-n$ to n. It follows that the number, ν, of elements of $\mathcal{M}^{(N)}$ is given by the equation

$$\nu = 10n + 5. \tag{11.3.6}$$

We denote the set of the ν self-adjoint operators that comprise the Hermitian and anti-Hermitian parts of the elements by $\mathcal{M}^{(N)}$ by $x^{(N)} = (x_1^{(N)}, ..., x_\nu^{(N)})$. These then are the macroscopic observables of the model. It is unneccessary, for our purposes, to specify the $x_j^{(N)}$'s individually.

By Eqs. (11.2.1), (11.2.2), (11.2.12), (11.2.13) and (11.3.1)–(11.3.3), $\mathcal{M}^{(N)}$ is a Lie algebra with respect to commutation: its nonzero Lie brackets are the following ones.

$$[s_l^{(N)}, s_m^{(N)\star}] = -(2N+1)^{-1} p_{[l-m]}^{(N)}, \quad [s_l^{(N)}, p_m^{(N)}] = 2(2N+1)^{-1} s_{[l-m]}^{(N)},$$

$$[s_l^{(N)\star}, p_m^{(N)}] = -2(2N+1)^{-1} s_{[l+m]}^{(N)\star}, \quad [\alpha_l^{(N)}, \alpha_m^{(N)\star}] = (2N+1)^{-1} I \delta_{lm},$$

$$\tag{11.3.7}$$

where $[l \pm m] = l \pm m \pmod{2n}$. Thus, in view of Eq. (11.3.5), the observables $\mathcal{M}^{(N)}$ intercommute and therefore become classical in the limit $N \to \infty$.

Our objective now is to extract the dynamics of $\mathcal{M}^{(N)}$ from the microscopic equation of motion (11.2.27), in a limit where $N \to \infty$ and n remains fixed and finite. For this purpose, the number N must evidently be treated as a variable parameter.

In order to analyse the macroscopic dynamics of the model, we first note that, by Eqs. (11.2.22), (11.2.25) and (11.3.1)–(11.3.3), the interaction Hamiltonian and the local radiation field $\phi_r^{(N)}$ are functions of the macro-observables only, as specified by the formulae

$$H_{int}^{(N)} = i(2N+1) \sum_{l=-n}^{n} \lambda_l (\alpha_l^{(N)\star} s_l^{(N)} - \alpha_l^{(N)} s_l^{(N)\star}) \tag{11.3.8}$$

and

$$\phi_r^{(N)} = -i \sum_{l=-n}^{n} \lambda_l \alpha_l \exp(2\pi i k_l r). \tag{11.3.9}$$

Likewise, by Eqs. (11.2.10), (11.2.11), (11.2.13), (11.2.20), (11.2.28), (11.3.1)–(11.3.3), (11.3.7) and (11.3.8), the action of the dynamical generator, $L^{(N)}$, on the elements of $\mathcal{M}^{(N)}$ yields functions of these observables only, that is,

$$L^{(N)} \alpha_l^{(N)} = -(i\omega_l + \kappa_l) \alpha_l^{(N)} + \lambda_l s_l^{(N)}, \tag{11.3.10a}$$

$$L^{(N)} s_l^{(N)} = -(i\epsilon + \gamma_\perp) s_l^{(N)} + \sum_{m=-n}^{n} \lambda_m p_{[l-m]}^{(N)} \alpha_m^{(N)} \tag{11.3.10b}$$

and

$$L^{(N)}p_l^{(N)} = -\gamma_{\parallel}(p_l^{(N)} - \eta_l^{(N)}I) - 2\sum_{m=-n}^{n} \lambda_m(\alpha_m^{(N)\star}s_{[l+m]}^{(N)} + \alpha_m^{(N)}s_{[m-l]}^{(N)\star}),$$

$$(11.3.10c)$$

where

$$\eta_l^{(N)} = \eta\delta_{l,0} + O(N^{-1}).\qquad(11.3.11)$$

We assume that the initial state, $\rho^{(N)}$, of $\Sigma^{(N)}$ satisfies the following conditions.

(I.1) $\rho^{(N)}$ lies in the domain $\mathcal{D}^{(N)}$, specified in Definition 11.2.1 (4).

(I.2) The mean number of photons carried by this state does not increase faster than N, that is, for some finite constant B,

$$\rho^{(N)}(\alpha_l^{(N)\star}\alpha_l^{(N)}) < B \quad \text{for } l = -n, .., n \text{ and } \forall N \in \mathbf{N}. \quad (11.3.12)$$

(I.3) The dispersions of the macroscopic observables $x^{(N)}$ in the state $\rho^{(N)}$ vanish in the limit $N \to \infty$.

We now remark that, since the observables $(x_1^{(N)}, ..., x_\nu^{(N)})$ intercommute in the limit $N \to \infty$, it follows from condition (I.3) that they behave initially as classical, dispersion-free variables in this limit. Further, since, by Eqs. (11.3.10), the action of the dynamical generator $L^{(N)}$ on these observables yields only algebraic functions of them, one might suspect that they subsequently evolve according to a classical, deterministic law. In order to prepare the way for a derivation of such a law, we now formulate a classical model, $\mathcal{K}$, whose dynamics corresponds to that of a natural formal limit of the equations of motion for the observables $x^{(N)}$.

The Classical Model $\mathcal{K}$

We assume that this is a dynamical system, whose variables form a set, $\mathcal{M}$, of complex numbers $\{s_l, p_l, \alpha_l \mid l = -n, ..., n\}$. We define $x = (x_1, ..., x_\nu)$ to be the set of real numbers given by the real and imaginary parts of these complex variables, which are ordered is such a way that the relationship of x_j to $\mathcal{M}$ is the canonical analogue of that of $x_j^{(N)}$ to $\mathcal{M}^{(N)}$. Thus, $\mathcal{M}$, s_l, p_l, α_l and x_j correspond to classical counterparts of $\mathcal{M}^{(N)}$, $s_l^{(N)}$, $p_l^{(N)}$, $\alpha_l^{(N)}$ and $x_j^{(N)}$, respectively.

We assume that the equation of motion for the evolute, x_t, of x, is of the autonomous form

$$\frac{dx_t}{dt} = F(x_t),\qquad(11.3.13)$$

and that, explicitly, it is the modification of Eqs. (11.3.10) obtained formally by inserting the subscript t to the variables $s_l^{(N)}$, $p_l^{(N)}$ and $\alpha_l^{(N)}$, removing the superscript N from these variables and $\eta_l^{(N)}$, and replacing $L^{(N)}$ by the time derivative d/dt. Thus,

$$\frac{d}{dt}\alpha_{l,t} = -(i\omega_l + \kappa_l)\alpha_{l,t} + \lambda_l s_{l,t}, \qquad (11.3.14a)$$

$$\frac{d}{dt}s_{l,t} = -(i\epsilon + \gamma_\perp)s_{l,t} + \sum_{m=-n}^{n} \lambda_m p_{[l-m],t}\alpha_{m,t} \qquad (11.3.14b)$$

and

$$\frac{d}{dt}p_{l,t} = -\gamma_\parallel(p_{l,t} - \eta_l) - 2\sum_{m=-n}^{n} \lambda_m(\overline{\alpha}_{m,t}s_{[l+m],t} + \alpha_m \overline{s}_{[m-l],t}), \qquad (11.3.14c)$$

where again $[l\pm m] = l\pm m \pmod{n}$.

We note that Eqs. (11.3.14) have been proved to have a unique, global solution [AlSe]. This corresponds to the solution of Eq. (11.3.13), and takes the form

$$x_t = \tau(t)x, \qquad (11.3.15)$$

where $\{\tau(t) \mid t \in \mathbf{R}_+\}$ is a one-parameter semigroup of transformations of $\mathbf{R}^\nu$.

We conclude that the classical system, $\mathcal{K}$, constructed here comprises the pair (X, τ), where X is the phase space $\mathbf{R}^\nu$ and τ is the one-parameter semigroup of transformations of X corresponding to the solution of Eq. (11.3.13), or equivalently of Eqs. (11.3.14).

Classical Limit of the Macrodynamics of $\Sigma^{(N)}$

Returning now to the quantum system $\Sigma^{(N)}$, we formulate the dynamics of its macroscopic observables $x^{(N)}$ in terms of the characteristic function $\mu_t^{(N)}$ on $\mathbf{R}^\nu$, defined by the formula

$$\mu_t^{(N)}(y) = \langle\rho_t^{(N)}; \exp(iy.x^{(N)})\rangle \quad \forall y \in \mathbf{R}^\nu, \qquad (11.3.16)$$

where the dot in the exponent denotes the $\mathbf{R}^\nu$ inner product and, as above, $\rho_t^{(N)}$ is the evolute of $\rho^{(N)}$ at time t. The function μ_t is evidently the quantum analogue of a classical characteristic function, and it carries the time-dependent statistics of the observables $x^{(N)}$. The following proposition was proved in [AlSe].

Proposition 11.3.1 *Assuming the initial conditions (I.1)–(I.3), $\mu_t^{(N)}$ converges pointwise, as N tends to infinity, to the characteristic function of the classical Dirac probability measure, δ_{x_t}, on X, with support at $x_t = \tau(t)x$.*

Comments

(1) This proposition signifies that, under the specified initial conditions, the macroscopic observables $x^{(N)}$ reduce, in the limit $N \to \infty$, to classical ones, x, that evolve according to the deterministic dynamics of the system $\mathcal{K}$. Hence, by Eq. (11.3.9), the local radiation field $\phi_r^{(N)}$ reduces to a classical field $\phi_{r,t}$ at time t, where

$$\phi_{r,t} = -i \sum_{l=-n}^{n} \lambda_l \alpha_{l,t} \exp(2\pi i k_l r). \tag{11.3.17}$$

(2) [AlSe] If the condition (I.3) is removed, the proposition merely becomes modified to the form where $\mu_t^{(N)}$ converges to the characteristic function of a probability measure, m_t, on $\mathbf{R}^\nu$, whose evolution is governed by the $\mathcal{K}$-dynamics according to the formula

$$\int_{\mathbf{R}^\nu} f(x) dm_t(x) = \int_{\mathbf{R}^\nu} f(\tau(t)x) dm_0(x))$$

for all continuous functions, f, on $\mathbf{R}^\nu$, whose support is compact.

11.4 THE DYNAMICAL PHASE TRANSITIONS

We now study the equations of motion (11.3.13) for the classical system $\mathcal{K}$ since, by Proposition 11.3.1, it is they that govern the dynamics of the macroscopic observables of $\Sigma^{(N)}$ in the limit $N \to \infty$.

The Normal Phase

By equations (11.3.14), the system $\mathcal{K}$ has a stationary phase point, $x^{(0)}$, corresponding to the values $\{s_l^{(0)}, p_l^{(0)}, \alpha_l^{(0)}\}$ of $\{s_{l,t}, p_{l,t}, \alpha_{l,t}\}$ given by

$$s_l^{(0)} = 0, \quad p_l^{(0)} = \eta \delta_{l0}, \quad \alpha_l^{(0)} = 0 \quad \forall l \in [-n, n]. \tag{11.4.1}$$

In order to investigate its stability, we introduce a "small" perturbation x'_t of $x^{(0)}$, as represented by the elements $\{\alpha'_{l,t}, s'_{l,t}, p'_{l,t} \mid l \in [-n, n]\}$ of $\mathcal{M}$, and infer from Eqs. (11.3.14) and (11.4.1) that its *linearised* equations of motion are

$$\frac{d\alpha'_{l,t}}{dt} = -(i\omega_l + \kappa_l)\alpha'_{l,t} + \lambda_l s'_{l,t} \tag{11.4.2a}$$

$$\frac{ds'_{l,t}}{dt} = -(i\epsilon + \gamma_\perp)s'_{l,t} + \lambda_l \eta \alpha'_{l,t} \tag{11.4.2b}$$

and

$$\frac{dp'_{l,t}}{dt} = -\gamma_\| p'_{l,t}. \tag{11.4.2c}$$

From the last equation, we see that $p'_{l,t}$ decays exponentially to zero, and so the stability of $p^{(0)}$ is guaranteed. To test for that of $\alpha^{(0)}$ and $s^{(0)}$, it suffices to look at the solutions of Eqs. (11.4.2a,b) for which $\alpha'_{l,t}$ and $s'_{l,t}$ are constant multiples of $\exp(-\zeta_l t)$, with ζ_l a complex constant. Thus, ζ_l is determined by the roots of the quadratic equation

$$(\zeta_l - \kappa_l - i\omega_l)(\zeta_l - \gamma_\perp - i\epsilon) - \lambda_l^2 \eta = 0. \tag{11.4.3}$$

The condition for stability of the fixed point $x^{(0)}$, that is, for exponential decay rather than amplification of the perturbation, is simply that the real parts of the roots of this equation are both positive. Thus, it follows from Eq. (11.4.3) that this condition is simply that, for each value of l, η is less than the critical value $\eta_l^{(c)}$, given by the equation

$$\eta_l^{(c)} = \frac{\kappa_l \gamma_\perp}{\lambda_l^2}\left[1 + \frac{(\epsilon - \omega_l)^2}{(\kappa_l + \gamma_\perp)^2}\right]. \tag{11.4.4}$$

Hence, the fixed point $x^{(0)}$ is stable provided that η is less than $\min_l\{\eta_l^{(c)}\}$. We denote this value by $\eta^{(1)}$.

We conclude that, for $\eta < \eta^{(1)}$, the system supports a phase corresponding to a stable fixed point for which, by Eq. (11.4.1), the α_l are all zero. This is therefore an optically quiescent phase.

The Coherent Phase

Noting now that, by Eqs. (11.2.19), (11.2.23) and (11.4.4),

$$\eta_l^{(c)} = \eta_{-l}^{(c)}, \tag{11.4.5}$$

we assume, for simplicity, that there is just one value, l_1, of $|l|$ at which $\eta_{\pm l}$ attains the value $\eta_1^{(c)}$. Then, if η is increased beyond $\eta^{(1)}$, the real part of a root of Eq. (11.4.3) becomes positive for $l = \pm l_1$ and so for these values of l, the l'th radiation mode and matter wave become unstable.

On the other hand, for $\eta > \eta^{(1)}$, the equations of motion (11.3.14) have periodic solutions in which *either* $s_{l_1,t}$, $p_{0,t}$ and $\alpha_{l_1,t}$ *or* $s_{-l_1,t}$, $p_{0,t}$ and $\alpha_{-l_1,t}$ are the only nonzero elements of $\mathcal{M}_c$. These solutions are of the following form, with $\tilde{l}_1 = \pm l_1$.

$$\alpha_{l,t} = \alpha_l \delta_{l,\tilde{l}_1}\exp(-i\Omega t), \quad s_{l,t} = s_l \delta_{l,\tilde{l}_1}\exp(-i\Omega t), \quad p_{l,t} = \eta^{(1)}\delta_{l,0}, \tag{11.4.6}$$

where

$$\Omega = \frac{\gamma_\perp \omega_{l_1} + \epsilon \kappa_{l_1}}{\gamma_\perp + \kappa_{l_1}}, \tag{11.4.7}$$

$$|\alpha_{\tilde{l}_1}| = \frac{1}{2}\left[\frac{\gamma_\perp(\eta - \eta^{(1)})}{\kappa_{l_1}}\right]^{\frac{1}{2}} \tag{11.4.8}$$

and

$$s_{\bar{l}_1} = -\frac{\left(\kappa_{l_1}(\gamma_\perp + \kappa_{l_1}) + i(\omega_{l_1} - \epsilon)\right)\alpha_{\bar{l}_1}}{\lambda_{l_1}(\gamma_\perp + \kappa_{l_1})}. \tag{11.4.9}$$

Further, by the Hopf bifurcation theory [Ho, JZ], these periodic solutions are stable for $\eta - \eta^{(1)}$ sufficiently small. Evidently, they corresponds to coherent laser light in the $\pm l_1$ modes, since the resultant radiation field reduces to a classical dispersion-free form and therefore satisfies the coherence condition of Section 3.3.2. Thus, the following result, obtained in [HL1] for the single mode case, prevails here too.

Proposition 11.4.1   *Under the specified conditions, there is a phase transition from a nonradiant state to a coherently radiating one, when η increases past the critical value $\eta^{(1)}$. Evidently, there is a breakdown of the gauge symmetry, specified at the end of Section 11.2, in the latter phase.*

Further Transitions: Polychromatic and Chaotic Laser Light

In the single mode case, it was proved in [HL1] that, in the particular case that $\gamma_\perp = \gamma_\| = \kappa$, the periodic orbit of $\mathcal{K}$, and hence the coherent monochromatic radiation of Σ, is stable for all $\eta > \eta^{(1)}$.

In general, however, it is clear from the theory of classical dynamical systems [RT, ER] that the equations of motion (11.3.14) for $\mathcal{K}$ provide the framework for further bifurcations, of the kind arising in hydrodynamics, as η is increased. Specifically, we have the following possibilities.

(A) There could be a bifurcation of the simply periodic orbit of $\mathcal{K}$ into a strange attractor [RT] at some value $\eta^{(2)}(> \eta^{(1)})$ of η. In this case, there would be a transition from monochromatic to chaotic radiation when η increased beyond the value $\eta^{(2)}$ (cf. [Ha3, HB, MSA]).

(B) There could be a succession of bifurcations, corresponding to the activation of different modes, according to the Landau mechanism for the onset of turbulence [LL2]. This would then correspond to polychromatic radiation, and would simulate optical chaos when the number of activated modes became very large.

In fact, we now adapt an argument of Haken [Ha3] to the present model in order to show that the scenario (A) can be achieved even when there is just one mode and when the resonance condition $\epsilon = \omega$ is fulfilled. Thus, assuming these conditions, we note first note that, by Eq. (11.4.4),

$$\eta^{(1)} = \frac{\kappa\gamma_\perp}{\lambda^2}. \tag{11.4.10}$$

Thus, introducing a transformation of variables, $(\alpha_t, s_t, p_t) \rightarrow (e_t, c_t, f_t)$, by the prescription

$$e_t = 2\left(\frac{\kappa}{\gamma_\|(\eta - \eta^{(1)})}\right)^{\frac{1}{2}} \alpha_t \exp(i\epsilon t), \qquad (11.4.11a)$$

$$c_t = 2\lambda(\kappa\gamma_\|(\eta - \eta^{(1)}))^{\frac{-1}{2}} s_t \exp(i\epsilon t), \qquad (11.4.11b)$$

and

$$f_t = \frac{p_t}{\eta^{(1)}}, \qquad (11.4.11c)$$

where the subscript $l (= 0)$ has been omitted, we see from Eqs. (11.4.10) and (11.4.11) that the equations of motion (11.3.14) take the following form.

$$\frac{de_t}{dt} = -\kappa(e_t - c_t), \qquad (11.4.12a)$$

$$\frac{dc_t}{dt} = -\gamma_\perp(c_t - e_t f_t), \qquad (11.4.12b)$$

and

$$\frac{df_t}{dt} = -\gamma_\|(f_t - g + (g - 1)\operatorname{Re}(e_t \bar{c}_t)), \qquad (11.4.12c)$$

where

$$g = \frac{\eta}{\eta^{(1)}}. \qquad (11.4.13)$$

Moreover, as proved in [AlSe], Eqs. (11.4.12) have a unique solution.
    We now infer from the first two of those equations that

$$\left(\frac{d}{dt} + \kappa + \gamma_\perp\right) \operatorname{Im}(e_t \bar{c}_t) = 0$$

which signifies that $\operatorname{Im}(e_t \bar{c}_t)$ is a transient quantity, which decays exponentially to zero. Hence, for the long-time dynamics, we may take $e_t \bar{c}_t$ to be real, that is, we may assume that e_t, c_t have the same phase. Thus,

$$e_t = e_t^{(0)} \exp(i\beta_t), \quad c_t = c_t^{(0)} \exp(i\beta_t), \qquad (11.4.14)$$

where $e_t^{(0)}$, $c_t^{(0)}$, β_t are real. Furthermore, on using these expressions for e_t and c_t in Eqs. (11.4.12) and taking the real and imaginary parts of the resultant formulae, we obtain the following equations of motion for the real variables $e_t^{(0)}$, $c_t^{(0)}$, f_t and β_t:

$$\frac{de_t^{(0)}}{dt} = -\kappa(e_t^{(0)} - c_t^{(0)}), \qquad (11.4.15a)$$

$$\frac{dc_t^{(0)}}{dt} = -\gamma_\perp(c_t^{(0)} - e_t^{(0)}f_t), \qquad (11.4.15b)$$

$$\frac{df_t}{dt} = -\gamma_\parallel\left(f_t - g + (g-1)(e_t^{(0)}c_t^{(0)})\right), \qquad (11.4.15c)$$

and

$$e_t^{(0)}\frac{d\beta_t}{dt} = c_t^{(0)}\frac{d\beta_t}{dt} = 0.$$

This last equation signifies that *either* β_t is constant *or* both $e_t^{(0)}$ and $c_t^{(0)}$ vanish. In fact there is no loss of generality in assuming the first alternative, since the value of β_t in Eq. (11.4.14) becomes irrelevant if the second one prevails.[3] Thus, the macroscopic evolution following the decay of transients is covered by the Maxwell–Bloch equations (11.4.15) and the constancy of β_t.

Now, as shown by Haken [Ha3], these equations transform, under a certain simple linear mapping, into the equations of motion for a Lorenz attractor, which is known to be undergo a transition to chaos for appropriate values of the control variables. As a result, the Maxwell–Bloch equations support such a transition if [Ha3]

$$\kappa > \gamma_\perp + \gamma_\parallel \quad \text{and} \quad \frac{\eta}{\eta^{(1)}} - 1 > \frac{(\gamma_\perp + \gamma_\parallel + \kappa)(\gamma_\perp + \kappa)}{\gamma_\perp(\kappa - \gamma_\perp - \gamma_\parallel)}.$$

Since these conditions are compatible with the demands of Eqs. (11.2.7) and (11.2.8), we have the following result.

Proposition 11.4.2 *The single-mode model, satisfying the resonance condition $\epsilon = \omega$, exhibits chaotic behavior when the control parameters $\gamma_\perp$, $\gamma_\parallel$, κ, λ lie in a certain domain.*

Comment The point to emphasise here is that, *even in the single-mode case,* there is a chaotic regime. In the multimode case, the prevalence of additional degrees of freedom would be expected to favor chaos still further [RT, ER].

11.5 THE MICROSCOPIC DYNAMICS

Our objective now is to obtain the microscopic dynamics of the model, in the limit $N \to \infty$. We take the local microscopic observables to be those of the

---

[3] In that case, which depends on the initial values of $e_t^{(0)}$ and $c_t^{(0)}$ being zero, Eqs. (11.4.12) simplify to the form

$$e_t = c_t = 0, \quad \frac{df_t}{dt} = -\gamma_\parallel(f_t - g).$$

matter only, since, as noted in the first comment following Proposition 11.3.1, the local radiation field of the model reduces to a classical macroscopic one, namely $\phi_{r,t}$, in this limit, and therefore carries no quantum observables.

We formulate the model, Σ_{mat}, of the infinitely extended system of atoms as follows. We assume that the $C^{\star}$–algebra, $\mathcal{A}_{\text{mat}}$, of the observables of this system is the standard quasi-local one given by the norm closure of the local algebra $\mathcal{A}_{\text{mat},L} := \bigcup_{N \in \mathbf{N}} \mathcal{A}_{\text{mat}}^{(N)}$, the norm being inherited from those of the local algebras $\mathcal{A}_{\text{mat}}^{(N)}$ in the usual way (cf. Section 2.5.2). We take the state space of Σ_{mat} to be the set, S_{mat}, of all states on $\mathcal{A}_{\text{mat}}$, and we represent the space translation group $\mathbf{Z}$ by the automorphisms V of this algebra, defined by the formula

$$V(m)\sigma_{r,u} = \sigma_{r+m,u} \quad \forall r, m \in \mathbf{Z},\ u = +, -, x, y, z. \quad (11.5.1)$$

We formulate the microscopic dynamics of Σ_{mat} as a limiting form, as $N \to \infty$, of that induced by $T^{(N)}$ on $\mathcal{A}_{\text{mat}}^{(N)}$, subject to the following assumptions concerning the initial states, ψ and $\rho^{(N)}$, of Σ_{mat} and $\Sigma^{(N)}$, respectively:

(I) ψ is primary and $\rho^{(N)}$ satisfies the conditions (I.1)–(I.3) of Section 11.3.

(II) The restrictions of ψ and $\rho^{(N)}$ to $\mathcal{A}_{\text{mat}}^{(N)}$ are the same, that is,

$$\psi_{|\mathcal{A}_{\text{mat}}^{(N)}} = \rho^{(N)}_{|\mathcal{A}_{\text{mat}}^{(N)}} \quad \forall N \in \mathbf{N}.$$

Under these assumptions, Bagarello and Sewell [BS] proved the following proposition, to the effect that the dynamics of Σ_{mat} is characterised by the independent motion of its atoms in the time-dependent classical radiation field ϕ, given by Eq. (11.3.17).

Proposition 11.5.1

(i) *Assuming the conditions (I) and (II), the microscopic dynamics of the model corresponds, in the limit $N \to \infty$, to a two-parameter family,*

$$\{\mathcal{T}(s,t \mid \phi) \mid 0 \leq s \leq t;\ \mathcal{T}(s,u \mid \phi)\mathcal{T}(u,t \mid \phi) = \mathcal{T}(s,t \mid \phi)\}$$

of CP transformations of $\mathcal{A}$, in that

$$\lim_{N \to \infty} \rho^{(N)}\left(T^{(N)}(t)A\right) = \psi(\mathcal{T}(0,t \mid \phi)A) \quad \forall A \in \mathcal{A}_{\text{mat},L},\ t \in \mathbf{R}_{+}.$$

$$(11.5.2)$$

Furthermore, the generator of $\mathcal{T}$ is

$$\mathcal{L}(t \mid \phi) := \frac{\partial}{\partial t} \mathcal{T}(s,t \mid \phi)_{|s=t} = L_{\text{mat}} + \sum_{r \in \mathbf{Z}} [\phi_{r,t}\sigma_{r,+} - h.c., \cdot]. \quad (11.5.3)$$

Thus,

$$\mathcal{T}(s,t \mid \phi) = \mathbf{T}_{a} \exp\left(\int_{s}^{t} du \mathcal{L}(u \mid \phi)\right), \quad (11.5.4)$$

where $\mathbf{T}_a$ is the antichronological operator. We define

$$\psi_t = \mathcal{T}^\star(0,t \mid \phi)\psi. \tag{11.5.5}$$

(ii) $\mathcal{T}$ factorises into a product of single site contributions according to the formula

$$\mathcal{T}(s,t \mid \phi) = \otimes_{r\in\mathbf{Z}} \mathcal{T}_r(s,t \mid \phi_r), \tag{11.5.6}$$

where

$$\mathcal{T}_r(s,t \mid \phi_r) = \mathbf{T}_a \exp\left(\int_s^t du \mathcal{L}_r(u \mid \phi)\right), \tag{11.5.7}$$

$$\mathcal{L}_r(t \mid \phi) = L_r + [\phi_{r,t}\sigma_{r,+} - h.c., \cdot], \tag{11.5.8}$$

and L_r is the copy of L_{at} for the site r.

Removal of Transient Effects

It follows immediately from this proposition that the microscopic dynamics of the model is completely determined by the transformations, $\mathcal{T}_r$, of the single site algebras $\mathcal{A}_r$. To investigate their action, we define

$$\sigma_{r,u}(t) := \mathcal{T}_r(0,t \mid \phi_r)\sigma_{r,u}, \quad \text{for } u = +,-,x,y,z, \tag{11.5.9}$$

and infer from equations (11.2.1), (11.2.2), (11.2.5), (11.5.7) and (11.5.8) that

$$\frac{d\sigma_{r,-}(t)}{dt} + (i\epsilon + \gamma_\perp)\sigma_{r,-}(t) - i\phi_{r,t}\sigma_{r,z}(t) = 0 \tag{11.5.10}$$

and

$$\frac{d\sigma_{r,z}(t)}{dt} + \gamma_\parallel\sigma_{r,z}(t) - 2i(\overline{\phi}_{r,t}\sigma_{r,-}(t) - h.c.) = \gamma_\parallel\eta I. \tag{11.5.11}$$

Therefore, since $\sigma_{r,\pm} = (\sigma_{r,x}\pm i\sigma_{r,y})/2$, Eqs. (11.5.10) and (11.5.11) signify that the spin vector $\sigma_r(t) = (\sigma_{r,x}(t), \sigma_{r,y}(t), \sigma_{r,z}(t))$ evolves according to a linear inhomogeneous equation of the form

$$\frac{d\sigma_r(t)}{dt} = b_r(t)\sigma_r(t) + cI, \tag{11.5.12}$$

where $b_r(t)$ is a linear transformation of $\mathbf{C}^3$, whose explicit form is obtained from the equivalence of Eq. (11.5.12) to the pair of equations (11.5.10) and (11.5.11), and

$$c = (0,0,\gamma_\parallel\eta). \tag{11.5.13}$$

The solution of Eq. (11.5.12) may be expressed in the form

$$\sigma_r(t) = g_r(t,0)\sigma_r(0) + \int_0^t dt' g_r(t,t')cI, \qquad (11.5.14)$$

where the Green function g_r is determined by the formula

$$\frac{\partial}{\partial t}g_r(t,t') = b_r(t)g_r(t,t') \quad \forall t \geq t' \geq 0, \ g_r(t,t) = I_3, \qquad (11.5.15)$$

and I_3 is the identity operator in $\mathbf{C}^3$. The following key lemma concerning g_r is proved in Appendix A.

Lemma 11.5.2 *For fixed, t', $g(t,t')$ decays to zero, with increasing t, at least as fast as* $\exp(-\gamma t)$, *where* $\gamma = \min(\gamma_\perp, \gamma_\parallel)$.

Comment This lemma signifies that the first term on the right-hand side of Eq. (11.5.14) is merely a transient, which decays in a time, γ^{-1}, of the order of the lifetime of an atomic excited state. Consequently, the asymptotic form for $\sigma_r(t)$, obtained by discarding the transients, is

$$\sigma_r(t) = \theta_r(t)I, \qquad (11.5.16)$$

where

$$\theta_r(t) = \int_0^t dt' g_r(t,t')(0,0,\gamma_\parallel \eta). \qquad (11.5.17)$$

Hence, since the state ψ_t is given by its action on the finite monomials in the components of the σ_r's, the following proposition is a simple consequence of these last two equations, together with Proposition 11.5.1.

Proposition 11.5.3 *The asymptotic form of the time-dependent state ψ_t carries no correlations between the observables at the different sites and is therefore of the form*

$$\psi_t = \otimes_{r \in \mathbf{Z}} \psi_{r,t}, \qquad (11.5.18)$$

where the atomic state $\psi_{r,t}$ is given by

$$\psi_{r,t}(\sigma_r) = \theta_r(t). \qquad (11.5.19)$$

Comment It follows immediately from this proposition that the asymptotic state of the model is completely determined by the form of θ_r. We denote the components of θ_r analogously with those of σ_r: thus $\theta_r = (\theta_{r,x}, \theta_{r,y}, \theta_{r,z})$ and $\theta_{r,\pm} = (\theta_{r,x} \pm i\theta_{r,y})/2$. The following proposition, which is proved in Appendix A, provides an explicit formula for θ_r in terms of the asymptotic forms of the macroscopic variables s and p.

Proposition 11.5.4 *The function θ_r is given by the equations*

$$\theta_{r,-}(t) = \sum_{l=-n}^{n} s_{l,t} \exp\left(\frac{2\pi i r l}{2n+1}\right) \tag{11.5.20}$$

and

$$\theta_{r,z}(t) = \sum_{l=-n}^{n-1} p_{l,t} \exp\left(\frac{2\pi i r l}{2n+1}\right). \tag{11.5.21}$$

Comments

(1) It follows from Eqs. (11.5.18)–(11.5.21) that the time-dependent state ψ_t is completely determined by the macroscopic variables, (s_t, p_t, α_t), that is, by the phase point x_t of the classical system $\mathcal{K}$. Hence, if Δ is an attractor of this system and if the initial value x_0 of the phase point lies in its domain of attraction, then x_t will subsequently move into Δ and, correspondingly, the microstate of Σ_{mat} will then be determined by the structure and dynamics of this attractor.

(2) The following corollary is an immediate consequence of Propositions 11.5.1 and 11.5.4.

Corollary 11.5.5 *The asymptotic state ψ_t is spatially periodic, with periodicity $(2n+1)$, that is,*

$$\psi_t = V^\star(2n+1)\psi_t, \tag{11.5.22}$$

where V is the group of spatial translational automorphisms of $\mathcal{A}$ defined by Eq. (11.5.1).

Explicit Asymptotic Forms of ψ_t

We are now in a position to obtain the explicit form of this state corresponding to the normal and coherent phases of the radiation. We have no means of doing the same thing for the chaotic phase, since we have no explicit solution for the macroscopic dynamics there.

(1) *The Normal Phase.* Here, by Eqs. (11.3.17), (11.4.1) and (11.5.18)–(11.5.21), ψ_t takes the stationary value $\psi^{(\text{normal})}$, defined by the formula

$$\psi^{(\text{normal})} = \bigotimes_{r \in \mathbf{Z}} \psi_r^{(\text{normal})} \tag{11.5.23}$$

where

$$\psi_r^{(\text{coh})}(\sigma_{r,\pm}) = 0 \quad \text{and} \quad \rho_r^{(\text{coh})}(\sigma_{r,z}) = \eta. \tag{11.5.23}$$

(2) *The Coherent Phase.* In this phase, it follows from Eqs. (11.4.6) and (11.5.16)–(11.5.21) that the asymptotic form, $\psi_t^{(\text{coh})}$, of ψ_t is given by the formula

$$\psi_t^{(coh)} = \bigotimes_{r \in \mathbf{Z}} \psi_{r,t}^{(coh)}, \qquad (11.5.24)$$

where

$$\psi_{r,t}^{(coh)}(\sigma_{r,-}) = s_{\tilde{l}_1}^{(0)} \exp\left(i(\frac{2\pi i \tilde{l}_1 r}{n} - \Omega t)\right), \qquad \psi_{r,t}^{(coh)}(\sigma_{r,z}) = \eta. \quad (11.5.25)$$

Comment By contrast with the thermal equilibrium states of pure phases, the stable states $\psi_t^{(coh)}$ are time-dependent. The same is true for the chaotic phase, since the macrosopic variables s_t and p_t are time-dependent there too. If they were constant, then, by Eq. (11.3.14a), α_t would also be constant, which would mean that the phase point, x_t, of $\mathcal{K}$ would be a fixed point, not a chaotically moving one.

11.6 A NONEQUILIBRIUM MAXIMUM ENTROPY PRINCIPLE

We now establish that, as a further consequence of the results of Section 11.5, ψ_t maximises the global entropy density of the matter, as defined on the states with spatial periodicity $(2n + 1)$, subject to the condition that the limiting value, as $N \to \infty$, of the expectation values of the macroscopic variables $s^{(N)}, p^{(N)}$ are s_t, p_t, respectively.

For this purpose, we start by defining S_n to be the set of spatially periodic states, ω, on $\mathcal{A}_{mat}$, with period $(2n + 1)$. We then define S_{n,s_t,p_t} to be the subset of these states for which

$$\lim_{N \to \infty} \omega(s_l^{(N)}) = s_{l,t} \quad \text{and} \quad \lim_{N \to \infty} \omega(p_l^{(N)}) = p_{l,t} \quad \forall l \in [-n, n], t \in \mathbf{R}_+.$$

Equivalently, by Eqs. (11.3.1) and (11.3.2), S_{n,s_t,p_t} consists of the states, ω, of spatial periodicity $(2n + 1)$ that satisfy the conditions

$$(2n + 1)^{-1} \sum_{r=-n}^{n} \omega(\sigma_{r,-}) \exp\left(-\frac{2\pi i l r}{2n + 1}\right) = s_{l,t} \qquad (11.6.1)$$

and

$$(2n + 1)^{-1} \sum_{r=-n}^{n} \omega(\sigma_{r,z}) \exp\left(-\frac{2\pi i l r}{2n + 1}\right) = p_{l,t}, \qquad (11.6.2)$$

for $l \in [-n, n]$ and $t \in \mathbf{R}_+$.

For any ω in S_{n,s_t,p_t}, we denote by $\hat{\omega}_N$ the density matrix in $\mathcal{H}_{mat}^{(N)}$ representing the restriction of the state ω to $\mathcal{A}_{mat}^{(N)}$. Further, recalling from Section 11.2 that the algebra $\mathcal{A}_{mat}^{(N)}$ and the Hilbert space $\mathcal{H}_{mat}^{(N)}$ are tensor products, $\bigotimes_{r=-N}^{N} \mathcal{A}_r$ and $\bigotimes_{r=-N}^{N} \mathcal{H}_r$, respectively, of their atomic components, we denote by $\hat{\omega}_{\{r\}}$ the density matrix in $\mathcal{H}_r$ representing the restriction of ω to $\mathcal{A}_r$.

We define the entropy functionals on S_n along the lines specified in Section 3.2.3. Thus we define the entropy induced by the state ω ($\in S_n$) on the atom at the site r and on the matter in the region $[-N, N]$ to be

$$S_{\{r\}}(\omega) = -\mathrm{Tr}_{\mathcal{H}_r}(\hat{\omega}_{\{r\}} \ln \hat{\omega}_{\{r\}}) \tag{11.6.3}$$

and

$$S_N(\omega) = -\mathrm{Tr}(\hat{\omega}_N \ln \hat{\omega}_N), \tag{11.6.4}$$

respectively. We then define the global entropy density of that state to be[4]

$$\sigma(\omega) = \lim_{N \to \infty} \frac{S_N(\omega)}{2N + 1}, \tag{11.6.5}$$

the convergence being guaranteed by the subadditivity of entropy and the periodicity of ω.

The following proposition constitutes a characterisation of the asymptotic nonequilibrium states of Σ_{mat} by a maximum entropy principle.

Proposition 11.6.1 *The asymptotic time-dependent state ψ_t maximises the functional σ, as restricted to S_{n,s_t,p_t}.*

Proof. We first note that, by Corollary 11.5.5, $\psi_t \in S_n$ and that, by equations (11.5.18)–(11.5.21), it satisfies the conditions (11.6.1) and (11.6.2). Hence, this state lies in S_{n,s_t,p_t}.

Next, we observe that, if ω is an arbitrary element of this latter set of states, then it follows from Eqs. (11.5.20), (11.5.21), (11.6.1) and (11.6.2), by elementary Fourier analysis, that $\omega(\sigma_r) = \theta_r(t)$, for $r \in [-n, n]$, and hence, by the periodicity of ω, for all $r \in \mathbf{Z}$. Therefore, by Eq. (11.5.19), the restrictions of ω and ψ_t to each single site algebra $\mathcal{A}_r$ are identical. In other words, *the elements of S_{n,s_t,p_t} all coincide on each of the single site algebras $\mathcal{A}_r$.*

Further, by the subadditivity of entropy,

$$S_N(\omega) \le \sum_{r=-N}^{N} S_{\{r\}}(\omega) \quad \forall \omega \in S_{n,s_t,p_t},$$

while, by Eq. (11.5.18),

$$\hat{\psi}_{N,t} = \otimes_{r=-N}^{N} \hat{\psi}_{r,t}$$

and therefore, by Eqs. (11.6.3) and (11.6.4),

---

[4] At this stage, the use of the symbol σ for the entropy functional defined by Eq. (11.6.5) should cause no confusion, since the spin observables will not occur in the remainder of this chapter.

$$S_N(\psi_t) = \sum_{r=-N}^{(N)} S_{\{r\}}(\psi_t).$$

Consequently, as ψ_t and ω coincide on the single site algebras,

$$S_N(\omega) \leq S_N(\psi_t),$$

and therefore, by Eq. (11.6.5),

$$\sigma(\omega) \leq \sigma(\psi_t) \ \forall \ \omega \in S_{n,s_t,p_t},$$

which proves the proposition. $\quad\square$

11.7 CONCLUDING REMARKS

The theory of this chapter embodies an interplay between the microscopic and macroscopic behavior of a dissipative quantum model that provides a realisation of the general scheme of Chapter 10. Since the model treated here is but a caricature of a physical system (specifically of a laser), the question naturally arises as to whether more realistic ones of dissipative quantum systems will possess similar features, in particular whether they will exhibit similar non-equilibrium phase structure and whether their microstates conform to a maximum entropy principle such as that of Proposition 11.6.1.

In this connection, an interesting model to consider would be a modification of the present one with a continuum of radiation modes. That could comprise the thermodynamical limiting version of a system of N atoms coupled to N radiation modes, all living on the same lattice; we envisage that, as in equilibrium statistical mechanics of non-mean field theoretic systems, the lattice should be three-dimensional in order that the model may support phase transitions with continuous symmetry breakdown. In fact, there is no difficulty in formulating such a model for finite N, since the validity of Proposition 11.2.3 depends neither on the dimensionality of the system nor on the values of N and n (cf. [AlSe]). However, in this case where n is equal or even just proportional to N, it is not clear whether the theory of quantum dynamical semigroups is sufficiently developed at present to permit a passage to the limit where N tends to infinity.

APPENDIX A: PROOF OF LEMMA 11.5.2 AND PROPOSITION 11.5.4

Proof of Lemma 11.5.2. Let $\xi_r = (\xi_{r,x}, \xi_{r,y}, \xi_{r,z})$ be an arbitrary element of $\mathbf{R}^3$, and, for fixed $t'(\in \mathbf{R}_+)$ and any $t > t'$, let

$$\xi_r(t) \equiv (\xi_{r,x}(t), \xi_{r,y}(t), \xi_{r,z}(t)) := g_r(t, t')\xi_r, \qquad (A.1)$$

where g_r is defined by Eq. (11.5.15). Then it follows from the latter formula and Eq. (A.1) that $\xi_r(t)$ satisfies the homogeneous linear equation obtained by replacing σ_r by ξ_r in Eq. (11.5.12) and discarding the term cI on the right-hand side. Hence, defining $\xi_{r,\pm} = (\xi_{r,x} \pm i\xi_{r,y})/2$, it follows from the equivalence between Eq. (11.5.12) and the pair of equations (11.5.10) and (11.5.11) that the equation of motion for $\xi_r(t)$ is also given by replacing σ by ξ in the latter equations and discarding the right-hand side of (11.5.11). Thus,

$$\frac{d}{dt}\xi_{r,-}(t) + (\gamma_\perp - i\epsilon)\xi_{r,-}(t) - i\phi_{r,t}\xi_{r,z}(t) = 0$$

and

$$\frac{d}{dt}\xi_{r,z}(t) + \gamma_\|\xi_{r,z}(t) - 2i(\overline{\phi}_{r,t}\xi_{r,-}(t) - c.c.) = 0.$$

Hence,

$$\frac{d}{dt}|\xi_r(t)|^2 \equiv \frac{d}{dt}(4|\xi_{r,-}(t)|^2 + \xi_{r,z}(t)^2)$$

$$= -8\gamma_\perp|\xi_{r,-}(t)|^2 - 2\gamma_\||\xi_{r,z}(t)^2$$

$$= -2[\gamma_\perp\xi_{r,x}(t)^2 + \gamma_\perp\xi_{r,y}(t)^2 + \gamma_\||\xi_{r,z}(t)^2]$$

$$\leq -2\min(\gamma_\perp, \gamma_\|)|\xi_r(t)|^2.$$

Consequently, defining γ to be $\min(\gamma_\perp, \gamma_\|)$,

$$\frac{d}{dt}(|\xi_r(t)|^2 \exp(2\gamma t)) \leq 0,$$

which implies that $|\xi_r(t)|$, and hence $g_r(t, t')$, decay to zero at least as fast as $\exp(-\gamma t)$, with increasing t.   □

Proof of Proposition 11.5.4. By Eqs. (11.5.10), (11.5.11), (11.5.15) and (11.5.17), the equations of motion for θ_r are formally the same as those for σ_r, namely

$$\frac{d\theta_{r,-}(t)}{dt} + (i\epsilon + \gamma_\perp)\theta_{r,-}(t) - i\phi_{r,t}\theta_{r,z}(t) = 0 \qquad (A.2)$$

and

$$\frac{d\theta_{r,z}(t)}{dt} + \gamma_\|\theta_{r,z}(t) - 2i(\overline{\phi}_{r,t}\theta_{r,-}(t) - c.c.) = \gamma_\|\eta. \qquad (A.3)$$

Further, defining

$$\hat{s}_r(t) = \sum_{l=-n}^{n} s_{l,t} \exp(2\pi i r l/2n + 1) \qquad (A.4)$$

and

$$\hat{p}_r(t) = \sum_{l=-n}^{n} p_{l,t} \exp(2\pi irl/2n + 1), \tag{A.5}$$

we see, by elementary Fourier analysis, that Eqs. (11.3.14b) and (11.3.14c) are equivalent to

$$\frac{d\hat{s}_r(t)}{dt} + (i\epsilon + \gamma_\perp)\hat{s}_r(t) - i\phi_{r,t}\hat{p}_r(t) = 0 \tag{A.6}$$

and

$$\frac{d\hat{p}_r(t)}{dt} + \gamma_\|\hat{p}_r(t) - 2i(\overline{\phi}_{r,t}\hat{s}_r(t) - c.c.) = \gamma_\|\eta, \tag{A.7}$$

respectively. Hence, the equations of motion (A.2) and (A.3) for $\theta_{r,-}$ and $\theta_{r,z}$ are equivalent to those given by Eqs. (A.6) and (A.7) for $\hat{s}_r$ and $\hat{p}_r$. Consequently, in the asymptotic regime, where the initial conditions are irrelevant, $\theta_{r,-} = \hat{s}_r$ and $\theta_{r,z} = \hat{p}_r$. By equations (A.4) and (A.5), this is the required result. $\square$

References

[Abr]   A. A. Abrikosov: *On the magnetic properties of superconductors of the second group*, Soviet Phys. JETP **32**, 1442–1452, 1957.

[An1]   P. W. Anderson: *Random phase approximation in the theory of superconductivity*, Phys. Rev. **110**, 1900–1916, 1959.

[An2]   P. W. Anderson: *The Theory of Superconductivity in the High-T_c Cuprates*, Princeton University Press, Princeton, NJ, 1977.

[Ar1]   H. Araki: *Relative entropy of states of Von Neumann algebras*, Publ. RIMS Kyoto Univ. **11**, 809–833, 1976; and *Relative entropy of states of Von Neumann algebras II*, Publ. RIMS, Kyoto Univ. **13**, 173–192, 1977.

[Ar2]   H. Araki: *Structure of some Von Neumann algebras with isolated discrete modular spectrum*, Publ. RIMS Kyoto Univ. **9**, 1–44, 1973.

[Ar3]   H. Araki: *On the equivalence of the KMS condition and the variational principle for quantum lattice systems*, Commun. Math. Phys. **38**, 1–10, 1974.

[Ar4]   H. Araki: *Multiple time analyticity of a quantum statistical state satisfying the KMS boundary condition*, Publ. RIMS Kyoto Univ. **4**, 361–371, 1968.

[Ar5]   H. Araki: *Positive cone, Radon–Nikodym theorems, relative Hamiltonian and the Gibbs condition in statistical mechanics. An application of the Tomita-Takesaki theory*, in: D. Kastler (editor), $C^{\star}$-Algebras and their Application to Statistical Mechanics and Quantum Field Theory, North-Holland, Amsterdam, pp. 64–100, 1976.

[AB]   Y. Aharonov and D. Bohm: *Significance of electromagnetic potentials in the quantum theory*, Phys. Rev. **115**, 485–491, 1959.

[AF]   R. Alicki and M. Fannes: *Quantum dynamical systems*, Oxford University Press, Oxford, 2001.

[AFL]   L. Accardi, A. Frigerio and J. T. Lewis: *Quantum stochastic processes*, Publ. RIMS **18**, 97–133, 1982.

[AHKT]   H. Araki, R. Haag, D. Kastler and M. Takesaki: *Extension of KMS states and chemical potential*, Commun. Math. Phys. **53**, 97–134, 1977.

[AM]   A. S. Alexander and N. F. Mott: *Bipolarons*, Rep. Prog. Phys. **57**, 1197–1288, 1994.

[AR]   A. S. Alexandrov and J. Ranninger: *Theory of bipolarons and bipolaronic bands*, Phys. Rev. B **23**, 1976–1801, 1981; and *Photoemission spectroscopy of the superconducting and normal state of polaronic systems*, Physica C **198**, 360–370, 1992.

[AS]   H. Araki and G. L. Sewell: *Local thermodynamical stability and the KMS conditions of quantum lattice systems*, Commun. Math. Phys. **52**, 103–109, 1977.

[AlSe]   G. Alli and G. L. Sewell: *New methods and structures in the theory of the Dicke laser model*, J. Math. Phys. **36**, 5598–5626, 1995.

[AW]   H. Araki and E. J. Woods: *Representations of the canonical commutation relations describing a nonrelativistic free Bose Gas*, J. Math. Phys. **4**, 637–662, 1963.

[Bl1]   F. Bloch: Z. Phys. **52**, 555, 1928; *ibid* **59**, 208, 1930.

[Bl2]   F. Bloch: *Generalised theory of relaxation*, Phys. Rev. **105**, 1206–1222, 1957.

[Bo]   N. Bohr: *Discussion with Einstein on epistomological problems in atomic physics*, in: P. A. Schilpp (editor), *Albert Einstein: Philosopher–Scientist*, The Library of Living Philosophers, Evanston, IL, pp. 200–241, 1949.

[Bog]   N. N. Bogoliubov: *Problems of a dynamical theory in statistical mechanics*, in: J. De Boer and G. E. Uhlenbeck (editors), *Studies in Statistical Mechanics*, Vol. 1, North-Holland, Amsterdam, 1962.

[Bol]   L. Boltzmann: *Lectures on Gas Theory*, University of California Press, Berkeley, CA, 1964.

[Bu]   H. A. Buchdahl: *The Concepts of Classical Thermodynamics*, Cambridge University Press, 1966.

[BC]   M. Born and K. C. Cheng: J. Phys. Radium **9**, 249, 1948.

[BCS]   J. Bardeen, L. N. Cooper and J. R. Schrieffer: *Theory of superconductivity*, Phys. Rev. **108**, 1175–1204, 1957.

[BFG]   L. Bugliaro, J. Fröhlich and G. M. Graf: *Stabilty of quantum electrodynamics with nonrelativistic matter*, Phys. Rev. Lett. **77**, 3494–3497, 1996.

[BM]   J. G. Bednorz and K. A. Müller: *Possible high T_c superconductivity in the Ba-La-Cu-O system*, Z. Phys. B **64**, 189–193, 1986.

[BR]   O. Bratteli and D. W. Robinson: *Operator Algebras and Quantum Statistical Mechanics*, Springer, Heidelberg, Vol. 1, 1979, Vol. 2, 1981.

[BS]   F. Bagarello and G. L. Sewell: *New structures in the theory of the laser model II: microscopic dynamics and a nonequilibrium entropy principle*, J. Math. Phys. **39**, 2730–2747, 1998.

[Ca]   H. Callen: *Thermodynamics and an Introduction to Thermostatistics*, Wiley, New York, 1985.

[Cas]   H. B. G. Casimir: *On Onsager's principle of microscopic reversibility*, Rev. Mod. Phys. **17**, 343–50, 1945.

[Ch]   G. Choquet: *Lectures on Analysis*, Vol. I, W. A. Benjamin, New York, 1969.

[Cha]   S. Chandrasekhar: *Hydrodynamic and Hydromagnetic Stability*, Oxford University Press, Oxford, 1961.

[Co]   L. N. Cooper: *Bound electron pairs in a degenerate Fermi gas*, Phys. Rev. **104**, 1189–90, 1956.

[CCO]   D. Capocaccia, M. Cassandro and E. Olivieri: *A study of metastability in the Ising model*, Commun. Math. Phys. **39**, 185–204, 1974.

[CF]   A. M. Chebotarev and F. Fagnola: *Sufficient conditions for conservativity of quantum dynamical semigroups*, J. Funct. Anal. **118**, 131–153, 1997.

[CW]   H. Callen and T. A. Welton: *Irreversibility and generalised noise*, Phys. Rev. **83**, 34–40, 1951.

[CNT]   A. Connes, H. Narnhofer and W. Thirring: *Dynamical entropy of $C^\star$-algebras and Von Neumann algebras*, Commun. Math. Phys. **112**, 691–719, 1987.

[CWWH] Y. H. Chen, F. Wilczek, E. Witten and B. Halperin: *On anyon superconductivity*, Int. J. Mod. Phys. B **3**, 1001–67, 1989.

[Da] E. B. Davies: *Markovian master equations*, Commun. Math. Phys. **39**, 91–110, 1974; and *Markovian master equations II*, Math. Ann. **219**, 147–158, 1976.

[DeG] P. G. De Gennes: *Superconductivity of Metals and Alloys*, W. A. Benjamin, New York, 1966.

[Di] P. A. M. Dirac: *Principles of Quantum Mechanics*, Clarendon Press, Oxford, 1958.

[Dic] R. H. Dicke: *Coherence in spontaneous radiation processes*, Phys. Rev. **93**, 99–110, 1954.

[Dix1] J. Dixmier: *Les Algebres des Operateurs dans l'Espace Hilbertien*, Gauthier–Villars, Paris, 1957.

[Dix2] J. Dixmier: *Les $C^{\star}$-Algebres et Leurs Representations*, Gauthier–Villars, Paris, 1964.

[Do] J. L. Doob: *Stochastic Processes*, Wiley, New York, 1953.

[Dob] R. L. Dobrushin: *The Gibbs state that describes the coexistence of phases for a three–dimensional Ising model*, Theory Prob. Appl. **17**, 582–600, 1972.

[Dor] T. C. Dorlas: *Statistical Mechanics*, Institute of Physics Publishing, Bristol, 1999.

[Duf1] N. G. Duffield: *Stability of Bose-Einstein condensation in Fröhlich's pumped phonon system*, Phys. Lett. A **110**, 332–334, 1985.

[Duf2] N. G. Duffield: *Global stability of continuum limit in Fröhlich's pumped phonon system*, J. Phys. A **21**, 625–641, 1988.

[deGM] S. R. de Groot and P. Mazur: *Non-equilibrium Thermodynamics*, North-Holland, Amsterdam, 1962.

[DDR] G.–F. Dell'Antonio, S. Doplicher and D. Ruelle: *A theorem on canonical commutation and anticommutation relations*, Commun. Math. Phys. **2**, 223–230, 1966.

[DF] B. Deaver and W. M. Fairbank: *Experimental evidence for quantised flux in a cylindrical superconductor*, Phys. Rev. Lett. **7**, 43–46, 1961.

[DL] F. J. Dyson and A. Lenard: *Stability of matter, I*, J. Math. Phys. **8**, 423–434, 1967; and *Stability of matter, II*, J. Math. Phys. **9**, 698–711, 1968.

[DLS] F. J. Dyson, E. H. Lieb and B. Simon: *Phase transitions in quantum spin systems with isotropic and non-isotropic interactions*, J. Stat. Phys. **18**, 335–383, 1978.

[DS] D. A. Dubin and G. L. Sewell: *Time-translations in the algebraic formulation of statistical mechanics*, J. Math. Phys. **11**, 2990–2998, 1970.

[Em1] G. G. Emch: *Algebraic Methods in Statistical Mechanics and Quantum Field Theory*, Wiley, New York, 1972.

[Em2] G. G. Emch: *The definition of states in quantum statistical mechanics*, J. Math. Phys. **7**, 1413–1420, 1966.

[EK] G. G. Emch and H. J. F. Knops: *Pure thermodynamical phases as extremal KMS states*, J. Math. Phys. **11**, 3008–3018, 1970.

[EKV] G. G. Emch, H. J. F. Knops and E. J. Verboven: *Breaking of Euclidean symmetry with an application to crystallisation*, J. Math. Phys. **11**, 1655–1667, 1970.

[ER]   J. P. Eckmann and D. Ruelle: *Ergodic theory of chaos and strange attrac-tors*, Rev. Mod. Phys. **57**, 617–656, 1985.

[Fa]   F. Fagnola: *On quantum stochastic differential equations with unbounded coefficients*, Prob. Th. Rel. Fields **86**, 501–516, 1990.

[Far]   I. E. Farqhuar: *Ergodic Theory in Statistical Mechanics*, Interscience, London, 1964.

[Fe]   E. Fermi: *Thermodynamics*, Dover, New York, 1936.

[Fi1]   M. E. Fisher: *The theory of condensation and the critical point*, Physics **3**, 255–283, 1967.

[Fi2]   M. E. Fisher: *The free energy of a macroscopic system*, Arch. Rat. Mech. Anal. **17**, 377–410, 1964.

[Fr1]   H. Fröhlich: *Long range correlations and energy storage in biological systems*, Int. J. Quantum Chem. **2**, 641–649, 1968.

[Fr2]   H. Fröhlich: *Theory of the superconducting state, I. The ground state at the absolute zero of temperature*, Phys. Rev. **79**, 845–856, 1950.

[Fr3]   H. Fröhlich: *Interaction of electrons with lattice vibrations*, Proc. R. Soc. A **215**, 291–298, 1952.

[Fr4]   H. Fröhlich: *Macroscopic wave functions in superconductors*, Proc. Phys. Soc. **87**, 330–332, 1966.

[Fr5]   H. Fröhlich: *The theory of the superconducting state*, Rep. Prog. Phys. **24**, 1–23, 1961.

[Fr6]   H. Fröhlich: *The biological effects of microwaves and related questions*, Adv. Elect. Electron Phys. **53**, 85–152, 1980.

[FKM]   G. W. Ford, M. Kac and P. Mazur: *Statistical mechanics of assemblies of coupled oscillators*, J. Math. Phys. **6**, 504–515, 1965.

[FV]   M. Fannes and A. Verbeure: *Correlation inequalities and equilibrium states. II*, Commun. Math. Phys. **57**, 165–171, 1977.

[Gi]   V. L. Ginzburg: *On the theory of superconductivity*, Nuov. Cim. **2**, 1235–1250, 1955.

[Gl]   R. J. Glauber: *The quantum theory of optical coherence*, Phys. Rev. **130**, 2529–2539, 1963.

[Go]   L. P. Gor'kov: *Microscopic derivation of the Ginzburg–Landau equations in the theory of superconductivity*, Sov. Phys. JETP **36**, 1364–1367, 1959.

[GH]   R. Graham and H. Haken: *Laser light–first example of a phase transition far away from equilibrium*, Z. Phys. **237**, 31–46, 1970.

[GK]   V. Gorini and A. Kossakowski: *N-level system in contact with a singular reservoir*, J. Math. Phys. **17**, 1298–1305, 1976.

[GL]   V. L. Ginzburg and L. D. Landau: Zu. Eksper. Teor. Phys. **20**, 1064, 1950 (in Russian).

[GMR]   T. Gerisch, R. Münzner and A. Rieckers: *Global $C^\star$-dynamics and its KMS states of weakly inhomogeneous bipolaronic superconductors*, J. Stat. Phys. **9**, 751–779, 1999.

[GP]   P. Glansdorff and I. Prigogine: *Thermodynamic Theory of Structure, Stabi-lity and Fluctuations*, Wiley-Interscience, London, 1971.

[GS]   I. I. Gihman and A. V. Skorohod: *The Theory of Stochastic Processes I*, Springer, Berlin, 1974.

[GV]   I. M. Gelfand and N. Ya. Vilenkin: *Generalised Functions*, Vol. 4, Academic Press, New York, 1964.

[GVV]   D. Goderis, A. Verbeure and P. Vets: *Noncommutative central limits*, Prob. Th. Rel. Fields **82**, 527–544, 1989.

[GW]   L. Garding and A. Wightman: *Representations of the anticommutation relations*, Proc. Natl. Acad. Sci. USA **40**, 617–621, 1954; and *Representations of the commutation relations*, Proc. Natl. Acad. Sci. USA **40**, 622–626, 1954.

[GWM]   J. C. Garrison, J. Wong and H. L. Morrison: *Absence of long range order in thin films*, J. Math. Phys. **13**, 1735–1742, 1972.

[Ha1]   H. Haken: *Handbuch der Physik*, Bd. XXV/2C, Springer, Heidelberg, 1970.

[Ha2]   H. Haken: *Advanced Synergetics*, Springer, Berlin, 1983.

[Ha3]   H. Haken: *Analogies between higher instabilities in fluids and lasers*, Phys. Lett. A **53**, 77–78, 1975.

[Haa1]   R. Haag: *Local Quantum Physics*, Springer, Berlin, 1992.

[Haa2]   R. Haag: *Canonical commutation relations in quantum field theory and functional integration*, in: W. E. Brittin, B. W. Downs and J. Downs (editors), *Lectures in Theoretical Physics*, III, Interscience, New York, pp. 353–381, 1961.

[Hah]   E. C. Hahn: *Spin echoes*, Phys. Rev. **80**, 580–594, 1950

[Hal]   P. R. Halmos: *Measure Theory*, Van Nostrand, New York, 1966.

[He]   K. Hepp: *Quantum theory of measurement and macroscopic observables*, Helv. Phys. Acta **45**, 237–48, 1972.

[Hei]   W. Heisenberg: Z. Naturforsch. **2a**, 185, 1947.

[Ho]   E. Hopf: *Abzweigung einer periodichen Lösung von einer stationären Lösung eines Differentialsystems*, Ber. Math-Phys. Kl. Sächs. Akad. Wiss. Leipzig **94**, 1–22, 1942.

[Hu]   J. Hubbard: *Electron correlations in narrow energy bands*, Proc. R. Soc. A **276**, 238–257, 1963.

[HB]   R. J. Harrison and D. J. Biswas; *Chaos in light*, Nature **321**, 394–401, 1986.

[HHW]   R. Haag, N. M. Hugenholtz and M. Winnink: *On the equilibrium states in quantum statistical mechanics*, Commun. Math. Phys. **5**, 215–236, 1967.

[HK]   R. Haag and D. Kastler: *An algebraic approach to quantum field theory*, J. Math. Phys. **5**, 848, 1964.

[HKT–P]   R. Haag, D. Kastler and E. B. Trych–Pohlmeyer: *Stability and equilibrium states*, Commun. Math. Phys. **38**, 173–193, 1974.

[HL1]   K. Hepp and E. H. Lieb: *Phase transitions in reservoir driven open systems, with applications to lasers and superconductors*, Helv. Phys. Acta **46**, 573–603, 1973.

[HL2]   K. Hepp and E. H. Lieb: *On the superradiant phase transitions for molecules in a quantised electromagnetic field: the Dicke maser model*, Ann. Phys. **76**, 360–404, 1973.

[HL3]   K. Hepp and E. H. Lieb: *A reversible quantum dynamical system with irreversible classical macroscopic motion*, in: J. Moser (editor), *Dynamical Systems, Theory and Applications*, Lecture Notes in Physics, 38, Springer, Heidelberg, pp. 178–208, 1975.

[HP]   R. L. Hudson and K. R. Parthaserathy: *Quantum Ito's formula and stochastic evolutions*, Commun. Math. Phys. **93**, 301–323, 1984.

[Ja]   J. M. Jauch: *Foundations of Quantum Mechanics*, Addison Wesley, Reading, MA, 1968.

[Jo]   B. D. Josephson: *Coupled superconductors*, Rev. Mod. Phys. **36**, 216–220, 1964.

[JZ]   R. Jost and E. Zehnder: *A generalisation of the Hopf bifuraction theorem* , Helv. Phys. Acta **45**, 258–276, 1972.

[Kh1]   A. I. Khinchin: *Mathematical Foundations of Information Theory*, Dover, New York, 1957.

[Kh2]   A. I. Khinchin: *Mathematical Foundations of Statistical Mechanics*, Dover, New York, 1949.

[Kho]   A. Kholevo: *On the structure of quantum dynamical semigroups*, J. Funct. Anal. **131**, 255–278, 1995.

[Ko1]   A. N. Kolmogorov: *Logical basis for information theory and probability theory*, IEEE Trans. Info. Theory **14**, 662–664, 1968.

[Ko2]   A. N. Kolmogorov: *Foundations of Probability*, Chelsea, New York, 1956.

[Kr]   K. Krauss: *General state changes in quantum theory*, Ann. Phys. **64**, 311–335, 1971.

[Ku]   R. Kubo: *Statistical mechanical theory of irreversible processes, I*, J. Phys. Soc. Japan **12**, 570–586, 1957.

[Kum]   B. Kümmerer: *Markov dilations on $W^\star$-algebras*, J. Funct. Anal. **63**, 139–177, 1985.

[KFGV]   A. Kossakowski, A. Frigerio, V. Gorini and M. Verri: *Quantum detailed balance and the KMS condition*, Commun. Math. Phys. **57**, 97–110, 1977.

[KL]   T. Kennedy and E. H. Lieb: *An itinerant electron model with crystalline or magnetic long range order*, Physica A **138**, 320–358, 1986.

[KRS]   S. A. Kivelson, D. S. Rokhsar and J. P. Sethna: *Topology of the resonating valence–bond state*, Phys. Rev. B, **35**, 8865–8868, 1987.

[La1]   L. D. Landau: Zeit. Phys. **64**, 629, 1930.

[La2]   L. D. Landau: *The theory of superfluidity of He II*, J. Phys. USSR **5**, 71–90, 1941.

[Lau]   R. B. Laughlin: *Superconducting ground state of noninteracting particles obeying fractional statistics*, Phys. Rev. Lett. **60**, 2677–80, 1988.

[Li1]   G. Lindblad: *Entropy, information and quantum measurements*, Commun. Math. Phys. **33**, 305–322, 1973.

[Li2]   G. Lindblad: *Completely positive maps and entropy inequalities*, Commun. Math. Phys. **40**, 147–151, 1975.

[Li3]   G. Lindblad: *On the generators of quantum dynamical semigroups*, Commun. Math. Phys. **48**, 119–130, 1976.

[Lo]   F. London: *Superfluids*, Vol. 1, Wiley, New York and Chapman and Hall, London, 1950.

[Lon]   F. London and H. London: Physica **2**, 341, 1935.

[LL1]   L. D. Landau and E. M. Lifshitz: *Fluid Mechanics*, Pergamon, Oxford, 1984.

[LL2]   L. D. Landau and E. M. Lifshitz: *Statistical Physics*, Pergamon, New York, 1959.

[LLS]   E. H. Lieb, M. Loss and J. P. Soloviev, *Stability of matter in magnetic fields*, Phys. Rev. Lett. **75**, 985–989, 1995.

[LaRo]   O. E. Lanford and D. W. Robinson: *Statistical mechanics of quantum spin systems III*, Commun. Math. Phys. **9**, 327–338, 1968.

[LaRu]   O. E. Lanford and D. Ruelle: *Observables at infinity and states with short range correlations in statistical mechanics*, Commun. Math. Phys. **13**, 194–215, 1969.

[LiRus]   E. H. Lieb and M. B. Ruskai: *Proof of the strong subadditivity of quantum mechanical entropy*, J. Math. Phys. **14**, 1938–1941, 1973.

[LT]   E. H. Lieb and W. Thirring: *Bound for the kinetic energy of fermions which proves the stability of matter*, Phys. Rev. Lett. **35**, 687–689, 1975.

[LTh]   J. T. Lewis and L. C. Thomas: *On the existence of a class of stationary quantum stochastic processes*, Ann. Inst. H. Poincare **22**, 241–248, 1975.

[LY]   T. D. Lee and C. N. Yang: *Statistical theory of equations of state and phase transitions II*. Phys. Rev. **87**, 410–419, 1952.

[Ma]   D. Mattis: *The Theory of Magnetism*, Harper and Row, New York, 1964.

[MM]   A. Messager and S. Miracle–Sole: *Equilibrium states of the two-dimensional Ising model in the two phase region*, Commun. Math. Phys. **40**, 187–198, 1975.

[MR]   S. Miracle–Sole and D. W. Robinson: *Statistical mechanics of quantum particles with hard cores II: the equilibrium states*, Commun. Math. Phys. **19**, 204–218, 1968.

[MS]   P. C. Martin and J. Schwinger: *Theory of many–particle systems I*, Phys. Rev. **115**, 1342–1373, 1959.

[MSA]   T. Milloni, M. Shih and J. R. Ackerhalt: *Chaos in laser-matter interactions*, World Scientific, Singapore, 1976.

[Na]   Y. Nambu: *Quasi-particles and gauge invariance in the theory of super-conductivity*, Phys. Rev. **117**, 648–663, 1960

[Ne]   E. Nelson: *Dynamical Theories of Brownian Motion*, Princeton University Press, Princeton, NJ, 1967.

[NS]   H. Narnhofer and G. L. Sewell: *Equilibrium states of gravitational systems*, Commun. Math. Phys. **71**, 1–28, 1980.

[NSZ]   H. T. Nieh, G. Su and B. H. Zhao: *Off-diagonal long range order: Meissner effect and flux quantisation*, Phys. Rev. B **51**, 3760–3771, 1995.

[On]   L. Onsager: *Reciprocal relations in irreversible processes, part I*, Phys. Rev. **37**, 405–426, 1931; and *Reciprocal relations in irreversible processes, part II*, Phys. Rev. **38**, 2265–2279, 1931.

[Pe]   O. Penrose: *On the quantum mechanics of He II*, Philos. Mag. **42**, 1373–1377, 1951.

[Pi]   D. Pines (editor): *The Many-body Problem*, W. A. Benjamin, New York, 1961.

[Pip]   A. B. Pippard: *An experimental and theoretical study of the relation between magnetic field and current in a superconductor*, Proc. R. Soc. A **216**, 547–568, 1953.

[Pl]   M. Planck: *Uber das Gesetz der Energieverteilung im Normalspectrum*, Annalen der Physik, **4**, 553–563, 1901.

[PL]   O. Penrose and J. L. Lebowitz: *Rigorous treatment of metastable states in the Van der Waals-Maxwell theory*, J. Stat. Phys. **3**, 211–236, 1971.

[PO]   O. Penrose and L. Onsager: *Bose-Einstein condensation and liquid helium*,
          Phys. Rev. **104**, 576–584, 1956.
[PR]   Y. V. Prohorov and Y. A. Rozanov: *Probability Theory*, Springer, Berlin,
          1969.
[PSSW]   E. Presutti, E.Scacciatelli, G. L. Sewell and F. Wanderlingh: *Studies in the
          $C^\star$-algebraic theory of non-equilibrium statistical mechanics: dynamics of
          open and mechanically driven systems*, J. Math. Phys., **13**, 1085–1098, 1972.
[Ra1]   C. Radin: *The dynamical instability of nonrelativistic many-body systems*,
          Commun. Math. Phys. **54**, 69–79, 1977.
[Ra2]   C. Radin: *The ground state for soft discs*, J. Stat. Phys. **26**, 365–373, 1981.
[Ra3]   C. Radin: *Classical ground states in one dimension*, J. Stat. Phys. **35**,
          109–117, 1984.
[Ri]   G. Rickayzen: *Collective excitations in the theory of superconductivity*,
          Phys. Rev. **115**, 795–808, 1959.
[Ro1]   D. W. Robinson: *Statistical mechanics of quantum spin systems II*, Commun.
          Math. Phys. **7**, 337–348, 1968.
[Ro2]   D. W. Robinson: *Return to equilibrium*, Commun. Math. Phys. **31**, 171–189,
          1973.
[Ro3]   D. W. Robinson: *The thermodynamical pressure in quantum statistical
          mechanics*: Lecture Notes in Physics, Vol. 9, Springer, Berlin, 1971.
[Ru1]   D. Ruelle: *Statistical Mechanics*, W. A. Benjamin, New York, 1969.
[Ru2]   D. Ruelle: *Symmetry breakdown in statistical mechanics*, in: D. Kastler
          (editor), *Cargese Lectures*, Vol. 4, Gordon and Breach, New York, pp.
          169–194, 1969.
[Ru3]   D. Ruelle: *Chaotic Evolution and Strange Attractors*, Cambridge University
          Press, Cambridge, 1989.
[RN]   F. Riesz and B. Sz-Nagy: *Functional Analysis*, F. Ungar, New York, 1955.
[RST]   F. Rocca, M. Sirugue and D. Testard: *On a class of equilibrium states under
          the Kubo-Martin-Schwinger conditions: II bosons*, Commun. Math. Phys.
          19, 119–141, 1970.
[RT]   D. Ruelle and F. Takens: *On the nature of turbulence*, Commun. Math. Phys.
          20, 187–192, 1971.
[RV]   A. W. Roberts and D. E. Varberg: *Convex Functions*, Academic Press,
          London, 1973.
[Sa]   S. Sakai: $C^\star$-*Algebras and* $W^\star$-*Algebras*, Springer, Berlin, 1971.
[Sch]   E. Schrödinger: *What is Life?* Cambridge University Press, 1967.
[Scha1]   M. R. Schafroth: *Superconductivity of a charged ideal Bose gas*, Phys. Rev.
          100, 463–475, 1955.
[Scha2]   M. R. Schafroth: *Remarks on Fröhlich's theory of superconductivity*, Helv.
          Phys. Acta **24**, 645–656, 1951.
[Scha3]   M. R. Schafroth: *Remarks on the Meissner effect*, Phys. Rev. **111**, 72–74,
          1958.
[Scho]:   D. Shoenberg: *Superconductivity*, Cambridge University Press, Cambridge,
          1952.
[Schw]   L. Schwartz: *Theorie des Distributions*, Tome 1, Hermann, Paris, 1950;
          Tome 2, Hermann, Paris, 1951.

[Se1]   G. L. Sewell: *Off-diagonal long range order and superconductive electro-dynamics*, J. Math. Phys. **38**, 2053–2071, 1997.

[Se2]   G. L. Sewell: *Quantum Theory of Collective Phenomena*, Oxford University Press, Oxford, 1989.

[Se3]   G. L. Sewell: *Unbounded local observables in quantum statistical mechanics*, J. Math. Phys. **11**, 1868–1884, 1970.

[Se4]   G. L. Sewell: $W^{\star}$-*dynamics of infinite quantum systems*, Lett. Math. Phys. **6**, 209–213, 1982.

[Se5]   G. L. Sewell: *Ergodic theory in algebraic statistical mechanics*, in: W. E. Brittin (editor), *Lectures in Theoretical Physics*, Vol. XIV B, Colorado University Press, pp. 511–538, 1973.

[Se6]   G. L. Sewell: *Nonequilibrium statistical mechanics: irreversibility and macroscopic causality*, in: A. O. Barut and W. E. Brittin (editors), *Lectures in Theoretical Physics*, Vol. X A, Gordon and Breach, New York, pp. 289–327, 1968.

[Se7]   G. L. Sewell: *KMS conditions and local thermodynamical stability of quantum lattice systems II*, Commun. Math. Phys. **55**, 53–61, 1977.

[Se8]   G. L. Sewell: *Stability, equilibrium and metastability in statistical mechanics*, Phys. Rep. **57**, 307–342, 1980.

[Se9]   G. L. Sewell: *Relaxation, amplification and the KMS conditions*, Ann. Phys. **85**, 336–377, 1974.

[Se10]   G. L. Sewell: *Quantum macrostatistics and irreversible thermodynamics*, in: L. Accardi and W. Von Waldenfels (editors), *Quantum Probability and Applications V, Lecture Notes in Mathematics*, Vol. 1442, Springer, Berlin, pp. 368–383, 1990.

[Se11]   G. L. Sewell: *Off-diagonal long range order and the Meissner effect*, J. Stat. Phys. **61**, 415–422, 1990.

[Se12]   G. L. Sewell: *Macroscopic quantum theory of superconductivity and the Higgs mechanism*, in: S. Albeverio, U. Cattaneo and D. Merlini (editors), *Stochastics, Physics and Geometry II*, World Scientific, Singapore, pp. 634–661, 1995.

[Seg]   I. E. Segal: *Postulates for general quantum mechanics*, Ann. Math. **48**, 930–948, 1947.

[Sei]   E. Seiler: *Gauge Theories as a problem of constructive field theories and statistical mechanics*, in: *Lecture Notes in Physics*, 159, Springer, Berlin, 1982.

[St]   R. F. Streater: *The Heisenberg ferromagnet as a quantum field theory*, Commun. Math. Phys. **6**, 233–247, 1967.

[Sta]   H. E. Stanley: *Introduction to Phase Transitions and Critical Phenomena*, Oxford University Press, Oxford, 1971.

[Sti]   W. F. Stinespring: *Positive functions on $C^{\star}$-algebras*, Proc. Am. Math. Soc. **6**, 211–216, 1955.

[Sto]   E. Stormer: *Positive linear maps of $C^{\star}$-algebras*, in: A. Harkämper and H. Neumann (editors), *Foundations of Quantum Mechanics and Ordered Spaces, Lecture Notes in Physics*, 29, Springer, Berlin, pp. 85–106, 1974.

[SS]   R. H. Schonmann and S. Shlosman: *Wulff droplets and the metastable relaxation of the kinetic Ising model*, Commun. Math. Phys. **194**, 389–462, 1988.

[SW]   C. E. Shannon and W. Weaver: *The Mathematical Theory of Communication*, University of Illinois Press, Urbana, IL, 1949.

[StWi]   R. F. Streater and A. S. Wightman: *PCT, Spin and Statistics, and All That*, W. A. Benjamin, New York, 1964.

[Ta]   M. Takesaki: *Tomita's theory of modular Hilbert algebras and its applications*, Lecture Notes in Mathematics, Vol. 128, Springer, Berlin, 1970.

[Th]   W. Thirring: *Quantum Mechanics of Large Systems*, Springer, New York, 1983.

[Tho]   D. J. Thouless: *The Quantum Mechanics of Many-body Systems*, Academic Press, New York, 1961.

[To]   P. L. Torres: *A classical Markov process in nonequilibrium quantum statistical mechanics*, J. Math. Phys. **18**, 301–305, 1977.

[TW]   M. Takesaki and M. Winnink: *Local normality in quantum statistical mechanics*, Commun. Math. Phys. **30**, 129–152, 1973.

[Uh]   G. E. Uhlenbeck: *An outline of statistical mechanics*, in: E. G. D. Cohen (editor), *Fundamental Problems in Statistical Mechanics* II, North-Holland, Amsterdam, pp. 1–29, 1968.

[Um]   H. Umegaki: *Conditional expectations in an operator algebra IV (entropy and information)*, Kodai Math. Sem. Rep. **14**, 59–85, 1962.

[Vh]   P. Vanheuverzwijn: *Generators for quasi-free completely positive semigroups*, Ann. Inst. H. Poincaré A **29**, 123–138, 1978; Erratum, *ibid*, **30**, 83, 1979.

[VB]   H. Van Beijeren: *Interface sharpness in the Ising system*, Commun. Math. Phys. **40**, 1–7, 1975.

[VK1]   N. G. Van Kampen: *Stochastic Processes in Physics and Chemistry*, North-Holland, Amsterdam, 1981.

[VK2]   N. G. Van Kampen: *Quantum statistics of irreversible processes*, Physica **20**, 603–622, 1954.

[VN1]   J. Von Neumann: *Mathematical Foundations of Quantum Mechanics*, Princeton University Press, Princeton, NJ, 1955.

[VN2]   J. Von Neumann: *Die Eindeutigkeit der Schrödingerschen Operatoren*, Math. Ann. **104**, 570–578, 1931.

[We]   H. Weyl: *The Theory of Groups and Quantum Mechanics*, Dover, New York, 1931.

[Weh]   A. Wehrl: *General properties of entropy*, Rev. Mod. Phys. **50**, 221–260, 1978.

[Wi]   E. P. Wigner: *Group Theory and its Application to Quantum Mechanics of Atomic Spectra*, Academic Press, New York, 1958.

[WE]   B. Whitten-Wolfe and G. G. Emch: *A mechanical quantum measuring process*, Helv. Phys. Acta **49**, 45–55, 1976.

[WH]   K. Weiss and H. Haug: *Bose condensation and superfluid hydrodynamics*, in: H. Haken and M. Wagner (editors), *Cooperative Phenomena*, Springer, Berlin, pp. 219–235, 1973.

[Ya]   C. N. Yang: *Concept of off-diagonal long range order and the quantum phases of liquid He and of superconductors*, Rev. Mod. Phys. **34**, 694–704, 1962.

[YL]   C. N. Yang and T. D. Lee: *Statistical theory of equations of state and phase transitions I*, Phys. Rev. **87**, 404–409, 1952.

[Za]   J. Zak: *Magnetic translation group*, Phys. Rev. A **134**, 1602–1606, 1964.

[Zu]   W. H. Zurek: *Algorithmic randomness and physical entropy*, Phys. Rev. A **40**, 4731–4751, 1989.

[Zw]   R. Zwanzig: *Ensemble method in the theory of irreversibility*, J. Chem. Phys. **33**, 1338–1341, 1960; and *Statistical mechanics of irreversibility*, in: W. E. Brittin, B. W. Downs and J. Downs (editors), *Lectures in Theoretical Physics*, III, Interscience, New York, pp. 106–141, 1961.

[ZA]   Z. Zhou and P. W. Anderson: *Neutral fermion, charge-e boson excitations in the resonating-valence-bond state and superconductivity in La_2CuO_4 compounds*, Phys. Rev. B **37**, 627–630, 1988.

Index